Lecture Notes in Electrical Engineering

Volume 739

The book series *Lecture Notes in Electrical Engineering* (LNEE) publishes the latest developments in Electrical Engineering - quickly, informally and in high quality. While original research reported in proceedings and monographs has traditionally formed the core of LNEE, we also encourage authors to submit books devoted to supporting student education and professional training in the various fields and applications areas of electrical engineering. The series cover classical and emerging topics concerning:

- Communication Engineering, Information Theory and Networks
- Electronics Engineering and Microelectronics
- Signal, Image and Speech Processing
- Wireless and Mobile Communication
- Circuits and Systems
- Energy Systems, Power Electronics and Electrical Machines
- Electro-optical Engineering
- Instrumentation Engineering
- Avionics Engineering
- Control Systems
- Internet-of-Things and Cybersecurity
- Biomedical Devices, MEMS and NEMS

For general information about this book series, comments or suggestions, please contact leontina.dicecco@springer.com.

To submit a proposal or request further information, please contact the Publishing Editor in your country:

China

Jasmine Dou, Editor (jasmine.dou@springer.com)

India, Japan, Rest of Asia

Swati Meherishi, Editorial Director (Swati.Meherishi@springer.com)

Southeast Asia, Australia, New Zealand

Ramesh Nath Premnath, Editor (ramesh.premnath@springernature.com)

USA, Canada:

Michael Luby, Senior Editor (michael.luby@springer.com)

All other Countries:

Leontina Di Cecco, Senior Editor (leontina.dicecco@springer.com)

** **This series is indexed by EI Compendex and Scopus databases.** **

More information about this series at http://www.springer.com/series/7818

Hyuncheol Kim · Kuinam J. Kim ·
Suhyun Park

Editors

Information Science
and Applications

Proceedings of ICISA 2020

 Springer

Editors
Hyuncheol Kim
Namseoul University
Cheonan-si, Korea (Republic of)

Kuinam J. Kim
Kyong-gi University
Suwon-si, Gyeonggi-do, Korea (Republic of)

Suhyun Park
Department of Computer Engineering
Dongseo University
Busan, Korea (Republic of)

ISSN 1876-1100   ISSN 1876-1119 (electronic)
Lecture Notes in Electrical Engineering
ISBN 978-981-33-6387-8   ISBN 978-981-33-6385-4 (eBook)
https://doi.org/10.1007/978-981-33-6385-4

This Springer imprint is published by the registered company Springer Nature Singapore Pte Ltd.
The registered company address is: 152 Beach Road, #21-01/04 Gateway East, Singapore 189721,
Singapore

Organizing Committee

General Chairs

Suhyun Park, Dongseo University, Republic of Korea
Hyeunchul Kim, Namseoul University, Republic of Korea

Steering Committee

Nikolai Joukov, New York University and modelizeIT Inc, USA
Borko Furht, Florida Atlantic University, USA
Bezalel Gavish, Southern Methodist University, USA
Kin Fun Li, University of Victoria, Canada
Kuinam J. Kim, Institute of Creative Advanced Technologies, Science and Engineering, Korea
Naruemon Wattanapongsakorn, King Mongkut's University of Technology Thonburi, Thailand
Xiaoxia Huang, University of Science and Technology Beijing, China

Publicity Chairs

Minsu Kim, Kyonggi University, Republic of Korea
Hongseok Jeon, ETRI, Republic of Korea
Suresh Thanakodi, National Defence University of Malaysia, Malaysia

Financial Chair

Jongmin Kim, Kyonggi University, Republic of Korea

Publication Chair

Hyeunchul Kim, Namseoul University, Republic of Korea

Program Chairs

Kuinam J. Kim, Institute of Creative Advanced Technologies, Science and Engineering, Korea
Nakhoon Baek, Kyungpook National University, Republic of Korea

Organizers and Supporters

Institute of Creative Advanced Technologies, Science and Engineering (iCatse)
Korean Industry Security Forum (KISF)
Korea Convergence Security Association (KCSA)
Hongik University, Republic of Korea
Kyonggi University, Republic of Korea
Kyungpook National University, Republic of Korea
Namseoul University Wellness Institute
Korea Institute of Science and Technology Information (KISTI)
Electronics and Telecommunications Research Institute (ETRI)
BPU Holdings Inc.

Program Committee

Dr. Mohd Ashraf Ahmad, University Malaysia Pahang, Malaysia
Dr. Siti Sarah Maidin
Wing C. Kwong, Hofstra University, USA
Dr. Christof EBERT, Vector Consulting Services
Marco Listanti, DIET Department, University of Roma Sapienza, Italy
Pr. Pascal LORENZ, University of Haute Alsace, France

Assistant Dr. Somlak Wannarumon Kielarova, Faculty of Engineering, Naresuan University, Thailand

Zeyar Aung, Khalifa University of Science and Technology, UAE

Xiaochun Cheng, Middlesex University, London, UK

Vasco N. G. J. Soares, Instituto de Telecomunicações / Instituto Politécnico de Castelo Branco, Portugal

Pavel Loskot, Swansea University, UK

Luca Piras, University of Brighton, UK

José Manuel Matos Ribeiro da Fonseca, NOVA University of Lisbon, Portugal

Sangseo Park, The University of Melbourne, Australia

Robert S Laramee, University of Nottingham, UK

Bok-Min Goi, Universiti Tunku Abdul Rahman, Malaysia

Tommaso Zoppi, University of Florence, Italy

Prof. Baojun MA, Shanghai International Studies University, School of Business and Management, China

Johann Marquez-Barja, University of Antwerp—imec, Belgium

Shimpei Matsumoto, Hiroshima Institute of Technology, Japan

Hyunsung Kim, Kyungil University, Korea

Zbigniew Leonowicz, Wroclaw University of Science and Technology, Poland

William Emmanuel Yu, Ateneo de Manila, Philippines

Dr. Shitala Prasad, Scientist, I2R, A*Star, Singapore

Dr. Chittaranjan Pradhan, KIIT University, Bhubaneswar, India

IICKHO SONG, Korea Advanced Institute of Science and Technology, Korea

Dr. Terje Jensen, Telenor, Norway

Pao-Ann Hsiung, National Chung Cheng University, Taiwan

Pr Naoufel Kraiem, RIADI Lab—Tunisia, North Africa

MAINGUENAUD Michel, Institut National des Sciences Appliquées—Rouen, France

Nguyen Dinh Cuong, Nha Trang University, Vietnam

Maicon Stihler, Centro Federal de Educação Tecnológica de Minas Gerais—CEFETMG, Brazil

Oscar Mortagua Pereira, University of Aveiro, Portugal

Dr. Bernd E. Wolfinger, University of Hamburg, Germany

Michalis Pavlidis, University of Brighton, UK

Ad Lalit Prakash Saxena, Combo Legal Consultancy, Obra India

Dr. James Braman, Community College of Baltimore County, USA

Ljiljana Trajkovic, Simon Fraser University, Canada

Bongkyo Moon, Dongguk University, Korea

Mauro Gaggero, National Research Council of Italy, Italy

Ong Thian Song, Multimedia University Malaysia, Malaysia

Yanling Wei, KU Leuven, Belgium

Harikumar Rajaguru, Bannari Amman Institute of Technology, Sathyamangalam India

Dr. Ahmad Kamran Malik, COMSATS University Islamabad (CUI), Pakistan

Etibar Seyidzade, Baku Engineering University, Azerbaijan

Lorenzo Ricciardi Celsi, Consel – Consorzio ELIS & Dipartimento di Ingegneria
Informatica, Automatica e Gestionale, Università di Roma "La Sapienza", Italy
Sira Yongchareon, Auckland University of Technology, New Zealand
David Liu, Purdue University Fort Wayne, USA
Wun-She Yap, Universiti Tunku Abdul Rahman, Malaysia
Mohd. Saifuzzaman, Daffodil International University, Dhaka, Bangladesh
Daniel Bo-Wei Chen, Princeton University, USA
Dr. Ng Hui Fuang, Universiti Tunku Abdul Rahman, Malaysia
Min-Shiang Hwang, Asia University, Taiwan
Maurantonio Caprolu, Hamad Bin Khalifa University, Doha, Qatar
Sandro Leuchter, Mannheim University of Applied Sciences, Germany
Azeddien M. Sllame, University of Tripoli, Tripoli, Libya
Wg. Cdr. Tossapon Boongoen, Mae Fah Luang University, Chiang Rai, Thailand

Preface

This LNEE volume contains the papers presented at the iCatse International Conference on Information Science and Applications (ICISA2020) which was held in Zoom event platform (virtual Conference) during December 16–18, 2020.

ICISA2020 will provide an excellent international conference for sharing knowledge and results in information science and application. The aim of the conference is to provide a platform to the researchers and practitioners from both academia as well as industry to meet the share cutting-edge development in the field.

The primary goal of the conference is to exchange, share, and distribute the latest research and theories from our international community. The conference will be held every year to make it an ideal platform for people to share views and experiences in information science and application-related fields.

On behalf of the Organizing Committee, we would like to thank Springer for publishing the proceedings of ICISA2020. We also would like to express our gratitude to the 'Program Committee and Reviewers' for providing extra help in the review process. The quality of a refereed volume depends mainly on the expertise and dedication of the reviewers. We are indebted to the Program Committee members for their guidance and coordination in organizing the review process and to the authors for contributing their research results to the conference.

Our sincere thanks to the Institute of Creative Advanced Technology, Engineering and Science for designing the conference web page and also spending countless days in preparing the final program in time for printing. We would also like to thank our organization committee for their hard work in sorting our manuscripts from our authors.

We look forward to seeing all of you next year at ICISA.

Suwon-si, Korea (Republic of)   Kuinam J. Kim
Busan, Korea (Republic of)   Suhyun Park
Cheonan-si, Korea (Republic of)   Hyuncheol Kim

Contents

Experience and Prospects of Tele-health Technologies

Valery Stolyar, Anastasiia Goy, and Maya Amcheslavskaya

Abstract The article describes the concept of modern telemedicine, and deals with basic directions of using tele-health technologies in clinical practice and the educational process. Particular attention is paid to such areas as remote video consultation and master classes, tele-mentoring. To conclude, the prospects of new technological solutions for telemedicine are estimated.

Keywords Informatics · Digital health · Telemedicine · Video consultation · Tele-lecture · Tele-mentoring

1 Introduction

An unprecedented leap in the development of information, multimedia and telecommunication technologies that took place at the turn of the twentieth and twenty-first centuries has had a significant impact on all spheres of our daily life. Computer technologies is so deeply rooted in everyday activities that human functioning without them seems impossible. The education and health care systems were also influenced by them, as a result of which the telemedicine direction was founded.

The subject of telemedicine is the transfer of medical information between points that are distant from each other (for example, between individual medical institutions). Telemedicine involves the use of telecommunications to connect healthcare professionals with clinics, hospitals, physicians, remote patients for diagnosis, treatment, consultation and continuing education.

V. Stolyar · A. Goy (✉) · M. Amcheslavskaya
Medical Informatics and Telemedicine Department, Peoples' Friendship University of Russia, RUDN University), Miklukho-Maklaya str.6, 117198 Moscow, Russia
e-mail: goy.telemed@yahoo.com

V. Stolyar
e-mail: v_stolyar@yahoo.com

M. Amcheslavskaya
e-mail: telemed@ntt.ru

The communication of doctors in the framework of medical consultations, medical education had reached a new qualitative level by telemedicine technologies. Such terms as remote video consultations «Doctor-Doctor» and «Doctor-Patient», interactive master classes and tele-mentoring are now in common parlance, in our vocabulary. However, there are unrealistic expectations in the medical community of using telemedicine. Therefore, we deemed it necessary to consider the possibilities, limitations and prospects of telemedicine technologies in this article.

2 Main "Vectors" of Modern Telemedicine

Put simply, telemedicine is medicine at a distance. It involves diagnostic consultations, managerial, educational, scientific and enlightenment activities in the field of health implemented with the use telecommunication technologies.

To date, the main application areas of telemedicine technologies are remote video consultations, interactive distance education, remote master classes, tele-mentoring, «Doctor-Patient» video consultations.

During a *remote video consultation* (scheduled or emergency) doctors from several clinics can jointly review and discuss patient information in order to clarify the diagnosis, treatment method, and indications for surgery. Patient participation and researching is also possible at the consultation. The attending physician reports on the patient's condition to the consultants, thereby turning for help to the experience and knowledge of senior colleagues from leading clinics. In some cases, remote video consultations are much more convenient than face-to-face consultations in these clinics, and help save the patient's time and energy (Fig. 1).

An up-to-date interactive distance education allow participate in lectures and seminars by the best professors from leading Russian and foreign universities, with

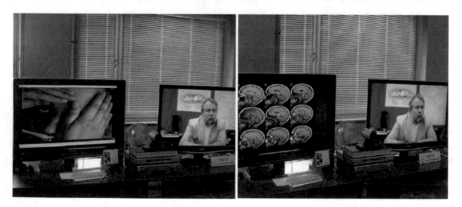

Fig. 1 Remote video consultation

studio audio and video quality, take a part in clinical discussions or scientific-practical conference. This technology allows to collect training courses for doctors and students from building blocks—from individual lectures by leaders in their fields [1]. Having extensive experience in this area, we regularly organize tele-lectures for RUDN University students with the participation of famous professors from Russia, India and Germany. Thus we are providing learners to receive knowledge from top professionals, which greatly improves the quality of education (Fig. 2).

Remote master classes based on modern video conferencing systems, and could get us into the operating room of leading clinics. A contemporary high-tech developments allow to connect to multiple cameras in the operating room simultaneously (to cameras on the surgeon's helmet, in the operating lamp, on the ceiling of the operating room, operational monitors, medical equipment), and display qualitative images on large monitors. Trainees have the opportunity to seeing an operation from different positions. Most cameras have remote control (for example, zoom, rotate), so that listeners can observe in detail the work of the operating surgeon and team members, to conduct a dialogue with them while in the classroom. A similar technology is adapted for interactive distance learning of diagnostic procedures and manipulations (ultrasound diagnostics of the heart, endoscopic procedures, etc.) when the full-time presence of students is undesirable (Fig. 3).

A *tele-mentoring* is the inverse task of remote master classes. A tele-mentoring make it possible to control the operation of young surgeons and diagnosticians with an option of remote consulting support in complex cases by more experienced colleagues based on video conferencing. This technology has already shown to be effective in the learning of new surgical and diagnostic methods and medical equipment (Fig. 4).

The stereoscopic complex can be used as an instrument of tele-mentoring. Stereoscopic technologies develop spatial thinking and make it possible to present a detailed model of any object. It is important to understand the implementation of one or another stereoscopic content is aimed at solving a narrow range of tasks for specific areas of activity. Great value stereoscopic technologies are presented in the task of

Fig. 2 The distance lecture from India

Fig. 3 Remote interactive master classes in ophthalmology and cardiac surgery

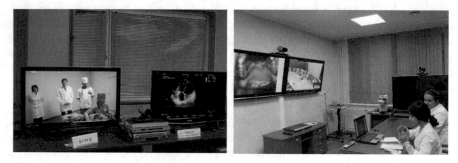

Fig. 4 Examples of tele-mentoring in ultrasound heart diagnosis and dentistry

improving the qualifications of already established surgeons in the course of interactive master classes and are successfully used in such areas as neurosurgery, oncology, maxillofacial surgery, laser surgery.

3 Future Developments and Expectations

When considering the possibility of using telemedicine technologies in the curative process and medical education, it is important to understand objectively both the possibilities and limitations (technical, organizational, economic, psychological etc.) of modern telemedicine. There will be fewer restrictions as technology develops, but they will never disappear in principle.

What should we expect from telemedicine in the next five years?

Firstly, it is dynamic transmission of a stereoscopic image of the operating field (for example, for maxillo-facial operations in addition to static images of 3D scanners) with spatial image processing to improve the effectiveness of remote interactive master classes, or issuing recommendations to the operating surgeon from an experienced colleague.

Secondly, remote transmission of tactile sensations during teleconsultation for palpation, percussion, assessment of skin areas surface problem using a remote «hand-glove».

Thirdly, remote transmission of smell during teleconsultation. Electronic dog nose created, and it is important to learn how correctly generate sets of smells on the receiving side. This technology will be in demand for a whole range of medical specialties (for example, dentistry, purulent surgery, etc.).

Fourthly, use of artificial intelligence (AI) technologies during emergency video consultations for rapid analysis of patient information and presentation of effective recommendations to the doctor based on available databases. At the current stage of development of digital medicine, it is extremely important to support medical decisions in emergency situations using intelligent systems located on the servers of large medical organizations or in "cloud" data centers. The appeal to these decision support systems (DSS) should be provided by calling the AI system by the doctor directly in the process of working in the electronic medical record system. The corresponding intelligent system will analyze the signs noted in the patient's card and form hypotheses of the differential diagnostic series, the possible prognosis of complications and recommendations for treatment.

Fifthly, use of augmented reality (AR) methods in conducting video conferences and educational process (tele-mentoring, master classes). AR considered one of the variations of virtual reality (VR). The main difference is that VR technology completely immerses the user in the virtual world. AR brings 3D virtual objects in real time into a 3D real environment. Thus, due to the possibility of generating any 3D-objects and their further integration into the real environment, augmented reality technologies are widely used in surgical practice, namely in training future surgeons. With the help of augmented reality, it has become possible to simulate various surgeries online, which greatly facilitates, for example, the practical part of the training of future surgeons [2].

Sixthly, the emergence of a whole range of implanted and pasted sensors connected by a single digital platform with pre-processing and analysis of data, for monitoring the condition of chronic patients and patients during long-term rehabilitation after complex operations. Currently available services, gadgets and medical devices for home use will remind the patient of the need to take medications on time, conduct research, fill out a diary about their condition, track the dynamics of indicators and automatically transmit them to the attending doctor. In this way, the doctor can fully assess the patient's condition by the time of the teleconsultation. Despite the fact that a remote appointment takes an amount of time equal to a face-to-face appointment, it is always less informative for the doctor, and the psychological burden for specialist is significantly higher.

4   Conclusion

Telemedicine services are actively developing all over the world, involving Телемедицинские сервисы активно развиваются во всем мире, involving the best minds of doctors and programmers, implementing artificial intelligence and the latest computer technologies. Success achieved only where they based on knowledge and experience of skilled doctors and diagnosticians, who should be involved as analysts at the stage of setting the task. We actively cooperate with Russian and foreign developers of new technologies and equipment for telemedicine, taking part in the search for ideas, formulating new tasks, creating and testing prototypes. We are optimistic about the prospects for the development and practical use of telemedicine, given the presence of strong medical schools in Russian.

Reference lists

1. At'kov O, Stolyar V (2001) Interaktivnoe teleobuchenie vrachej na osnove sistem videokonfer-encsvyazi. Vizualizaciya V Med 18:34–40
2. Namiot E (2019) Augmented reality in medicine. Int J Open Info Technol 7(11):94–98

Exploring the Effects of Network Topology Layers on Quality of Service Mechanisms in the Context of Software-Defined Networking

Josiah Eleazar T. Regencia and William Emmanuel S. Yu

Abstract The separation of the control plane and the data plane in the Software-Defined Networking (SDN) architecture makes it easier and more feasible to implement per-flow Quality of Service (QoS) provisioning in the network which is beneficial for flows that require special QoS handling such as VoIP and video streaming. Applying QoS mechanisms to a distributed single-layered set of switches in the context of SDN has shown promise in previous studies. Using a test framework that was previously developed for QoS algorithms and networking topologies in SDN, this study explored how layers in the network topology affect the performance of QoS in the network. This is in order to replicate real-world networking scenarios where packets travel through layers of networking equipment. Generally, results have shown that having less layers in the network means better performance. However, this does not mean that adding layers creates a material disadvantage. Results also show that despite adding layers to the networking topology, the SDN distributed Leaf-enforcement of QoS performed better or as good as compared to the traditional centralized Core-enforcement of QoS. This is despite having leaf-enforced environments have more points of enforcement as layers are added.

Keywords Software-defined networking · Quality of service · Class-based queueing

1 Introduction

The emergence of Software-Defined Networking (SDN) provides dynamic, flexible and scalable—all of which are weaknesses of traditional networks—control and management for networks. The separation of the control plane and the data plane

J. E. T. Regencia (✉) · W. E. S. Yu
Ateneo de Manila University, Loyola Heights, 1108 Quezon City, Philippines
e-mail: josiah.regencia@obf.ateneo.edu

W. E. S. Yu
e-mail: wyu@ateneo.edu

© The Author(s), under exclusive license to Springer Nature Singapore Pte Ltd. 2021
H. Kim et al. (eds.), *Information Science and Applications*, Lecture Notes in Electrical Engineering 739, https://doi.org/10.1007/978-981-33-6385-4_2

enables flow-management and resources to be dynamically optimized. In addition, this makes it easier and more feasible to implement per-flow Quality of Service (QoS) provisioning in the network which is beneficial for flows that require special QoS handling such as VoIP and video streaming [7].

Multiple studies have explored QoS in the context of SDN. Chato and Yu [2, 3] explored Class-Based Queuing (CBQ) QoS algorithms (Basic CBQ and Source CBQ) in a virtual network topology implemented inside the Mininet emulator [4] using a custom POX OpenFlow Controller to in terms of throughput, bandwidth, and latency. The Basic CBQ algorithm was a protocol-based classification queuing and the Source CBQ algorithm was a classification queuing based on source IP addresses. Results showed that QoS algorithms applied to the distributed nature of SDN have better performance in terms of bandwidth and decreased latency. In 2020, Regencia and Yu [8] introduced a testing framework for QoS algorithms and network topologies in the context of Software-Defined Networking. As an anchor to test the framework, Destination IP address based CBQ and Source-Destination IP addresses CBQ algorithms are introduced and compared against the CBQ algorithms in the Chato and Yu [2, 3] inside the test framework. The test framework has two test cases for each QoS algorithm: the traditional centralized Core-enforcement where QoS is only enforced in a centralized core switch and the SDN distributed nature Leaf-enforcement where QoS is distributedly enforced to switches that are closer to the edge called the client leaf switches. The test framework introduced by Regencia and Yu [8] has been submitted to the CANDAR'20 Conference and is currently awaiting paper acceptance notification.

In the previous 2020 study by Regencia and Yu [8], Leaf-enforcement is only applied to a single-layered set of client-leaf switches where results showed that Leaf-enforcement generally showed better performance compared to Core-enforcement. In this study, the test framework introduced by Regencia and Yu [8] is used to test how layers in the network topology affect QoS in both Core-enforced and Leaf-enforced test cases of QoS algorithms. The addition of layers is an experiment to replicate a real-word scenario networking topologies.

2   Related Literature

The flexibility and programmability of SDN has been seen as an answer to the issues traditional networks continue to face when it comes to QoS especially as standard QoS policies have become no longer capable of supporting today's complex networks due to the rapid increase of network applications and devices [7].

Applying proper QoS mechanism is a difficult task due to different needs of different types of flows. For example, multimedia flows from applications such as video conferences or video streaming require steady network resources and as much as possible, little to no packet drop and delay. In designing frameworks for such flows, prioritization and classification are both key factors [7]. Queuing is a common QoS mechanism. QosFlow [6] provides flexible control mechanisms by manipulating multiple packet schedulers such as Hierarchical Token Bucket (HTB), Random Early Detection (RED), and Stochastic Fairness Queuing (SFQ) schedulers. Chato and Yu [2, 3] also explored use of queuing QoS mechanisms by applying two (2) algorithms Class-Based Queuing (CBQ): Basic CBQ and Source CBQ. Each algorithm is tested on both centralized Core-enforcement and distributed Leaf-enforcement.

In 2020, Regencia and Yu [8] introduced a testing framework for QoS algorithms and network topologies in the context of Software-Defined Networking. As an anchor to test the framework, Destination IP address based CBQ and Source-Destination IP addresses CBQ algorithms are introduced and compared against the CBQ algorithms in the Chato and Yu [2, 3] inside the test framework. As shown in Fig. 1, the key component of the test framework is the separation of the SDN Controller, the Topology which is software simulated using Mininet, the Test Simulator, and the implementations of QoS algorithms. In addition, generator scripts create configuration files that define features of networking devices in the virtual topology. The test framework has two test cases for each QoS algorithm: Core-enforcement where QoS is only enforced in a centralized core switch and Leaf-enforcement where QoS is distributedly enforced to switches that are closer to the edge called the client leaf switches. In this study, the test framework introduced by Regencia and Yu [8] is used to test how layers in the network topology affect QoS in both Core-enforced and Leaf-enforced test cases of QoS algorithms. The addition of layers is an experiment to replicate a real-word scenario networking topologies.

3 Framework

This study was performed using an Amazon Web Services (AWS) EC2 m5.xlarge Ubuntu 18.04 environment with 40GB storage size. The implementation of the network topology was performed using Mininet version 2.3 [5] along with a custom Ryu OpenFlow 1.3 Controller [1]. The type of network topology used in this study is a fat-tree topology. Figure 2 shows two examples of how the topology looks when the number of client leaf switch layers is 2(a) and 3(b). Switches encompassed within the bracket are all called the client leaf switches. Quality of Service (QoS) is enforced in these switches if the test case is Leaf-enforcement. If the test case is Core-enforcement, QoS is enforced in the core switch. Switch 7 in the case of (a) and Switch 15 in the case of (b).

In the experiments performed, the network also had a total of 70 client hosts divided as equally as possible across the client leaf switches in the first layer. All client hosts run both Apache Bench and VLCt. The network also has six (6) server hosts connected to server aggregate switch, Switch 8 in the case of (a) and Switch 16 in the case of (b). Three (3) servers function as HTTP servers running Python3 simple.http server inside them. The other three (3) function as Video on Demand (VOD) servers running VLC Streaming Media using VLC version 3.8. All links in the network run at the default 10Gbps of Mininet.

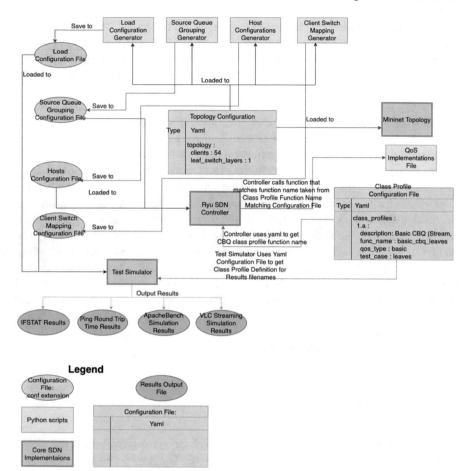

Fig. 1 Test framework architecture

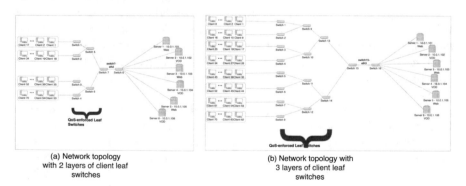

(a) Network topology with 2 layers of client leaf switches	(b) Network topology with 3 layers of client leaf switches

Fig. 2 Sample virtual network topology used

4 Methodology

In this study, Quality of Service (QoS) mechanisms are tested in order to answer how layers in the topology affect their performance. QoS mechanisms used are all listed in Table 1. The experiments in this study focused on testing from 2 layers up to 6 layers of client leaf switches. This is mainly due to computational constraints. For every layer test case, the number of first layer switches is always 2^{layers}. The number of switches in the following layers will then only be half of the number of switches from its previous layer.

The simulations in this study follow the same simulations from Chato and Yu [3]. This simulation is also adapted in the test framework introduced by Regencia and Yu [8]. Each Class Profile listed in Table 1 is executed one at a time through the process shown in Fig. 3. Each Class Profile in this study was given a 5-min window to perform test simulations. The following tools were used to perform the simulations and get data results concurrently:

- Ifstat Bandwidth results—This tool is used to get Bandwidth In and Bandwidth Out for every second during the 5-min window. This was performed and executed at the Core Switch for all Class Profiles
- Ping Round Trip Time—Round Trip Time (RTT) was measured in milliseconds (ms) by sending Ping packets to all destination servers from each of the 70 client hosts. Each destination server has its own measurement but the overall result was calculated by getting the mean results of all destination servers.
- ApacheBench HTTP Results—ApacheBench was used to simulate the HTTP traffic to web servers. The data taken from this tool were the transfer rate, mean time per request in ms, and total data transferred.
- VLC Streaming Media statistics results—VLC Streaming Media software client take both De-multiplexer (Demux) Bytes Read (KB) and Demux Bitrate (in Kbps). The De-multiplexer takes feeds from disparate and separate streams and assembles them into a single coherent media stream for playing.

Table 1 Class profiles

CBQ class profile	CBQ class grouping	Switch QoS
Basic at leaf	Traffic protocol	Client leaf switches
Basic at core	Traffic protocol	Core switch
Source at leaf	Source IP address	Client leaf switches
Source at core	Source IP address	Core switch
Destination at leaf	Destination IP address	Client leaf switches
Destination at core	Destination IP address	Core switch
Source-destination at leaf	Source-destination IP address	Client leaf switches
Source-destination at core	Source-destination IP address	Core switch

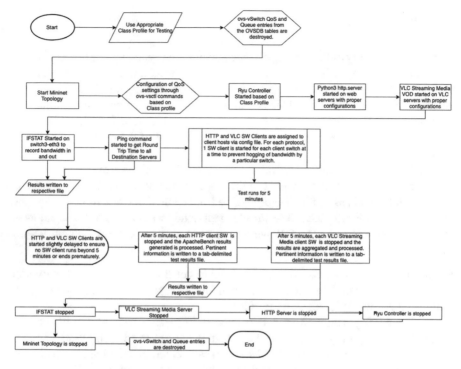

Fig. 3 Methodology flowchart based on Chato and Yu [3]

5 Results

The mean Quality of Service (QoS) performance results from layer 2 up to layer 6 can be found in Figs. 4, 5, 6, 7 and 8 respectively. Generally, it can be observed from the charts that for both Leaf-enforcement and Core-enforcement test cases, having less layers in the topology results to better performance. However, with regards to video streaming, the charts from Figs. 4, 5, 6, 7 and 8 show that layering does not have substantial impact to its performance. This is likely because the protocol is quite heavy and the additional switching enforcement is not too material.

For Apache Bench Transfer Rate, Figs. 4, 5, 7, and 8 show that when the network has 2, 3, 5, or 6 layers respectively, Leaf-enforcement of QoS performed better than Core-enforcement where we can conclude that in this aspect, the distributed nature of Software-Defined Networking is better than the traditional centralized enforcement of QoS. IFSTAT Bandwidth In and Bandwidth Out can also be generally observed to have similar results.

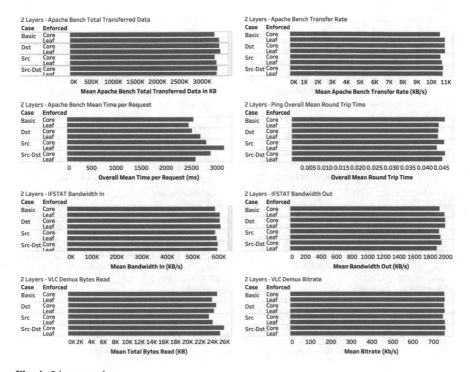

Fig. 4 2 layer results

For the latency tests, Destination Class-Based Queueing (CBQ) at the Leaf showed to have performed best as can be seen where 3 out 5 layer cases Destination CBQ at the Leaf performed best latency while in the other 2 layer cases, Destination CBQ at the Leaf performed second best latency.

However, it should also be noted that some tests indicate that the Core-enforcement algorithm performed better than the Leaf-enforcement algorithm. In order to test the significance of the result, a student's t-test was performed comparing both algorithms with significant level is set to $\alpha = 0.05$. If the p-value is less than $\alpha = 0.05$, then the difference will be considered as statistically significant. Otherwise, the difference will be considered as statistically insignificant.

The following instances show examples when Core-enforcement performed better than Leaf-enforcement and what the t-tests conclude with regards to the comparison.

- For VLC Demux Bitrate in 3 layer test case, Source CBQ at the Core (3rd) performed better than Source CBQ at the Leaf (5th). However, the t-test shows that its p-value is 0.9312 which is greater than $\alpha = 0.05$. Hence, the performance advantage of Source CBQ at the Core over Source CBQ at the Leaf for VLC Demux Bitrate in the 3 layer test case is statistically insignificant.

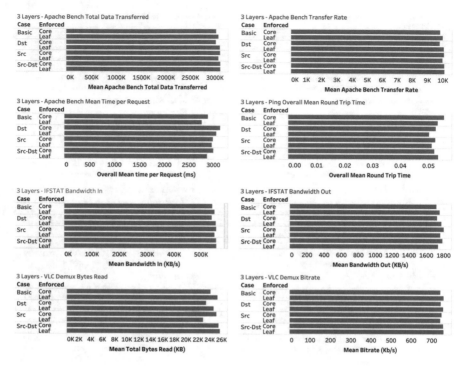

Fig. 5 3 layer results

- For VLC Demux Bitrate in 5 layer test case, Source CBQ at the Core (best) performed better than Source CBQ at the Leaf (3rd). However, the t-test shows that its p-value is 0.9522 which is greater than $\alpha = 0.05$. Hence, the performance advantage of Source CBQ at the Core over Source CBQ at the Leaf for VLC Demux Bitrate in the 5 layer test case is statistically insignificant.
- For Ping Round Trip Time latency test in 4 layer test case, Basic CBQ at the Core (4th) performed better than Basic CBQ at the Leaf (5th). However, the t-test shows that its p-value is 0.8851 which is greater than $\alpha = 0.05$. Hence, the latency advantage of Basic CBQ at the Core over Basic CBQ at the Leaf for VLC Demux Bitrate in the 4 layer test case is statistically insignificant.
- For Ping Round Trip Time latency test in 5 layer test case, Basic CBQ at the Core (3rd) performed better than Basic CBQ at the Leaf (7th). However, the t-test shows that its p-value is 0.4169 which is greater than $\alpha = 0.05$. Hence, the latency advantage of Basic CBQ at the Core over Basic CBQ at the Leaf for VLC Demux Bitrate in the 5 layer test case is statistically insignificant.

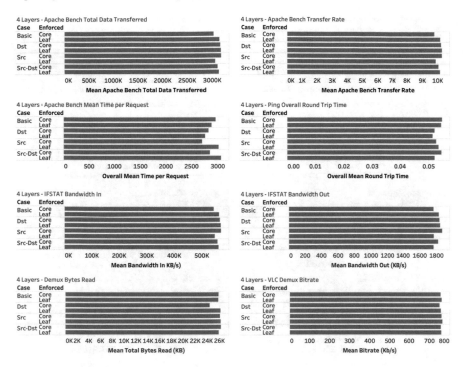

Fig. 6 4 layer results

- For Ping Round Trip Time latency test in 6 layer test case, Basic CBQ at the Core (3rd) performed better than Basic CBQ at the Leaf (7th). However, the t-test shows that its p-value is 0.1008 which is greater than $\alpha = 0.05$. Hence, the latency advantage of Basic CBQ at the Core over Basic CBQ at the Leaf for VLC Demux Bitrate in the 6 layer test case is statistically insignificant.

6 Conclusion

This study uses the testing framework for Quality of Service (QoS) algorithms and network topologies in order to test how layers in a network affect the performance of QoS which was introduced by Regencia and Yu [8] and is submitted to the CAN-DAR'20 Conference awaiting paper acceptance notification. The purpose for testing layers in the topology is to replicate a real-world networking scenario. Two types of QoS enforcements were tested and compared: Leaf-enforcement which is a distributed enforcement of QoS and Core-enforcement which is a centralized enforcement of QoS at the core switch. Having less layers in the network topology resulted to better performance in QoS for both Leaf-enforcement and Core-enforcement. How-

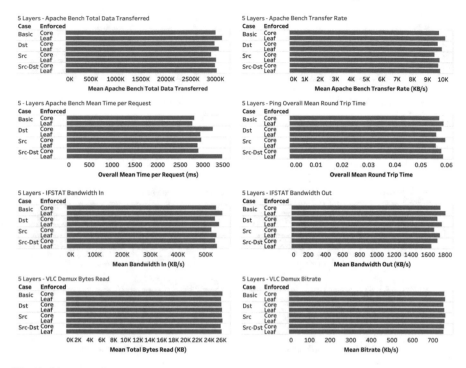

Fig. 7 5 layer results

ever, adding layers to the network topology does not create material disadvantage as can be seen from the results where Leaf-enforcement of QoS performed better or as good as Core-enforcement of QoS despite adding layers to the topology. Therefore, Software-Defined Networking logic has to be applied in many devices. This is despite the fact that enforcement in an SDN topology adds logic in more switches as the layers increase. Although it is noted that there were instances were Core-enforcement performed better than Leaf-enforcement, the student t-test results have shown that the difference in performance is not material and considered as insignificant.

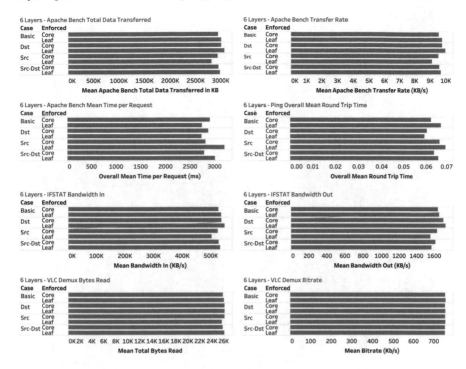

Fig. 8 6 layer results

References

1. https://github.com/faucetsdn/ryu
2. Chato O, Yu WES (2016) An exploration of various quality of service mechanisms in an open-flow and software defined networking environment in terms of latency performance. In: 2016 international conference on information science and security (ICISS). IEEE, New York, pp 1–7
3. Chato O, Yu WES (2016) An exploration of various QoS mechanisms in an Openflow and SDN environment. In: Accepted for presentation in the international conference on systems and informatics (ICSAI-2016)
4. De Oliveira RLS, Schweitzer CM, Shinoda AA, Prete LR (2014) Using Mininet for emulation and prototyping software-defined networks. In: 2014 IEEE Colombian conference on communications and computing (COLCOM). IEEE, New York, pp 1–6
5. Huang TY, Jeyakumar V, Lantz B, Feamster N, Winstein K, Sivaraman A (2014) Teaching computer networking with Mininet. In: ACM SIGCOMM
6. Ishimori A, Farias F, Cerqueira E, Abelém A (2013) Control of multiple packet schedulers for improving QoS on Openflow/SDN networking. In: 2013 Second European workshop on software defined networks. IEEE, New York, pp 81–86
7. Karakus M, Durresi A (2017) Quality of service (QoS) in software defined networking (SDN): A survey. J Network Comput Appl 80:200–218
8. Regencia JET, Yu WES, Introducing destination class-based queueing algorithm to ensure quality of service in a software-defined networking environment. Submitted to: CANDAR'20

Image Background Subtraction and Partial Stylization Based on Style Representation of Convolutional Neural Networks

Edwin Kurniawan, Bok-Deuk Song, Yeon-Jun Choi, and Suk-Ho Lee

Abstract Image background subtraction refers to the technique which eliminates the background in an image or video to extract the object (foreground) for further processing like object recognition in surveillance applications or object editing in film production, etc. The problem of image background subtraction becomes difficult if the background is cluttered, and therefore, it is still an open problem in computer vision. For a long time, conventional background extraction models have used only image-level information to model the background. However, recently, deep learning based methods have been successful in extracting styles from the image. Based on the use of deep neural styles, we propose an image background subtraction technique based on a style comparison. Furthermore, we show that the style comparison can be done pixel-wise, and also show that this can be applied to perform a spatially adaptive style transfer. As an example, we show that an automatic background elimination and background style transform can be achieved by the pixel-wise style comparison with minimal human interactions, i.e., by selecting only a small patch from the target and the style images.

Keywords Style transfer · Deep learning · Background subtraction

1 Introduction

Many background subtraction techniques rely on the statistical properties of the image intensity values. These techniques choose a statistical model for the background image intensity values and then train the model with the intensity values to find the right parameters for the background model. After that, a statistical test

E. Kurniawan · S.-H. Lee (✉)
Dongseo University, Busan, South Korea
e-mail: petrasuk@gmail.com

B.-D. Song · Y.-J. Choi
ETRI, Busan, South Korea

© The Author(s), under exclusive license to Springer Nature Singapore Pte Ltd. 2021
H. Kim et al. (eds.), *Information Science and Applications*, Lecture Notes
in Electrical Engineering 739, https://doi.org/10.1007/978-981-33-6385-4_3

is performed for every pixel in the image, and the pixels which fit to the background model are considered as background pixels [1–4]. Meanwhile, deep convolutional neural networks (Deep-CNNs) have been applied with success to background subtraction as they extract deep features of the image [5–8]. Recently, it has been also shown that Deep CNNs can well capture the style of an image which property has been applied to transfer a style of an artistic image to a photo image [9], resulting in the stylized image. This technique is known as style transfer. In style transfer, the style of the style image is transferred to the generated stylized image by matching the Gram matrices of the style and the stylized images, where the elements in the Gram matrices are the inner products of the feature maps.

It has been shown in [10] that the style loss function can be rewritten as the maximum mean discrepancy (MMD) with second order polynomial kernel. The second order polynomial kernels are the products of the feature vectors for different positions in the feature map, where the feature vectors can be accessed in a pixelwise manner, a fact which has not been further utilized in [10]. This property makes it possible to apply different styles from different sub-regions to different regions in the target image, i.e., makes the spatially adaptive style transfer possible. In this paper, we utilize this property also to measure the distance of a feature vector of a certain pixel to the distribution of the feature vectors in the background to determine whether the pixel has the same style as the background or not. Thus, we can perform a background subtraction based on the pixel-wise style representation. Furthermore, we also show how to partially stylize only the background region by adjusting the distribution of the feature vectors in the background region to that of a partial region in the style image.

2 Related Works

2.1 Neural Style Transfer

The original neural style transfer method makes use of the fact that the correlation between the activations of the layers can be used as a measure of the style similarity. Let x^c, x^*, and x^s denote the content image, the stylized image, and the style image, respectively. The feature maps of x^*, x^c, and x^s in the layer 1 of a CNN are denoted by

$$\mathbf{F}^l \in \mathbb{R}^{N_l \times M_l}, \ \mathbf{P}^l \in \mathbb{R}^{N_l \times M_l}, \ \mathbf{S}^l \in \mathbb{R}^{N_l \times M_l} \tag{1}$$

respectively, where N_l is the number of the feature maps in the layer l and M_l is the height times the width of the feature map. The traditional neural style transfer iteratively updates x^* by minimizing the following loss function,

$$L = \alpha L_{content} + \beta L_{style} \tag{2}$$

where $L_{content}$ and L_{style} are the content and the style losses, respectively, and α and β are the weights for the content and the style losses. The content loss $L_{content}$ is defined by the squared error between the feature maps of a specific layer l for x^* and x^c:

$$L_{content} = \frac{1}{2} \sum_{i=1}^{N_l} \sum_{j=1}^{M_l} \left(F_{ij}^l - P_{ij}^l \right)^2 \tag{3}$$

The style loss L_{style} is the sum of several style losses L_{style}^l in different layers:

$$L_{style} = \sum_l w_l L_{style}^l \tag{4}$$

where w_l is the weight of the loss in the layer l and L_{style}^l is defined by the squared error between the features correlations expressed by Gram matrices of x^c and x^s:

$$L_{style}^l = \frac{1}{4N_l^2 M_l^2} \sum_{i=1}^{N_l} \sum_{j=1}^{N_l} \left(G_{ij}^l - A_{ij}^l \right)^2 \tag{5}$$

where the Gram matrix $G^l \in \mathbb{R}^{N_l \times M_l}$ is the inner product between the vectorized feature maps of x^* in layer l:

$$G_{ij}^l = \sum_{k=1}^{M_l} F_{ik}^l F_{jk}^l \tag{6}$$

and similarly A^l is the Gram matrix corresponding to S^l. Here, both the content loss and the style loss are taking into account the correlation between feature maps, and therefore, the style transfer is performed in a channel-wise manner. Therefore, a spatially adaptive style transfer is not possible with the traditional approach.

2.2 Maximum Mean Discrepancy

Maximum mean discrepancy (MMD) is a popular test statistic for the two-sample testing problem, which is used to determine whether two sets of sample points are generated from the same distribution. Let $X = \{x_i\}_{i=1}^n$ and $Y = \{y_j\}_{j=1}^m$ be two different sets where x_i and y_j are generated from two different distributions. Let $\phi(\cdot)$ be an explicit feature mapping function of MMD. Applying the associated kernel function $k(x, y) = \phi(x), \phi(y)$, the MMD can be expressed in the form of kernels:

$$\text{MMD}^2[X, Y] = \frac{1}{n^2} \sum_{i=1}^{n} \sum_{i'=1}^{n} k(\boldsymbol{x}_i, \boldsymbol{x}_{i'}) + \frac{1}{m^2} \sum_{j=1}^{m} \sum_{j'=1}^{m} k(\boldsymbol{y}_j, \boldsymbol{y}_{j'}) - \frac{2}{nm} \sum_{i=1}^{n} \sum_{j=1}^{m} k(\boldsymbol{x}_i, \boldsymbol{y}_j)$$

$$(7)$$

A larger value of the MMD indicates a larger similarity between the two different distributions from which the sample points have been drawn.

2.3 Equivalence of Style Loss and Maximum Mean Discrepancy

It has been derived in [10], that the style loss function is equivalent to the Maximum mean discrepancy(MMD) between the distribution of the feature vectors of the stylized image and the distribution of the feature vectors of the style image. Again, let F_{ik}^l and S_{ik}^l denote the k-th pixel in the i-th feature map in layer l of x^* and x^s, respectively. Then, the style loss function is re-written in [10] as follows:

$$L_{style}^l = \frac{1}{4N_l^2 M_l^2} \sum_{i=1}^{N_l} \sum_{j=1}^{N_l} \left(\sum_{k=1}^{M_l} F_{ik}^l F_{jk}^l - \sum_{k=1}^{M_l} S_{ik}^l S_{jk}^l \right)^2$$

$$= \frac{1}{4N_l^2 M_l^2} \sum_{k_1=1}^{M_l} \sum_{k_2=1}^{M_l} \left(\left(\sum_{i=1}^{N_l} F_{ik_1}^l F_{ik_2}^l \right)^2 + \left(\sum_{i=1}^{N_l} S_{ik_1}^l S_{ik_2}^l \right)^2 \right)$$

$$- \frac{1}{2N_l^2 M_l^2} \sum_{k_1=1}^{M_l} \sum_{k_2=1}^{M_l} \left(\sum_{i=1}^{N_l} F_{ik_1}^l S_{ik_2}^l \right)^2 = \frac{1}{4N_l^2 M_l^2} \sum_{k_1=1}^{M_l} \sum_{k_2=1}^{M_l} (\mathbf{f}_{.k_1}^l {}^T \mathbf{f}_{.k_2}^l)^2$$

$$+ (\mathbf{s}_{.k_1}^l {}^T \mathbf{s}_{.k_2}^l)^2 - 2(\mathbf{f}_{.k_1}^l {}^T \mathbf{s}_{.k_2}^l)^2$$

$$(8)$$

where N_l is the number of the feature maps in the layer l and M_l is the height times the width of the feature map. Here $\mathbf{f}_{.k_1}^l$ is a feature vector for a fixed position k_1 which expands to the axis of i, i.e., across the feature maps $F_{i.}^l, i = 1, 2, \dots N_l$ in layer l, and likewise for $\mathbf{f}_{.k_2}^l$, $\mathbf{s}_{.k_1}^l$, and $\mathbf{s}_{.k_2}^l$. The last line in (8) is equivalent to

$$\frac{1}{4N_l^2 M_l^2} \text{MMD}^2[F^l, S^l]$$

$$(9)$$

which is the MMD between the distributions from which the feature vectors in F^l and S^l are derived.

3 Proposed Method

3.1 Spatially Adaptive Style Transfer

In this section, we investigate the derivation of the equivalence of the style loss function and the MMD as derived in [10], and then see how this derivation can be utilized to perform spatially adaptive style transfer. By observing (8), it can be seen that the summation is now with respect to the pixels, i.e., k_1 and k_2 are the indices of the pixel positions. The individual terms in the summation in (8) reflect the influence of the feature vector at k_1 on the feature vector at k_2. The term of $\left(s^l_{.k_1}{}^T s^l_{.k_2}\right)^2$ in (8) has no effect on the stylization as the minimization is with respect to the features in the target image, which makes this term a constant. The minimizing of the term of $\left(f^l_{.k_1}{}^T f^l_{.k_2}\right)^2$ works as a kind of regularization of the feature vectors in the target image, and has nothing to do with the style image. Therefore, the main term which effects the stylization is the last term $-2\left(f^l_{.k_1}{}^T s^l_{.k_2}\right)^2$. Minimizing this term is equal to maximizing $\left(f^l_{.k_1}{}^T s^l_{.k_2}\right)^2$, which tries to correlate the feature vectors of the target image and the style image in a pixel-wise manner, i.e., tries to correlate the feature vector $f^l_{.k_1}$ at the position k_1 in the target image with the feature vector $s^l_{.k_2}$ at the position k_2 in the style image.

As a result of this observation, we can guess that if we perform pixel-wise feature vector correlation maximization in partial regions we can perform a spatially adaptive style transfer, i.e., apply the style of a partial region in image I_1 to the partial region in image I_2. For example, in Fig. 1, we are applying the style in Ω_3 in the 'Style1' image and the style in Ω_4 in the 'Style2' image to the regions Ω_1 and Ω_2 in the

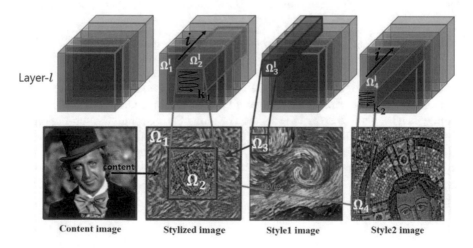

Content image   Stylized image   Style1 image   Style2 image

Fig. 1 Spatially adaptive style transfer with multiple styles. The style in Ω_3 in the 'Style1' image and the style in Ω_4 in the 'Style2' image are applied to the regions Ω_1 and Ω_2 in the stylized image, respectively

stylized image, respectively. This can be done by defining the region adaptive style loss function in the MMD form over the regions $\Omega_3^l \rightarrow \Omega_1^l$ and $\Omega_4^l \rightarrow \Omega_2^l$, where Ω_1^l, Ω_2^l, Ω_3^l, and Ω_4^l are the regions in layer l that correspond to the regions Ω_1, Ω_2, Ω_3, and Ω_4 in the images, respectively:

$$
\begin{aligned}
L_{style}^l = &\sum_{k_1 \in \Omega_1^l} \sum_{k_2 \in \Omega_3^l} \left(\mathbf{f}_{.k_1}^l{}^T \mathbf{f}_{.k_2}^l\right)^2 + \left(\mathbf{s}_{.k_1}^l{}^T \mathbf{s}_{.k_2}^l\right)^2 - 2\left(\mathbf{f}_{.k_1}^l{}^T \mathbf{s}_{.k_2}^l\right)^2 \\
+ &\sum_{k_1 \in \Omega_2^l} \sum_{k_2 \in \Omega_4^l} \left(\mathbf{f}_{.k_1}^l{}^T \mathbf{f}_{.k_2}^l\right)^2 + \left(\mathbf{s}_{.k_1}^l{}^T \mathbf{s}_{.k_2}^l\right)^2 - 2\left(\mathbf{f}_{.k_1}^l{}^T \mathbf{s}_{.k_2}^l\right)^2
\end{aligned} \tag{10}
$$

where in the last line, $\text{MMD}^2\left[F^l(\Omega_1^l), S_1^l(\Omega_3^l)\right]$ is the MMD with respect to the feature vectors in Ω_1^l and Ω_3^l, and likewise for $\text{MMD}^2\left[F^l(\Omega_2^l), S_2^l(\Omega_4^l)\right]$. The regions in layer l corresponding to the regions in the image are extracted by calculating the relative positions and sizes in the feature maps with respect to those in the image. Let define the spatially adaptive style loss for the l-th layer of the k-th style image as follows:

$$
L_{S_{\Omega_{m_k}^l \rightarrow \Omega_n^l}}^k = \text{MMD}^2\left[F^l(\Omega_n^l), S_k^l(\Omega_{m_k}^l)\right] \tag{11}
$$

where k denotes the fact that we apply the style from the k-th style image onto the target image, and Ω_m^l is the sub-region in the k-th style image from which we take the style, and Ω_n^l is the sub-region in the target image to which we apply the style. The total loss for all the style images, all the layers, and all the sub regions is defined as

$$
L_{style} = \sum_k \sum_{m_k} \sum_n \sum_{l \in S_l} L_{S_{\Omega_{m_k}^l \rightarrow \Omega_n^l}}^k \tag{12}
$$

where S_l is the set of layers used in the style transfer. The sum for k is over all the style images and the sum for m_k is over all the sub-regions in the k-th style image from which we want to extract the styles. The sum of n is over all the sub-regions in the target image to which we want to apply the styles from the style image. Figure 1 shows an example of getting the spatially adaptive stylized image with two different styles from two different style images by minimizing loss function in (12). Using the loss in (12), it is possible to apply any number of styles from any style image on any region in the stylized image.

3.2 Image Background Subtraction with Pixel—Wise Style Comparison

Utilizing the spatially adaptive style extraction method explained in the previous section, we show how to eliminate the background region. The background elimination is performed by manually selecting a small region from the background, and then eliminating all the regions that have similar styles to that of the selected region. As explained in the previous section, the pixels have the same styles if their high representation feature vectors are derived from the same distribution. Therefore, to estimate whether a pixel in the image has the same style as the background, we can estimate whether the feature vector in high representation of the pixel comes from the same distribution as the pixels in the selected background or not. We can use many sophisticated methods which decide whether or not a sample comes from a specific distribution, and also we can use a sophisticated model to model the distribution. However, in this paper, we use only the simple distance calculation and model the distribution only with the first order statistics: the mean feature vector. It is obvious that if we use a more sophisticated model the result improves. Figure 2 shows the progress of background elimination and stylizing with the proposed method. We put the same input image into two CNNs (CNN_1 and CNN_2) which have the same

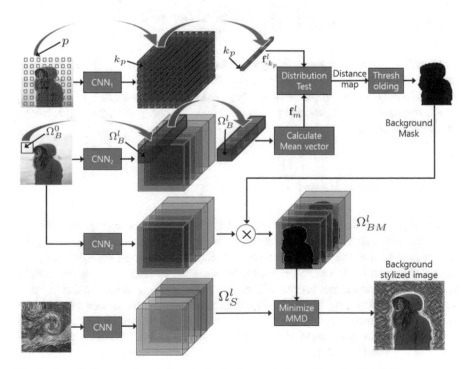

Fig. 2 Overall diagram of spatially adaptive background subtraction and stylization

structure and trained parameters. From CNN1 we extract the feature vectors from the domain Ω_B^l that correspond to the selected small background region (Ω_B^0) in the input image. Then, the mean of the feature vectors in Ω_B^l are calculated to yield the mean vector $\mathbf{f}_m^l$. Then, for every pixel p in the input image, we compute the following distance,

$$D(p) = \sum_{l \in S_l} \left\| \mathbf{f}_{\cdot k_p}^l - \mathbf{f}_m^{l2} \right\| \tag{13}$$

where S_l denotes the set of layers that are used in the background elimination process and $\mathbf{f}_{\cdot k_p}^l$ denotes the feature vector in layer l corresponding to the pixel p, where k_p denotes the spatial index in the feature maps corresponding to the position of the pixel p. The simple distance measure in (13), which measures the distance of the feature vector for a specific pixel to the mean feature vector in the selected background region for all selected layers, serves as a measure for the distribution test, i.e., whether the feature vector of a pixel comes from the same distribution as the feature vectors of the pixels in the selected background region. If the distance for a pixel is relatively small, we label the pixel as a background pixel. Then, we perform a thresholding on the distance map D to obtain the final background mask BM^l:

$$BM^l(p) = \begin{cases} 1 & \text{if } D(p) \leq \text{Threshold} \\ 0 & \text{otherwise} \end{cases} \tag{14}$$

If we further want to style the background in the image we can use this background mask to filter out the domain Ω_{BM}^l which contains only the background feature vectors. Finally, we minimize the MMD between the feature vectors in Ω_{BM}^l and Ω_S^l which results in the stylization of the image. Here, Ω_S^l can be the whole style image or a selected part of it.

4 Experimental Results

Figures 3, 4 and 5 show the experimental results with the proposed method for three different input images with different backgrounds. We took a small patch of the background of the images in (a), constructed the distance maps according to (13), and took a threshold to obtain the background masks. We dilated the masks a little to compensate for the boundary artifacts due to the pooling operations. After that we applied the styles of the images in column (b) onto the background regions by the spatially adaptive style transfer in (9) which resulted in the images shown in column (d). All the user has to do is to select a small patch of the background in the content image and a small patch in the style image, then, the background elimination

Fig. 3 Experimental result: **a** original image **b** style image **c** showing the subtracted background in black color **d** background stylized image

and the stylization is automatically made by the proposed method. It is possible to apply multiple of styles to different regions in the target image by selecting more patches from the style image and the target image. Even though the results are not state-of-the-art, they can be improved by using more sophisticated measures of the distribution distance. Furthermore, the results verify the fact that the style can be accessed spatially adaptively if we use the feature vectors mentioned above, which is different from the normal style transfer which uses the feature maps. We believe that further studies in the direction of utilizing the feature vectors will produce a variety of new results in the future.

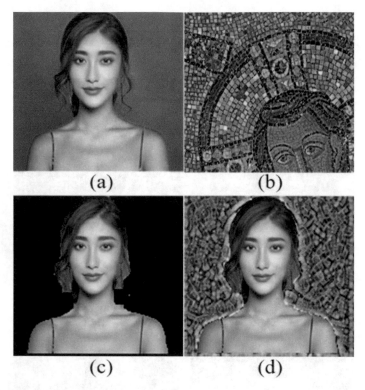

Fig. 4 Experimental result: **a** original image **b** style image **c** showing the subtracted background in black color **d** background stylized image

5 Conclusion

In this paper, we showed that the style transfer can be applied spatially adaptively and also showed how to compare the style point-wisely. Based on this property we performed a spatially adaptive background elimination and stylization. The proposed method can be improved if better measures for the distribution model and test are used, which are the topics of future works.

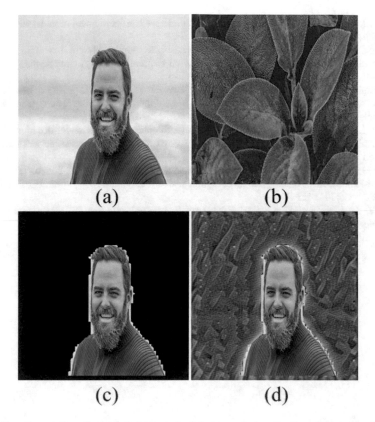

Fig. 5 Experimental result: **a** original image **b** style image **c** showing the subtracted background in black color **d** background stylized image

Acknowledgements This work was supported by the Basic Science Research Program through the National Research Foundation of Korea under Grant NRF-2019R1I1A3A01060150 and the Electronics and Telecommunications Research Institute (ETRI) grant funded by the Korean government. [19AH1200, Development of programmable interactive media creation service platform based on open scenario]

References

1. Bouwmans T (2014) Traditional and recent approaches in background modeling for foreground detection: an overview. Comput Sci Rev 11:31
2. Barnich O, Droogenbroeck MV (2011) ViBe: a universal background subtraction algorithm for video sequences. IEEE Trans Image Process 20:1709
3. St-Charles PL, Bilodeau GA, Bergevin R (2014) Subsense: A universal change detection method with local adaptive sensitivity. IEEE Trans Image Process 24:359
4. Rother C, Kolmogorov V, Blake A (2004) GrabCut: interactive foreground extraction using iterated graph cuts. ACM Trans Graph 23:309

5. Dou J, Qin Q, Tu Z (2019) Background subtraction based on deep convolutional neural networks features. Multimedia Tools Appl 78:14549
6. Braham M, Droogenbroeck MV (2016) Deep background subtraction with scene-specific convolutional neural networks. In: International conference on systems, signals and image processing, IWSSIP, Bratislava, Slovakia, pp 1–4
7. Campilani M, Maddalena L, Alcover GM, Petrosino A, Salgado L (2017) A benchmarking framework for background subtraction in RGBD videos. In: International conference on image analysis and processing, ICIAP, Catania, Italy, pp 219–229
8. Lim K, Jang W, Kim C (2017) Background subtraction using encoder-decoder structured convolutional neural network. In: IEEE International conference on advanced video and signal based surveillance, IEEE, Lecce, Italy, pp 1–6
9. Gatys LA, Ecker AS, Bethege M (2019) Image style transfer using convolutional neural networks. In: IEEE Conference on computer vision and pattern recognition, IEEE, Lasvegas, USA, pp 2414–2423
10. Li Y, Wang N, Liu J, Hou X (2017) Demystifying neural style transfer. In: International joint conference on artificial intelligence,Melbourne, Australia, pp 2230–2236

Multi-source Log Clustering in Distributed Systems

Jackson Raffety, Brooklynn Stone, Jan Svacina, Connor Woodahl, Tomas Cerny, and Pavel Tisnovsky

Abstract Distributed systems are seeing wider use as software becomes more complex and cloud systems increase in popularity. Preforming anomaly detection and other log analysis procedures on distributed systems have not seen much research. To this end, we propose a simple and generic method of clustering log statements from separate log files to perform future log analysis. We identify variable components of log statements and find matches of these variables between the sources. After scoring the variables, we select the one with the highest score to be the clustering basis. We performed a case study of our method on the two open-source projects, to which we found success in the results of our method and created an open-source project log-matcher.

Keywords Distributed system · Log clustering · Multi-source logs

1 Introduction

A distributed system is a complex set of machines that exchanges information and provides computational resources [14]. These systems got significant attention in recent years due to the increasing demand for conducting large computations that a single machine cannot manage. The distributed nature of the systems makes it particularly challenging to analyze them for anomalies, security intrusions, and attack patterns [14]. Traditionally, researchers proposed methods for detecting these errors by analyzing a log file, which are messages that the system outputs during its executions [9]. Nowadays, distributed systems became heterogeneous, with each machine producing differently structured messages [14]. The heterogeneity makes analysis of multiple log files challenging as there are no obvious elements that would con-

J. Raffety · B. Stone · J. Svacina · C. Woodahl · T. Cerny (✉)
Baylor University, Waco, TX 76701, USA
e-mail: tomas_cerny@baylor.edu

P. Tisnovsky
Red Hat, FBC II Purkyňova 97b, 61200 Brno, Czech Republic

H. Kim et al. (eds.), *Information Science and Applications*, Lecture Notes
in Electrical Engineering 739, https://doi.org/10.1007/978-981-33-6385-4_4

nect events from machines across the distributed system [16, 26]. Analyzing and connecting events across the distributed system is critical for finding anomalies or intrusions [6]. We propose a new generic method for finding common attributes in the log messages of any distributed system.

Our method proposes a simple way to cluster log statements from separate files with each other. Each log line contains static elements-which are the same text in all log statements (ie. in "`time=1:34:00`", time is the static element)- and variable elements-which dynamically change throughout the log statements (ie. in "`orgID=148253052`", "`148253052`" is the variable element). We begin by extracting the variable elements from each line of the log files and organize them by their regular expression (an example would be an element "IP Address" with a regular expression of "`1(\d{1,3}\.){4}`" and a variable element of 1`1127.0.0.1.`"). We then cross-compare the variable elements between the two log files, only considering those whose regular expressions are the same. We record variable elements mapping to each static element across the files and find a match between files - if two files share a common variable element mapping to a static element. We then send the plain text static elements, and their match counts to a scoring function based on their uniqueness, frequency, and commonality from our matches. We output two static elements with the highest score as the variable component for which the two separate log files can be clustered.

We tested our proposed method on a case study using log files provided by Red Hat from their two open-source projects called *ccx-data-pipeline* (pipeline) [20] and *ccx-data-aggregator* (aggregator) [19]. With the highest score, we detected static element `clusterID` and used the values to successfully cluster statements across both systems. We manually validated our results with the quality assurance team at Red Hat for its correctness.

This paper is organized as follows. Section 2 provides background and discusses related works. Section 3 describes our method. We share a case study in Sect. 4 and conclude the paper in Sect. 5.

2 Background and Related Work

We aim to analyze and match logs originating from an overall distributed system. Log are output messages generated by the system during a process to describe actions, errors, warnings, and other events during execution. Each log statement can hold information about the event such as an ID, timestamp, file name, etc. [1]. Log statements can then be analyzed to find patterns of logs that describe specific processes, and then those patterns can be used for anomaly detection, performance testing, security issues, and more [9]. However, log statements can originate from many different places within a distributed system (a large-scale complex set of uncoupled and unrelated nodes that interact asynchronously across the network [14]) is where multi-source log analysis becomes important. With the complexity and scale of a distributed system, many different log types are generated, such as system, event, security, and user logs. It can be difficult to trace issues and analyze problems manually.

Our research's motivation is the absence of a general method for multi-file log matching in a distributed system, as stated by Landauer et al. in the survey [9]. Log clustering has traditionally only been applied to the scope of a single file, thus making our research novel and necessary [9]. Multi-file log analysis is relatively new, few papers discuss it, and many of these papers focus solely on multi-file log analysis algorithms to solve problems related to cybersecurity [21]. However, logs can be utilized to solve issues outside of security like finding resource bottlenecks [8], monitoring system performance [7, 24], finding event sequences [5], diagnosing failures [3], and checking system upgrade results [8]. The growing number of distributed systems and log types over the past years necessitates a log clustering algorithm that would work across multiple log files for arbitrary log types [9].

Before any clustering can be done, the log statements must be converted into a structured form and potentially compressed. Some of the most popular approaches include log parsing [27], log compression [11], and natural language processing [2]. One study covers 13 log parsing techniques applied to 16 different data sources [27]. Since the data sources are quite large, a technique to compress the logs into a form that is easier to work with is also necessary [11]. The log parsing technique (Drain) chosen by this study uses heuristics to separate log messages into groups by analyzing constant and variable tokens. This technique is also applied to distributed systems, so the applicability of a heuristics approach on a multi-source log file system is supported. On the other hand, a natural language processing approach could be better suited to parse many types of logs, but since this technique has not been applied to a distributed system, there is no evidence that this would be more effective.

Log clustering survey [9] describes different methods of static and dynamic clustering and the ways in which they are applied in cybersecurity. This survey results give the best algorithms and techniques for outlier detection, anomaly detection, and log parsing. Together with Landauer et al. article on dynamic log file analysis [10], it shows how the combination of static and dynamic clustering proves effective. However, both tend to be exclusively applicable to cybersecurity and anomaly detection by analyzing outliers and sequences of logs. Our generic approach for log clustering applies beyond cybersecurity; it can help to build clusters between files, which is not a focus in either of these works.

Other log clustering approaches use signature extraction, statistics, and pattern recognition to cluster and perform analysis on logs [1, 4, 12, 16, 25]. Some of these methods generate templates for logs and use those templates to find log statement patterns [4, 25], while others group logs together with the same template [1, 16]. However, a better algorithm would take into account the variable pieces of logs like any ID's, event codes, and file names. An approach that takes this into account finds the frequent log constants, maps those constants to a specific number, then generates structured sessions based on execution order [13]. This approach targets security with the clusters based on execution order, still it provides a very useful method for grouping logs using a key value pairs.

Although the research on clustering log statements is extensive, few studies show interest in multi-source log clustering. The few studies on multi-source logs or logs from distributed systems are most often used to solve problems with anomaly detection, malicious requests, password guessing, brute-force cracking, and attack patterns [6, 15, 21, 22]. Since these studies' focus is mainly cybersecurity, the case-study log files primarily consist of security-formatted logs, rather than a variety of log types.

3   Proposed Method

In order to achieve multi-source logs clustering, there must be some connecting metric between groups of log statements from one file to the other. Finding this connector is the most challenging part, as every set of log data differs from on another in a variety of ways [27]. To begin, we need to find a variable (or a combination of variables) that uniquely identifies a set of one or more statements in a log file. It is possible that variables are not present across multiple files; however, we permit occurrences of the same variable in the two distinct files. Once this variable is identified, we can connect log statements from multiple files, identify their order, and perform analysis.

To provide an example, an IP address of a customer will be unique from all other customers, and if you find the same IP address in separate log files within the same distributed environment, you can say with some certainty, there is a correlation between these statements. Our goal then is to search for this IP address in all the log statements across the log files and compare them to each other until all matches are found. We then perform a statistical analysis to determine which match is the best. We cannot simply use the time to cluster, as being from different sources, we cannot know the discrepancies in time zones and communication delays. Including time in our algorithm would also cause unrelated events occurring synchronously would be clustered.

Our method is highlighted on Fig. 1. Firstly, we extract the log signature of the log messages to find the candidate variable (see Fig. 2). Extracting the log signature involves finding the static elements of a given log statement, which are the elements of log output that remain constant every time the line of code that prints the log output is executed [9]. This performs two things for us. First, it gives us a simple clustering of the log messages, which is generally the primary usage of log signature extraction [16]. With that, we have a template of the different messages found in a log file. The second thing finding the log signature does for us is to give us all the variable

Fig. 1 Overview of our proposed method

Fig. 2 Example signature extraction process to extract log statement variables

elements of a log statement. If we know what the static elements are, the reverse yields variable elements [9]. With both the log template and variable components of each log statement, we can proceed to the next step of the method .

We do not wish to compare every variable in one file with every variable of another, so we need to find some way to narrow down the possibilities. The straight forward approach is only to consider variables with the same length, however, this has some issues. For example, an IP address can be represented with different lengths, but should still be considered the same variable type. Our proposed method of doing this is to consider the regular expression of the variables. A regular expression is a string used to describe a text pattern in another string (ie. [a-z]+ describes "hello"), and would not describe "12345" [23]. These regular expressions need not be defined the same as traditional regular expressions are, but capture enough information to distinguish variables from each other. The generation of these regular expressions is beyond the paper's scope, so the length can be used as an alternative if needed. If two variables are to have matches, they must have the same regular expression. With this being the case, we do not have to check between variables that do not have the same regular expression. The data structure we used is a nested map to store the information for a single file's variables. We mapped the log templates (the key of which could simply be the static component of the message) to each of the variables in that template, which are then mapped to each of the possible regular expressions that that variable can take on (see Fig. 3). We then map the regular expressions to the actual values found in the data, along with what line(s) the values were found on.

With this, we can now begin matching. We wish to create matches between variables that share the same regex. Since some variables may have more than one regex, only one regex match is required to become a candidate match. When we find two

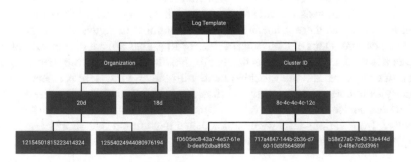

Fig. 3 Tree representation of our mapping structure

variables with the same regex, we create a match object, which also stores some relevant information regarding the two variables. We compare every instance of a variable to every value-instance of the other log file and create an entry for each variable regardless if there was a match or not. For each value, we store the total number of occurrences in each file. For example, if it was in each file once, then we would store 2. We also store the number of matches. A match is the number of occurrences of the value in one file multiplied by the number of occurrences in the other. This is because we can not know what the true match is without more information for each duplicate value in one file, so we record all possibilities. A variable will likely appear in multiple different log templates across a file, so there will likely be duplicate match groups as well. These can be avoided by either performing intra-file matches to identify repeating variables or considering the variable's key name, if available. Provided is an example in pseudo-code describing the process of matching the log templates to each other.

```
1  for each log template a and b in log files A and B:
2      if regex(a) == regex(b):
3          create match
4          for every value in a, b:
5              match.value_count += 1
6              if a == b:
7                  match.match_count += 1
```
Listing 1.1 Primary matching code example

With all of the matches throughout the files found, we can narrow down the candidate list. For each item in the match list, we can find the number of unique values by counting the number of distinct values in the match list. The number of unique matches can be calculated by counting the number of values in the match list with a match count greater than zero. From there, we can find the ratio of unique matches to total matches and unique values to total values. Variable ratios help give us an indication of how well the match is. Those with a low ratio have many repeated values and indicate no real correlation between two log statements with the same variable value. Those with high ratios indicate that they are more likely to be a strong indicator of a connection between statements. We also compare the number of unique matches with the number of unique values. Having a high ratio of unique to total matches is not enough if there is only one match over a thousand values, so we ensure that there is not an insignificant amount of matches. Along with this point, we also want to ensure that there is not an insignificant number of values. A match with 10 total values across tens of thousands of lines of logs does not cut it, so a ratio of total values to total number of lines is taken. All these variable ratios are then weighted to produce a score, and the variable match with the highest score is selected as the primary candidate to cluster between files. Definition 1 defines the score function in a more concise manner. The values of the weights depend on the data set given, however, from various tests we conducted, we have found that simply leaving the weights at 1.0 provided adequate results.

Definition 1 *(Score function)*

$$Score(A, G) = (w_m + (A_{um}/A_{tm})) + (w_v \times (A_{uv}/A_{tv})) + (w_g \times (A_{tv}/G))$$

where A is the match object, and each of it's subvalues, A_{um}, A_{tm}, A_{uv}, and A_{tv} represent the unique matches, total matches, unique values, and total values respectively. G is the global value count. w_m, w_v, w_g represent the weights for each of the ratios.

4 Case Study

We performed a blind case study on RedHat's logs from their distributed production system [17, 18]. The system is based on multiple heterogeneous machines together with Kafka. We selected two machines for our analysis called *ccx-data-pipeline* [20] and *ccx-data-aggregator* [19]. We include a Fig. 4 of the whole data pipeline from which the files originate. The pipeline file contained 133,000 log lines and was stored in plain text, and we identified 4 dynamic variables: Time, ClusterID, Organization, and partition. The aggregator file contained 196,000 lines of logs, was stored in JSON format, and we identified 4 dynamic variables: time, cluster_id organization, and partition. In the pipeline, file were logs of sent messages through the Kafka cloud system. These messages would have two primary attributes, organization and cluster ID. Both organization and cluster ID combined would contribute to a unique identifier for a given message, but both on their own had the possibility to be repeated. The organization had numerous repeating values, whereas cluster ID had very few repeating across different organizations. The other file, aggregator, recorded log messages of the messages being received. The attributes organization and cluster were also recorded here.

We then created tests for small portions of each log file at each phase of the method. After this, we performed a full test on entire log files with line counts in the hundreds of thousands. Our tests' results were that the match between the cluster ID variables in aggregator and pipeline was the primary cluster candidate. This is what the answer should have been, as cluster ID is the most unique variable of each message.

Provided are the specific results of one of our tests. Cluster ID scored the highest, greater than the next highest candidate, organization, by a factor of 10^6. Another variable we previously did not account for found via our algorithm was partition. Partition was a 1–2 digit variable which appeared semi-frequently throughout both files, but was only specific to the machine which created the log file, and had no correlation between the two files. Since the variable was only a few digits long and was repeated regularly over the course of hundreds of thousands of lines, it received a very low score.

Table 1 provides our results. The name of the match candidates is each variable's names from the pipeline and aggregator files, respectively. The score represents how close to a perfect match the candidate is. As can be seen, the cluster match performed

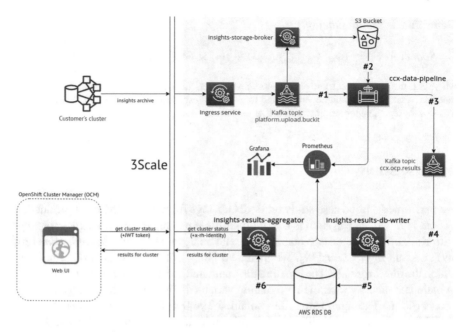

Fig. 4 RedHat data pipeline flow [18]

Table 1 Case study match candidate scores

Match candidate	Score
ClusterID, cluster_id	0.108
Organization, organization	9.834×10^{-6}
partition, partition	1.131×10^{-15}

with a very high score, while the others had very low scores, indicating that the cluster match is the best for clustering.

Threats to Validity: We expected the score for both cluster ID and organization to be as higher than the actual reported value, but after examination of the log files, it became clear why. The aggregator file would print multiple messages for each received message, with each including the organization and cluster ID. This meant that the aggregator would print organization and pipeline at least 5 times for every message sent from the pipeline. These repeated values skewed the ratios for unique values and matches. This ultimately posed no problem to our data set; however, another data set may be afflicted with false-positives from such a situation. Our proposed method always provides an output, so whether there is a real connection between two files cannot be determined, only if there are any matches, which match appears to be the most relevant. However, if the weighted score for the primary output is exceptionally low, then there's a chance there is no true correlation. Of course, to what extent the score must be to be considered exceptionally low depends on the data set. While intended to work on generic data, some knowledge of the data set would be useful to accurately understand the results.

5 Conclusion

This paper proposed a unique method for multi-file log matching in a distributed system. It provides researchers with a general method for further analysis in distributed systems using data mining techniques. This method enables to analyze security leaks, patterns, and intrusions across the distributed system using a centralized approach. We demonstrated the approach on a case study involving two production systems and proved that our method produces correct results on a generic system. Our long term goal is to provide methods for message tracing and automated security checks in the distributed system.

Our test cases set the weights ratio to 1.0; however, this requires deeper knowledge about the data set to set weights. Machine learning would be optimal for training the weights prediction for a particular data set.

The method we propose attempts to cluster log statements together with a single variable. In future work, we will cluster log statements based on multiple variables. This would prove useful for data sets where events have more than a single identifier. In our data set, each pair of organization and cluster ID were unique, whereas when only considering the individual values, organization and cluster ID was not unique. Even after clustering based on cluster ID, there may be a few incorrect cluster points. A combination of both organization and cluster ID would be the most accurate. If cluster ID were not unique, the single variable clustering algorithm would not be sufficient. Our approach could also cluster more files, and performance needs to be assessed.

Acknowledgements This material is based upon work supported by the National Science Foundation under Grant No. 1854049 and a grant from Red Hat Research.

References

1. Aharon M, Barash G, Cohen I, Mordechai E (2009) One graph is worth a thousand logs: uncovering hidden structures in massive system event logs. In: Machine learning and knowledge discovery in databases. Springer, Berlin Heidelberg, pp 227–243
2. Amato Flora C, Giovanni M, Antonino M (2019) Francesco: detect and correlate information system events through verbose logging messages analysis. https://doi.org/10.1007/s00607-018-0662-1
3. Chuah E, Kuo S, Hiew P, Tjhi W, Lee G, Hammond J, Michalewicz MT, Hung T, Browne JC (2010) Diagnosing the root-causes of failures from cluster log files. In: International conference on high performance computing, pp 1–10

4. Debnath B, Solaimani M, Gulzar MAG, Arora N, Lumezanu C, Xu J, Zong B, Zhang H, Jiang G, Khan L (2018) Loglens: a real-time log analysis system. In: 2018 IEEE 38th international conference on distributed computing systems (ICDCS), pp 1052–1062

5. He P, Zhu J, Zheng Z, Lyu MR (2017) Drain: an online log parsing approach with fixed depth tree. In: 2017 IEEE international conference on web services (ICWS), pp 33–40

6. Jia Z, Shen C, Yi X, Chen Y, Yu T, Guan X (2017) Big-data analysis of multi-source logs for anomaly detection on network-based system. In: 2017 13th IEEE conference on automation science and engineering (CASE), pp 1136–1141

7. Juvonen A, Sipola T, Hämäläinen T (2015) Online anomaly detection using dimensionality reduction techniques for http log analysis. Comput Networks 91:46–56. https://doi.org/10.1016/j.comnet.2015.07.019

8. Kubacki M, Sosnowski J (2019) Exploring operational profiles and anomalies in computer performance logs. Microprocessors Microsyst 69:1–15. https://doi.org/10.1016/j.micpro.2019.05.007

9. Landauer M, Skopik F, Wurzenberger M, Rauber A (2020) System log clustering approaches for cyber security applications: a survey. Comput Security 92:101739. https://doi.org/10.1016/j.cose.2020.101739

10. Landauer M, Wurzenberger M, Skopik F, Settanni G, Filzmoser P (2018) Dynamic log file analysis: An unsupervised cluster evolution approach for anomaly detection. Comput Security 79:94–116. https://doi.org/10.1016/j.cose.2018.08.009

11. Liu J, Zhu J, He S, He P, Zheng Z, Lyu MR (2019) Logzip: extracting hidden structures via iterative clustering for log compression. In: Proceedings of the 34th IEEE/ACM international conference on automated software engineering. ASE '19, IEEE Press, pp 863–873. https://doi.org/10.1109/ASE.2019.00085

12. Lu J, Liu C, Li F, Li L, Feng X, Xue J (2020) Cloudraid: detecting distributed concurrency bugs via log-mining and enhancement. IEEE Trans Software Eng p 1

13. Lu S, Wei X, Li Y, Wang L (2018) Detecting anomaly in big data system logs using convolutional neural network. In: 2018 IEEE 16th international conference on dependable, autonomic and secure computing, 16th international conference on pervasive intelligence and computing, 4th international conference on big data intelligence and computing and cyber science and technology Congress, pp 151–158

14. Nagaraj K, Killian C, Neville J (2011) Structured comparative analysis of systems logs to diagnose performance problems

15. Pei K, Gu Z, Saltaformaggio B, Ma S, Wang F, Zhang Z, Si L, Zhang X, Xu D (2016) Hercule: attack story reconstruction via community discovery on correlated log graph. In: Proceedings of the 32nd annual conference on computer security applications. ACSAC '16, ACM, New York, pp 583–595. https://doi.org/10.1145/2991079.2991122

16. Pokharel P, Pokhrel R, Joshi B (2019) A hybrid approach for log signature generation. Appl Comput Inform. https://doi.org/10.1016/j.aci.2019.05.002

17. RedHat: Logscraper. https://cloudhubs.github.io/logscraper/ (2020) online; accessed 30 July 2020

18. RedHat: Logscraper pipeline. https://cloudhubs.github.io/logscraper/Pipeline.html (2020), online; accessed 30 July 2020

19. RedHat: Redhatinsights aggregator. https://github.com/RedHatInsights/insights-results-aggregator (2020), online; accessed 30 July 2020

20. RedHat: Redhatinsights pipeline. https://github.com/RedHatInsights/ccx-data-pipeline-monitor (2020), online; accessed 30 July 2020

21. Shu X, Smiy J, Daphne Yao D, Lin H (2013) Massive distributed and parallel log analysis for organizational security. In: 2013 IEEE Globecom workshops (GC Wkshps), pp 194–199

22. Sun Y, Guo S, Chen Z (2019) Intelligent log analysis system for massive and multi-source security logs: Mmslas design and implementation plan. In: 15th international conference on mobile Ad-Hoc and sensor networks, pp 416–421

23. Tania KD, Tama BA (2017) Implementation of regular expression (regex) on knowledge management system. In: 2017 international conference on data and software engineering (ICoDSE), pp 1–5

24. Wurzenberger M, Skopik F, Landauer M, Greitbauer P, Fiedler R, Kastner W (2017) Incremental clustering for semi-supervised anomaly detection applied on log data. In: Proceedings of the 12th international conference on availability, reliability and security. ARES '17, ACM, New York, NY. https://doi.org/10.1145/3098954.3098973
25. Xu W (2010) System problem detection by mining console logs. PhD thesis, USA
26. Zeng Q, Duan H, Liu C (2020) Top-down process mining from multi-source running logs based on refinement of petri nets. IEEE Access 8:61355–61369
27. Zhu J, He S, Liu J, He P, Xie Q, Zheng Z, Lyu MR (2019) Tools and benchmarks for automated log parsing. In: Proceedings of the 41st international conference on software engineering: software engineering in practice. ICSE-SEIP '19, IEEE Press, pp 121–130. https://doi.org/10.1109/ICSE-SEIP.2019.00021

A Study on Classification and Integration of Research on Both AI and Security in the IoT Era

Ryoichi Sasaki, Tomoko Kaneko, and Nobukazu Yoshioka

Abstract With the arrival of the Internet of Things era, research combining AI and security is increasing. Because there are various types of research on both AI and security, the authors examined the classification of this research. There are four types of research related to both AI and security: (a) attacks using AI, (b) attacks by AI, (c) attacks on AI, and (d) security measures using AI. The research status and issues for each type were clarified. To realize highly secure measures using AI, we proposed combining the measures using AI with the three types of attacks to create a counter-measure that anticipates future attacks.

Keywords Security · AI · Machine learning · Integrated research · IoT

1 Introduction

With the advent of the Internet of Things (IoT) era, research on artificial intelligence (AI), centering on machine learning, has regained attention in various fields. On the other hand, in this era, ensuring security is more important than ever. Therefore, research related to both AI and security is increasing. A Google Scholar search using the keywords "Cyber Security AND Artificial Intelligence" revealed that the searches increased from 1870 in 2007 to 17,000 in 2019, about a ninefold increase.

Because there are various types of research on both AI and security, the authors classified the types. The authors determined that the types of research related to both AI and security are the following:

R. Sasaki (✉)
Research Institute, Tokyo Denki University, 5 Senju-Asahi-Cho, Adachi-Ku, Tokyo 1208551, Japan
e-mail: r.sasaki@mail.dendai.ac.jp

T. Kaneko · N. Yoshioka
National Institute of Informatics, 2-1-2 Hitotsubashi, Chiyoda-Ku, Tokyo 1018430, Japan
e-mail: t-kaneko@nii.ac.jp

N. Yoshioka
e-mail: nobukazu@nii.ac.jp

H. Kim et al. (eds.), *Information Science and Applications*, Lecture Notes in Electrical Engineering 739, https://doi.org/10.1007/978-981-33-6385-4_5

(a)   Attacks using AI
(b)   Attacks by AI
(c)   Attacks on AI
(d)   Security measures using AI

In addition, the status and issues for each research type were clarified. Then, to realize highly secure measures using AI, we proposed combining the measures using AI with the three types of attacks. Unfortunately, security measures have always reacted to attackers. We believe that an integrated approach that anticipates future attacks is indispensable for solving this problem and achieving security measures that are at least partially beyond the abilities of attackers.

There is a great deal of research related to each of the above types, but no approach has attempted to classify and to integrate them all for achieving higher security.

Section 2 presents the classification of the four types of research and the status and issues for each type. Section 3 describes a proposed integrated approach that combines these types.

2 Classification of Research on Both AI and Security

2.1 Attacks Using AI

In the future, incidents of cyber attackers using AI are expected to increase. The most basic type of attack would be to use AI to automate the attacks that humans once did. Consider the following recent example.

Bots have been used in attempts to purchase tickets for events such as concerts. To prevent access by bots, image authentication, which is difficult for computers to read, is used (see Fig. 1). However, it seems that AI has been used to improve the decoding ability of bots to perform image authentication. For example, in the case of the eplus Inc. website in August 2018, more than 90% of the access for ticket purchases was by bots.

Malware with AI functions will surely be used for attacks in the future. We believe that various types of minor malware with AI functions will invade and cooperate with each other to attack the most vulnerable environment. At a minimum, on the research level, it is important to consider future measures to counter such attacks.

There has been much research on attacks using AI (See [1–3]). A method of using AI, instead of simple automation, to evade security measures and to attack a target has been proposed recently. One example is DeepLocker. IBM Research developed it and announced it at Black Hat USA in August 2018 [3]. DeepLocker adopts the "AI-embedded attack" approach, which means the attack is embedded in the AI mechanism itself. Therefore, the attack is difficult to analyze because the process is black boxed, and it can avoid detection by security countermeasure products. Various attack methods that improve upon this have been studied recently.

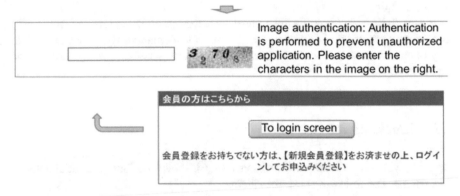

http://atom.eplus.jp/sys/main.jsp?prm=U=82:P6=001:P1=0375:P2=000376:P5≠0
001:P7=1:P0=GGWC01:P3=0144&_ga=2.43604771.767253417.1535088889-
1294490883.1535088889

Fig. 1 Example of image authentication screen

2.2 *Attacks by AI*

One of the biggest problems facing humanity is that an AI entity with the ability to surpass humans may have malicious intent and endanger human lives or society as a whole.

Google researcher Ray Karlwitz has said that an AI rebellion due to transcendental singularities could occur by 2045 [4]. In addition, famous physicist, Stephen Hawking has said, "The development of full artificial intelligence could spell the end of the human race" [4].

On the other hand, Japanese researchers have strong opinions that an AI rebellion will not occur [5].

First, current research is focused on "weak AI (dedicated AI)" rather than "strong AI (general-purpose AI)." It is difficult for weak AIs to have general-purpose capabilities and they typically cannot create higher-level AIs spontaneously. Even if a rebellion occurs, it can be suppressed by giving the AI system a constraint such as the "Three Laws of Robotics" devised by the science fiction author Isaac Asimov.

Former University of Tokyo professor Toru Nishigaki has said, "Fear of an artificial intelligence rebellion in the West is rooted in the fear of becoming a creator on behalf of God under the influence of monotheism" [5].

The present authors think that the possibility of an AI rebellion is extremely low. However, as seen with the Fukushima nuclear power plant accident, which occurred after the 2011 earthquake and tsunami in Japan, the ability to correctly perceive risks to human society can be inadequate. There is also a strong possibility that a rebellion will be irreversible if it occurs. Therefore, it is still important to monitor the evolution of AI carefully.

Research is needed to apply meta-constraints (e.g., the "Three Laws of Robotics") to prevent a rebellion. Additionally, if AI evolves spontaneously, it will be necessary to conduct research that confirms via simulations whether the rebellion can be suppressed.

2.3 Attacks on AI

Figure 2, which was created with reference to [6], shows the four types of attacks on AI systems. The following describes them.

The first type of attack is stopping the machine learning system and stealing file and communication path information. Since this is basically the same as an attack on a conventional system, it is excluded here.

The next type of attack is the noise-added attack that induces misclassification by the trained model. If noise is added to the judgment/prediction data, the accuracy of judgment/prediction may be reduced, possibly inducing misjudgment. A well-known example is an attack on a system that determines the name of an animal. In that attack, minute noise is added to the image of a panda to cause it to be misjudged as a gibbon [7].

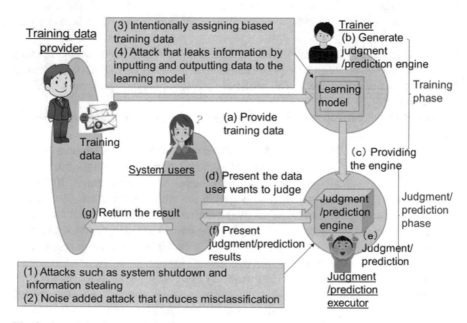

Fig. 2 Overview of machine learning usage patterns and attack methods

The third type of attack is creating inappropriate judgment by intentionally giving biased training data for machine learning. An infamous example is Microsoft's chatbot "Tay." Microsoft intended for Tay to learn how to chat by crowd sourcing; however, malicious users cooperated and repeatedly input discriminatory remarks. Tay began repeating those discriminatory remarks.

The last type of attack is causing information leakages by inputting/outputting data to/from a learning model. In a machine learning system, information related to training data may be leaked from the input/output of the judgment/prediction engine. For example, facial recognition systems use facial images as training data and as information to identify individuals. The facial images of individuals used as training data will likely have a large influence on the input/output of judgment/prediction engine (See [8]).

These are important subjects and fields in which various types of research are being conducted, but it is thought that research in the second and third types should be accelerated.

2.4 Measure Using AI

2.4.1 Overview of Methods

In this section, we discuss the approaches that use AI for security measures. Renowned security expert Bruce Schneier has said that introducing AI security measures can improve situations in which the attacker has an overwhelming advantage [9]. Therefore, research in this field is important. A Google Scholar search on the approaches of using AI for security measures produces many papers (e.g. [10–13]).

Applications of machine-learning-based AI that are currently being used for security measures include the following:

- Malware detection
- Log monitoring/analysis
- Continuous certification
- Traffic monitoring and analysis
- Security diagnosis
- Spam detection
- Information leaking

Various companies have released security measure tools that use AI, and those companies describe the merits on their respective websites. However, there is a lack of verifiable experimental results acquired using these tools, and it is thus unclear what their actual effectiveness is in practice. Accordingly, research on security measures using AI is expected to be highly important, both at present and in the future.

2.4.2   Research of Authors

The present authors are currently conducting the following research regarding the use of AI for security measures. Namely, we are researching the use of machine learning to create an automatic identification system that prevents targeted attacks by C&C servers by identifying and restricting communication with suspected servers before they attempt to communicate with a server (see Fig. 3).

In the targeted attacks that have become common in recent years, various communications are performed between a C&C server and a server infected with malware. Once a C&C server has been identified, it is put on a blacklist which restricts future communications with that server to prevent damage.

However, such a blacklist only has information about C&C servers that have attacked a server, so there is a possibility that an unknown C&C server can gain access. Therefore, we used information from DNS, WHOIS, and VirusTotal (a virus database), along with tools such as neural networks, to create a discriminant model that distinguishes between normal and C&C servers. Suspected C&C servers are put on a blacklist and communication with them is restricted before they try to communicate with a server.

When the actual data is applied to this method, a detection rate of about 99.3% was obtained [12]. It was also revealed that the detection rate could be 98.9% by using only information that is difficult for an attacker to detect [13].

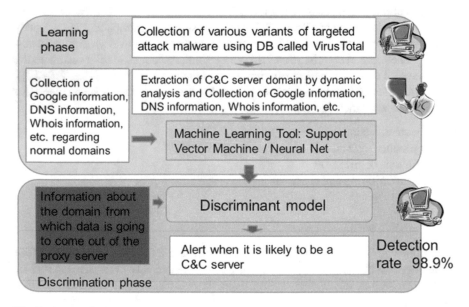

Fig. 3 Automatic C & C server identification system for targeted attacks

2.4.3 Problems in AI Application and Countermeasures

Below are some of the problems that occur when using AI for security measures. Solving these issues is crucial for the development of a successful AI-based security system.

First, it is desirable for an AI system, especially a machine learning system, to obtain a properly classified large-scale dataset, but obtaining such a dataset in this field is generally difficult. In particular, the characteristics of cyberattacks often change over time, so it is difficult to obtain a large enough amount of data in each period of time.

Furthermore, machine learning systems can typically improve accuracy only at the cost of increasing false positives. However, antiviruses should not incorrectly classify and block good applications, and the false positive rate should ideally be less than 1%. Therefore, if the false positive rate of an AI security system is high and cannot be reduced on its own, it will be necessary to combine the AI with a mechanism that can sufficiently reduce this rate.

Additionally, in the security field, it is desirable that a result be "explainable," but advanced machine learning algorithms, such as neural networks and deep learning, are often difficult to grasp in an intuitive manner.

3 Consideration on Integrated Research Approach

We have described the status and remaining issues for the four categories of research concerning attacks and preventative measures involving AI. Although the depth of each category is important, it is difficult to link them to results that increase the probability of advancing beyond the abilities of attackers. Therefore, as shown in Fig. 4, we think that it is necessary to create a model that can evolve to protect against new malware by combining the attacks discussed in previous sections with security measures using AI.

By combining the concept of attacks using AI with that of security measures using AI, it may be possible for the AI to analyze prior attacks and to estimate future attacks. The AI would then be able to respond to new attacks. To achieve this, the present paper proposes the following approaches.

First, a security expert conceives of new attack methods, selects the likely ones, devises efficient countermeasures, and incorporates them into the system. At the same time, it is necessary to consider the establishment of a system for responding properly and a mechanism for preventing unauthorized use.

Next, an AI method that uses a multi-agent is needed. The authors' collaborators, Professor Hiroshi Yamaki and colleagues, have developed a co-evolution model using multi-agents, as shown in Fig. 5 [14]. Here, the model for attacking malware and the defending system are evolving and weeding each other out. By implementing an evolving countermeasure that predicts likely future attacks, the possibility of advancing beyond the abilities of attackers increases.

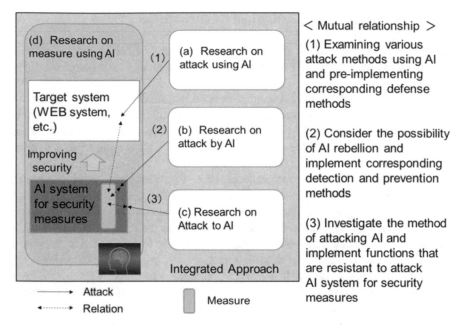

Fig. 4 Integrated approcah to both AI and security

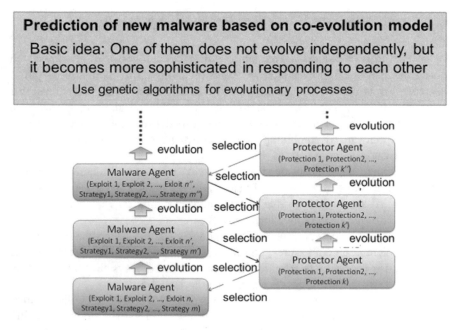

Fig. 5 Prediction method for similar malware with new functions

Every time there is a new attack, more data is collected, and, by using the analogical process, future attacks can be inferred. Measures against anticipated malware can be incorporated in advance, and can pre-emptively defend against attackers.

Also, by considering the combination of attacks by AI and security measures using AI, similar to the multi-agent method described above, an early detection method for an AI rebellion could be developed. When an AI begins to take wrong actions due to accidents or fraud, there is a possibility that a security countermeasure system using AI can detect these anomalies in advance.

Finally, by combining attacks on AI and security measures using AI, it is possible to develop a security system that counters the attacks on AI.

Thus, by combining the three types of attacks with security measures using AI, it is possible to deal with various cyberattacks, including anticipated attacks, and it will be easier to construct an AI system for security measures that is resistant to attacks on and by AI. Although prediction of every future attack is impossible, there is a greater possibility of advancing beyond the ability of attackers.

We think these methods and their effects will be important research topics in the future. A common platform for cyberattack/defense experiments and risk assessment is required to judge whether a given method is a good countermeasure.

4 Conclusions and Future Work

In this paper, we first analyzed research on AI and security and clarified the following types:

(a) Attacks using AI
(b) Attacks by AI
(c) Attacks on AI
(d) Security measures using AI

Next, the research status and issues of each research type were clarified. Then, to realize highly secure measures that use AI and to improve the possibility of advancing beyond the abilities of attackers, we clarified how to combine security measures using AI with the three types of attacks.

In the future, we plan to refine these approaches, to select the integrated methods that are considered highly effective, and to evaluate them by experiments.

Acknowledgements This research was mainly conducted with the support of the Ministry of Education, Culture, Sports, Science and Technology as part of a private university research branding project at Tokyo Denki University titled "Formation of a Secure and Advanced Biomedical Engineering Base in the Age of Global IoT" [15]. We thank the project members for their help with our research.

References

1. Vaithianathasamy S (2019) AI versus AI: fraudsters turn defensive technology into an attack tool. Comput Fraud and Secur 2019(8):6–8
2. Falco G, Viswanathan A, Caldera C, Shrobe H (2018) A Master attack methodology for an ai-based automated attack planner for smart cities. IEEE Access 6:48360–48373
3. "DeepLocker-Concealing Targeted Attacks with AI Locksmithing" https://www.blackhat.com/us-18/briefings/schedule/#deeplocker---concealing-targeted-attacks-with-ai-locksmithing-11549
4. "What are the problems that AI brings?" https://souspeak.jp/ks/ai-problem/
5. Nishigaki T (2016) In: Big data and artificial intelligence, Chuko Shinsho
6. Inoue S, Une M (2019) Utilization of machine learning system and security measures in the financial field" 2019 https://www.boj.or.jp/research/wps_rev/rev_2019/data/rev19j02.pdf
7. Barreno M et al (2010) The security of machine learning. Machine Learn 121–148
8. Tramèr F et al (2016) Stealing machine learning models via prediction APIs. In: Proceedings of USENIX security symposium, advanced computing systems association
9. Schneier B (2018) Artificial intelligence and the attack/defense balance. IEEE Secur Priv 16(02):96–96
10. Xiao L, Wan X, Lu X, Zhang Y, Wu D (2018) IoT Security techniques based on machine learning: how do IoT devices use AI to enhance security? IEEE Signal Process Mag 35(5):41–49
11. Sasaki R et al (2018) Development and evaluation of intelligent network forensic system LIFT using bayesian network for targeted attack detection and prevention. Int J Cyber-Secur Digital Forensics (IJCSDF) 7(4):344–353
12. Kuyama M, Kakizaki Y, Sasaki R (2016) Method for detecting a malicious domain by using only well-known information. Int J Cyber-Secur Digital Forensics (IJCSDF) 5(4):166–174
13. Kuyama M, Kakizaki Y, Sasaki R (2017) Proposal and evaluation of C&C server detection method using information hard to be detected by attacker.Trans Info Process Soc Japan 58(9):1410–1418
14. Ishikawa H, Yamaki H (2016) Co-progressive simulation of attack and defense in cyberspace. IPSJ, CSS2016
15. Private University Research Branding Project (2020) https://www.dendai.ac.jp/about/tdu/activities/branding/TokyoDenkiUniversityResearchBrandingProjectEnglishVer..pdf

Safe Internet: An Edutainment Tool for Teenagers

Han-Foon Neo, Chuan-Chin Teo, and Chee Lim Peng

Abstract Commonly, most people especially teenagers nowadays do not have a high awareness of their Internet privacy. This is due to lack of cybersecurity knowledge to protect their data and sensitive information. Although there are existing security systems to protect user's online information, there are many attacks to bypass in order to steal user information. Teenagers have become an easy target on the Internet. In this paper, an educational tool is designed to provide necessary basic cybersecurity and privacy knowledge targeted at teenagers. The aim is to create a mobile learning platform to educate the teenager towards Safe Internet. To increase user retention in using the learning materials, the app design has to be attractive, colorful and fun. There are three features available, namely Learning Materials, Quizzes and Games. The learning materials comprises three fundamental issues for instance cybersecurity, network and program security, network privacy and data protection. There are two quizzes available currently to evaluate user's understanding after learning. Total scores will be computed for the number of correct answers to indicate user proficiency in Safe Internet. Last but not least, hands-on games are designed to assess user's practical skills in using the Internet. The experimental results of 35 respondents show that more than 50% of them are satisfied and think that the proposed educational tool is effective, learnable and interactive.

Keywords Internet · Privacy · Teenagers · Cybersecurity · Edutainment

H.-F. Neo (✉) · C.-C. Teo · C. L. Peng
Faculty of Information Science and Technology, Multimedia University, 75450 Melaka, Malaysia
e-mail: hfneo@mmu.edu.my

C.-C. Teo
e-mail: ccteo@mmu.edu.my

C. L. Peng
e-mail: jaydenchee97@gmail.com

1   Introduction

Nowadays it is often reported that Internet privacy awareness among teenagers is very low. The reason being teenagers do not care about keeping their personal information safe and private. This is a critical problem worldwide as the number of teenagers using the Internet has amplified significantly. In a survey, 22% of teenagers used their social networking sites at least ten times daily while more than 50% of them accessed to social networking sites at least once a day. 75% of them have their own smart phones for the purposes of social networking, texting and instant messaging [1].

The Internet and smart devices have become an essential part of our lifestyle. Lack of awareness towards cybersecurity and privacy is one of the biggest issues today. Many efforts had been implemented to prevent cybersecurity issues online, but cyber attackers still managed to hack the system and access to the information and data. According to the data from the FBI, since 2016, there were more than 4000 cases of ransomware attacks everyday which was a 300% increase as compared to a thousand cases daily in 2015. The issue has not been solved mainly due to lack of knowledge, low awareness and do not know what to do after attacked by the malicious people. Many people use and rely on their digital devices daily but do not know how to strengthen the security of the devices. Every day, many teenagers are new to the Internet and technology but unfortunately, they are not taught the basic knowledge about cybersecurity and privacy.

Cybersecurity is one of the most critical issues in the Internet era [2]. It is an important issue due to the increasing number of risks and technological devices that we possess. It was reported that there was a correlation between user knowledge and online privacy behavior. Having a greater sense of privacy awareness was positively associated with better controlling of social networking privacy profiles [3]. In a study of 236 teenagers, 60 of them had affected by online scams and fraud [4]. Online scammers preferred to exploit teenagers explicitly to steal their identity because they could open a new credit account easily with their clean slate of credit [5].

The traditional mode of delivering cybersecurity and privacy training is through a trainer at a dedicated schedule in a physical classroom. In contrast, the idea of this paper is to design a mobile app where students can learn at their own pace and anytime they want to. The learning contents are designed to be educational and interesting which consists of three aspects, namely cyber security, network and program security, network privacy and data protection. The materials are designed to be simple to understand, easy to read and suffice for novice Internet users especially teenagers.

Subsequently, quiz is another highlight of the application. Various questions are designed based on the learning materials in order to assess user's cognitive learning process. In addition, challenges of real life scenarios are embedded into the games feature so that users can apply what they have learnt. This function is practical, hands-on and state-of-the-art due to its distinctiveness compared to other existing related applications. In a nutshell, the application is user friendly and flexible to provide clear and sufficient information so that users can have enjoyable and hedonic experience. The concept of edutainment should rouse the user's interest to interact with the application to learn of the Internet threats, and to protect their online privacy as their top priority.

2 Cybersecurity and Privacy

The Internet and technology have become an integral part of today's generation of people, from communicating through instant texts and emails to banking, shopping, studying and travelling, it connects every aspects of life. As the cases of identity theft, internet fraud and cybercrimes are constantly on the rise, education is the pathway in addressing the issue.

Cybersecurity is defined as technologies and processes used to protect computers, hardware, technology devices, software, networks and data from unauthorized access and vulnerabilities. It is supposed to protect the Internet and Web from unauthorized alteration and access. However, it is being abused all over the Internet for instance hackers attacking someone's digital devices by accessing through their personal data. There are various types of attacks and it has extended to incorporate individual's assets and resources [6] for example malware, account hijacking, DDoS, malicious script injection, malvertising, DNS hijacking and defacement.

Privacy is a personal private information that should be kept as secret while information privacy refers to the right to have control over how our own personal information is used and collected. It is our responsibility to know who collects what information and how it is going to be used [7]. Most of the teenagers did not know how to protect their privacy and the importance of privacy when they often give up their personal data freely to join various social networks or to obtain free gifts. In addition, it is perceived a common lifestyle to share photos and locations to the social network. It has now become a major concern in the Internet due to enormous quantity of data available publicly [8]. Little did the user know that private information of every Internet user can be acquired by anybody at any place and at any time through the network. Most importantly, the risk of sharing sensitive information enabled the possibilities to use this information to predict confidential data such as password [9].

Thus, practicing a safe internet approach is vital and parents should have taken parental control for teenagers who use the computers, Internet and smart devices. This is because at times, teenagers do not recognize the risk of certain actions such as pornography and malicious links when they surf the Web [10]. On the other hand, usage of public Wi-Fi will also make them an easy target for cybercriminal.

Before designing the conceptual framework, an analysis had been carried out with other state-of-the-art related mobile cybersecurity and privacy applications. For instance, "CyberSecurity" is a learning platform that includes cybersecurity materials, news, chat and quiz. The materials are comprehensive which covers many cybersecurity aspects but it is purely text-based and not attractive. On the other hand, "Cybrary" is an open sourced cybersecurity learning application with a purely

content feed platform. User can enroll the courses and obtain a certificate after they
have successfully completed the learning process. However, in order to get certified,
user needs to pay and attempt for the test.

"WebMe" is a 5-stages game application which is interactive and easy for user to
learn to raise the awareness of cybersecurity and privacy protection. The application
shows clearly the rules and processes to complete a set of tasks without falling prey
to cyber criminals. Whether or not the user had answered the question correctly, a
reasoning of the topic would be reinforced to the user. Last but not least, "TechSafe
Privacy" is an information learning application about online privacy for adults and
children. User can choose to learn the knowledge of online reputation, privacy, and
identity theft. User also can choose to take quiz to test knowledge. Help feature is
available to facilitate and provide guidance.

3   Methodology

Figure 1 shows the research conceptual framework. There are altogether three
processes applicable to the users. User can choose to learn, do the quiz or play
the games. Learn section consists of three learning modules, namely cyber security, network and program security, network privacy and data protection. On the
other hand, there are two distinctive quizzes available in the Quiz section. After the
user chooses the quiz, the application will display the question. The application will
display the score to the user after the quiz is completed. Once click on the Games
button, the application will show the rules to the user. The user can read and learn the
rules before starting the games. If the user fails on one of the levels of the game, the
application will display the reason of the failure. It helps the user to replay the games
in order to complete the session. After the user has passed all levels, the application
will display a congratulatory page to the user.

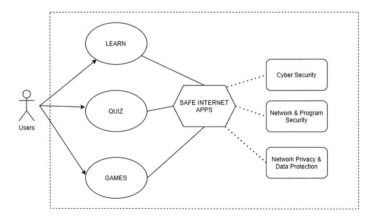

Fig. 1 Safe internet conceptual framework

This research adopts the experimental design which include pre-test and post-test experiments to determine user's perception of the Safe Internet tool using convenience sampling approach. In the pre-test, a survey was conducted to a total of 30 respondents. Data collection was conducted from 30th July 2018 to 19th August 2018. Post-test experiment focused in the usability and survey method, 35 volunteers were required to explore and test the features of the tool followed by filling up the survey form. The survey form consisted of 15 statements to evaluate their perception on the tool. 5-point Likert Scale was used to evaluate their level of agreement with 1 represents "Strongly Disagree" to 5 which refers to "Strongly Agree". The whole testing process for each volunteer took around 30 min to 45 min. The testing has been carried out from 28 January 2019 to 8 February 2019.

4 System Design and Interfaces

Figures 2 and 3 show the main menu scene which consists of three options which are "Learn", "Quiz" and "Game" and three types of learning materials related to cybersecurity and Internet privacy. The materials are designed to be simple, easy-to-learn and attractive targeted at teenagers aged 11 to 18. Specifically, the topics are "Introduction to Cybersecurity", "Network and Program Security" and "Privacy and Data Protection".

Figures 4, 5, 6 and 7 show the samples of learning materials. The topic has been chosen and curated according to the needs and suitability to teenagers. Figure 8 and 9 illustrate the samples of quizzes. There are a total of two quizzes with multiple choice. and user needs to answer all the questions. If the user has picked the wrong answer, a dialog box would appear to provide simple explanations.

The games are designed to be hands-on and practical to improve user's psychomotor skill in practicing the Safe Internet strategies in real life. The games require the user to complete a set of tasks when surfing the internet, but the user need to be mindful of cybercriminal attack. The game consists three levels with different set of tasks. Examples of games are shown in Figs. 10, 11, 12 and 13. In Fig. 10 a stranger initiates a conversation with the user and the user needs to react by configuring the "Setting" at the top right corner of the chatroom to block the stranger from appearing anymore. If the player attempts any conversation with the stranger, he/she fails this task.

Figure 12 is another illustration of what action should the user performed when meeting with strangers. It shows that after blocking the stranger, a hint light bulb button appears. The user will need to click the button and a message box will pop out asking the user the consequence of strangers appearing and attempting conversation in a chatroom. The user should click "Yes" to complete Level 1 of Safe Internet. Hence, it demonstrates that the layer has learnt to protect themselves by rejecting any communication with suspicious stranger while resorting to report to their parents.

Fig. 2 Main menu

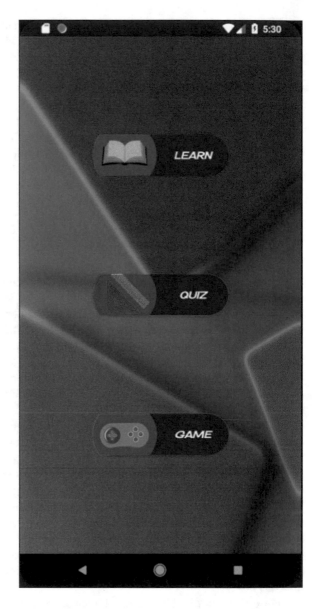

5 Results

In the pre-test, 50% of respondents were females and another 50% were males. In terms of age, the highest percentage of 18 years old was 30% while 20 and 21 years' old both were 20%. Age 19 had 16.7% while age of 17 occupied 10%. 22 year-olds had only 3.3%.

Table 1 summarizes some cybersecurity and privacy issues affected by the respon-

Fig. 3 Learning materials
topics

dents. 53% of the respondents were aware of their identity would be stolen and 46.7% did not. In assessing the method to set passwords, 36.7% would use numbers only, 36.7% used the combination of numbers and letters and the remaining used a more secure method, which consisted of symbols, numbers and letters. Undoubtedly, a high percentage of respondents used the same password for every online account that they own i.e. 70% and only 36.7% of respondents had taken into consideration the setting of the privacy function of their online accounts. 63.3% of respondents

Fig. 4 Main menu

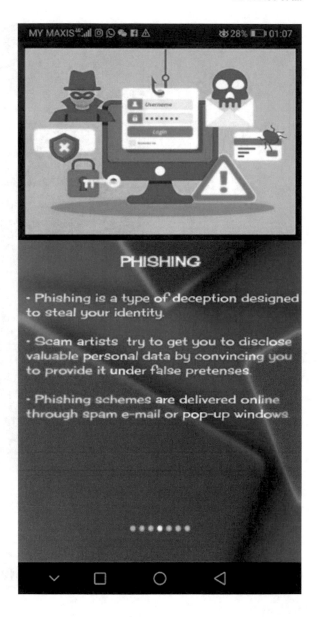

admitted that they did not have basic cybersecurity and privacy knowledge. Majority would just use the public Wi-Fi and did not aware if it was malicious. In terms of accessing unknown links, 70% would click on it without checking its authenticity. Phishing is not widely recognized by the respondents as 60% of them did not have any idea about it and did not recognize malicious emails.

Fig. 5 Learning materials topics

As for the post-test results, out of 35 respondents, one of them aged 10, 2 of them aged 11, 12 and 13, 5 of them aged 14, 8 of them aged 15, 6 of them aged 16, 3 of them aged 17, 5 aged 18 and 1 aged 19. 57.1% are males while 42.9% are females.

71.4% of them agreed that after using the app, they had learnt some basic cyber-security and privacy knowledge and there was also 14.3% of respondents strongly agreed while 2.9% of respondents disagreed. On the other hand, 57.1% agreed that they had learnt the importance of the online privacy with the support of another

Fig. 6 Main menu

28.6% of the respondents strongly agreed. More than half of them strongly agreed that Internet privacy is of utmost important while 74.3% agreed that they were now more aware of identity theft. 65.7% of respondents would set stronger passwords for all their accounts and 68.8% were more aware of their Internet privacy when they are surfing Internet after using the app.

Generally, the respondents felt that the Safe Internet application was interactive, easy to learn to use, and useful. They also suggested that the notes were easy to read, the quizzes were interesting, and the games were fun. In summary, they would recommend their friends to use this app.

Fig. 7 Learning materials topics

6 Conclusion

The app is created as one of the mobile educational tools to learn the basic knowledge of cybersecurity and Internet privacy. It provides various innovative features such as learning materials, quizzes and games to engage teenagers' attention to focus and increase their learning motivation. The app is just in time as Internet is becoming our daily lifestyle and a low awareness of cybersecurity and privacy exposed the users to easily become a target of identity theft or other cybercrimes.

The learning materials consists of three topics, namely Introduction to Cybersecurity, Network and Program Security and Privacy and Data Protection. The materials are designed to be colorful, easy to read and precise while the contents had been curated to include the fundamental theme for instance types of hackers, threats, password setting mechanism, protecting the privacy and others. Besides learning,

Fig. 8 Main menu

users could assess their level of understanding through quizzes. Currently, there are two sets of quizzes which comprise of 10 questions each. The quizzes are easy to play whereby users just need to tap on the correct answer. Finally, users could enjoy themselves by playing the games. The games are designed to be hands-on where users need to complete a set of tasks related to cybersecurity and privacy.

Fig. 9 Learning materials topics

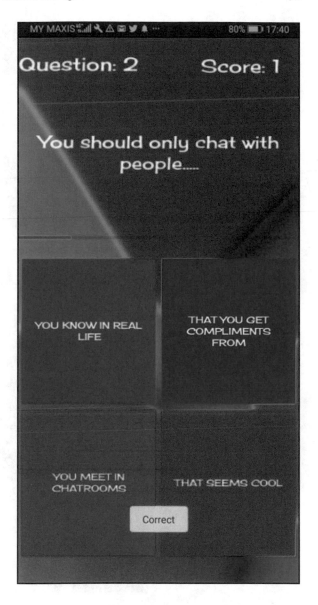

The experimental results show positive feedback from the respondents. They agree that Safe Internet is an effective tool to learn and increase their understanding on cybersecurity and Internet privacy.

Fig. 10 Main menu

Fig. 11 Learning materials
topics

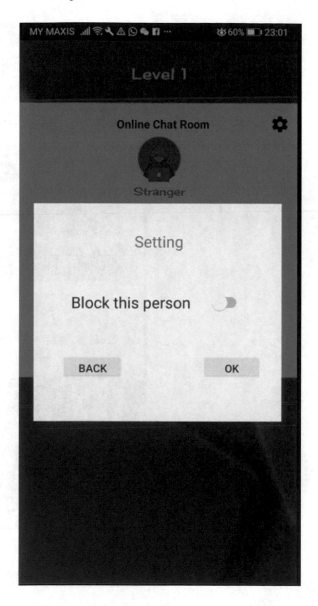

Fig. 12 Main menu

Fig. 13 Learning materials topics

Table 1 Cybersecurity versus privacy issues affected by respondents

Cybersecurity	Privacy
Using the same passwords for all applications	Stolen identity / spoofing
Using public Wi-Fi anytime available	Not adjusting privacy setting for all applications
Do not know what is phishing	Less awareness on the need to have privacy
Unable to recognize malicious emails	

References

1. O'Keeffe GS, Clarke-Pearson K (2011) The impact of social media on children, adolescents, and families. Pediatrics, 127(4):800–804. https://doi.org/10.1542/peds.2011-0054
2. de Bruijn H, Janssen M (2017) Building Cybersecurity awareness: the need for evidence-based framing strategies. Govern Inf Quarterly 34(1):1–7. https://doi.org/10.1016/J.GIQ.2017.02.007
3. Kang R, Dabbish L, Fruchter N, Kiesler S (2015) My data just goes everywhere: user mental models of the internet and implications for privacy and security. In: Symposium on usable privacy and security (SOUPS), pp 39–52
4. Cheung R (2018) More teens roped in for online scams. The Standard (2018). Retrieved from: https://www.thestandard.com.hk/section-news.php?id=197949&sid=4
5. The Star Online, Police: Beware of mule account scam (2018). Retrieved from: https://www.thestar.com.my/news/nation/2018/12/29/police-beware-of-mule-account-scam
6. Newman RC, C R.. Cybercrime identity theft, and fraud. In: Proceedings of the 3rd annual conference on information security curriculum development-InfoSecCD '06. New York, USA, ACM Press, pp 68. https://doi.org/10.1145/1231047.1231064
7. Barnes SB (2006) A privacy paradox: social networking in the United States. First Monday 11(9). https://doi.org/10.5210/fm.v11i9.1394
8. Maple C (2017) Security and privacy in the internet of things. J Cyber Policy 2(2):155–184. https://doi.org/10.1080/23738871.2017.1366536
9. Jabee R, Afshar AM (2016) Issues and challenges of cyber security for social networking sites (Facebook). Int J Comput Appl 144(3):36–40. https://doi.org/10.5120/ijca2016910174
10. Willard NE (2007) Cyber-safe kids, cyber-savvy teens: helping young people learn to use the Internet safely and responsibly. Jossey-Bass . Retrieved from https://books.google.com.my/books?hl=en&lr=&id=fDo7f-ldz_EC&oi=fnd&pg=PR10&dq=safe+internet&ots=nuZmSYKBf3&sig=EGXjQSmJ1rzNZECmKjI1PYzxJ6w&redir_esc=y#v=onepage&q=safe internet&f=false

Predictive Analytics Service for Security of Blockchain and Peer-to-Peer Payment Solutions

Svetlana Boudko, Habtamu Abie⬤, Mirna Boscolo, and Davide Ferrario

Abstract The blockchain and Peer-To-Peer Payment solutions become adopted by financial institutions. While these changes bring significant service benefits they also increase the risks and vulnerabilities of the financial services. In this paper, we investigate, develop, and evaluate machine learning (ML) algorithms for predicting attacks on blockchain nodes and a Peer to Peer payment system. We have evaluated a set of machine learning algorithms that include classification ML algorithms from the scikit-learn library. We demonstrate that the proposed solution is able to predict cyber-physical attacks close to 100% accuracy. We have implemented a service prototype as a proof of concept. The prediction is done based on the collected data of the blockchain and peer-to-peer payment nodes. For the evaluation of the algorithms, a set of highly reputable classification metrics has been selected and applied.

Keywords Predictive analytics · Security · Machine learning · Deep learning · Blockchain · Peer-to-Peer payment

1 Introduction

The financial sector develops into a highly digitised and interconnected industry. The real-time payments infrastructure changes drastically the ways the financial world operates. The recent advances in the ICT technologies like BigData, Internet of

S. Boudko (✉) · H. Abie
Norwegian Computing Center, Oslo 0314, Norway
e-mail: Svetlana@nr.no

H. Abie
e-mail: Abie@nr.no

M. Boscolo
SIA S.p.A, Milan 20147, Italy
e-mail: Mirna.Boscolo@sia.eu

D. Ferrario
Zanasi and Partners, Via Giardini, 45, 41124 Modena, Italy
e-mail: Davide.Ferrario@zanasi-alessandro.eu

© The Author(s), under exclusive license to Springer Nature Singapore Pte Ltd. 2021   71
H. Kim et al. (eds.), *Information Science and Applications*, Lecture Notes
in Electrical Engineering 739, https://doi.org/10.1007/978-981-33-6385-4_7

Things (IoT), Artificial Intelligence (AI), blockchains, mobile Apps, Cloud services and web infrastructures connected with the financial technology innovations have caused an explosion of the financial transactions. While this development brings significant benefits for the customers and financial institutions, it also drastically increases the risks and vulnerabilities of the financial services. This expands the attack surface and complicates attack detection. The financial sector is reported as one of the infrastructures exposed to highest cyber risks [1, 2]. To address this challenge, there is a need for intelligent predictive analytics for immediately predicting complex attacks and allowing mitigation actions before the attacks occur.

Blockchain solutions become widely adopted by financial institutions to secure the payment system. Once considered as a very reliable solution, the attacks on blockchain are now also reported to be a serious threat. Peer-to-Peer (P2P) Payment is another solution of electronic money transfers that has been successfully deployed due to the rapid technological development of mobile devices and applications. While the number of the customers grows rapidly, the attackers follow the development by detecting and exploiting vulnerabilities in blockchain solutions [3].

To predict and prevent these attacks, financial institutions require an advanced cyber-defence infrastructure that facilitates the prediction and identification of complex attack patterns. For this purpose, we have designed and developed a predictive analytics service (PAS) as a component of the FINSEC (https://www.finsec-pro ject.eu/) platform. A set of machine learning algorithms has been selected, evaluated and applied for training ML models. These models are utilized by the PAS to predict attacks.

The major contributions of the paper are: (1) The definition of the architecture and the workflow of the system; (2) The selection of a set of evaluation metrics and the evaluation of a set of machine learning models for attack prediction; (3) The design and development of a proof-of-concept, a prototype of the predictive analytics service and the successful testing and validation of it; (4) The prediction of cyber-physical attacks close to 100% accuracy; and (5) The outline of a roadmap for the future work.

2   Related Work

Machine Learning (ML) and Deep Learning (DL) are used to predict security attacks, threats, and anomalies [4–6]. The analysis in this section, focuses on ML and DL applications to blockchain and P2P payment solutions.

A number of work in the area of ML and blockchain exist. Dey [7] proposes a methodology for using intelligent software agents to monitor the activity of stakeholders in the blockchain networks to detect anomaly such as collusion, using supervised machine learning algorithms and algorithmic game theory. The author argues that the methodology is capable of stopping the majority of attacks from taking place. Dinh and Thai [8] argue that a disruptive integration of AI and Blockchain will reshape how we live, work, and interact, and summarize existing efforts and discuss the promising future of their integration. Scicchitano et al. [9] define an encoder-decoder deep learning model trained exploiting aggregate information extracted by monitoring blockchain activities to detect anomalies in the usage of blockchain systems. Their main contributions are the identification of a relevant set of features computed on blockchain logs describing network status in determined time steps, and the usage of a sequence-to-sequence neural network model to recognize anomalous changes in the blockchain network. Podgorelec et al. [10] propose a machine learning-based method for automated signing of blockchain transactions, including also a personalized identification of anomalous transactions. They envision the method to be included within the software that operates on top of blockchain technology and with so-called blockchain-based user wallets, and the anomaly detection system to operate and store the data (i.e., anomaly detection model) on a user device.

In the area of P2P payment services, Lara-Rubio et al. [11] identify the factors affecting the intention to use P2P mobile payment. They analyse consumers' adoption of P2P mobile payment services, review of previous literature in this field, identify the main factors that determine the adoption of mobile payments, and then perform a logistic regression (LR) analysis and propose a neural network to predict this adoption. From this analysis, they conclude that six variables significantly influence intentions to use P2P payment: ease of use, perceived risk, personal innovativeness, perceived usefulness, subjective norms and perceived enjoyment. The authors argue that with respect to the nonparametric technique, the multilayer perceptrons (MLP) prediction model for the use of P2P payment obtains higher AUC (area under the ROC (Receiver Operating Characteristic) curve values, and thus is more accurate, than the LR model. Soni and Bhushan [12] present a comprehensive survey on blockchain, working of blockchain, security analysis on blockchain, privacy threats for blockchain and potential applications of blockchain, in a decentralized blockchain as well as distributed public ledger technology in P2P networks.

Our work is similar to most of the above work in the application of ML, DL, and AI to predict cyber security attacks, threats and anomalies in blockchain and P2P payment environments. However, our work differs in the prediction of combined cyber-physical security attacks, threats, and anomalies, and our service operates in real-time.

3   System Overview

The PAS is a part of the FINSEC platform. In this paper, we limit the overview of the platform operation to the PAS and its communication with the involved FINSEC components. Our work uses a proper data model that provides an integrated representation of physical and cyber assets and their relationships, to operate on data and to define the scope of the prediction algorithms. This data model named FINSTIX extends the Structured Threat Information eXpression (STIX) 2 standard [13] combining information coming from both physical and logical worlds, thus supporting predictions and defences against both cyber and physical threats.

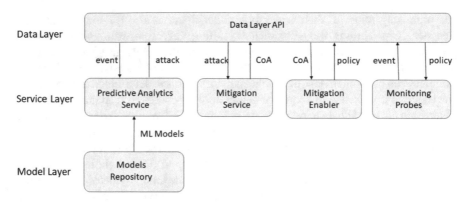

Fig. 1 A view of the PAS showing the data flow between the involved components

3.1 Predictive Analytics Service

The operation of the PAS is depicted in Fig. 1. We outline the system architecture and define how the workflow is implemented. The main purpose of the PAS is to predict cyber-physical attacks and report these results to the Data Layer. The service considers various types of attacks that include both cyber and physical attacks. The workflow is implemented as follows. The probes of the P2P Payment and blockchain nodes monitor different parameters of the system, and collect the observations.

The results are translated into FINSTIX objects representing events and inserted in the Data Layer. The Data Layer implements an OpenAPI end point that streams the received data in a real-time manner. The PAS operates in real-time. It listens to the data stream. When a new event is received it is processed and analysed by the prediction methods. For the analysis, the ML and DL models have been pre-trained and stored in the Models Repository. If an attack is predicted the PAS generates an attack object and inserts it in the Data Layer. The Mitigation Service listens to the Data Layer. When a new attack is received a new course-of-action (CoA) object is generated and inserted in the Data Layer. It triggers the generation of a policy object by the Mitigation Enabler. This object is inserted in the Data Layer and will be received by the probe. The probe applies the measures in compliance with the received policy.

3.2 Predictive Analytics Service in the Big Picture

The FINSEC Reference Architecture (RA) as shown in Fig. 2 is structured as a layered view with a clear separation of functions between the different levels. The Field Tier represents the lowest level where the FINSEC platform gets raw data and events through the cyber-physical probes. The Edge Tier represents the level of the RA gathering the data sent in real time from the probes in the form of raw

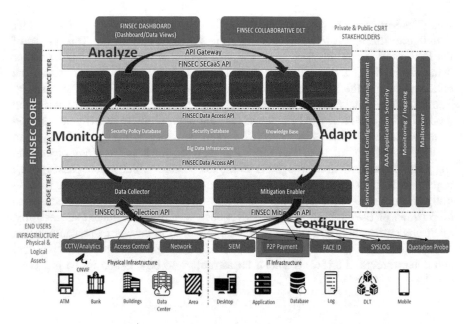

Fig. 2 Implementation of the predictive analytics service in the FINSEC RA

events to be analysed and detected as attack patterns by the upper layers of the FINSEC system. It is composed of Data Collector and Mitigation Enabler. The Data Tier represents the FINSEC Big Data Infrastructure with databases dedicated to the storage of the events coming from the Field Tier, to the knowledge base rules for the predictive and anomaly detection services, and to the security policies. The Service Tier represents the applications that the FINSEC project has developed and integrated in the FINSEC platform. It is composed of Anomaly Detection, Predictive Analytics, Risk Assessment Engine, Audit and Certification, Collaborative Risk Management, Mitigation and TLS Assistant. The last Upper Layer is the level of data sharing and presentation and is composed of FINSEC Dashboard and Collaborative DLT (distributed ledger technology).

The FINSEC RA thus provides capability to foster new, intelligent, collaborative and more dynamic approaches to detect, prevent, and mitigate integrated (cyber & physical) security incidents, intelligent monitoring, and data collection of security-related information; predictive analytics (the topic of this paper) over the collected data; triggering of preventive and mitigation measures in advance of the occurrence of the attack; and allowing all stakeholders to collaborate in vulnerability assessment, risk analysis, threat identification, threat mitigation, and compliance. All data exchange between the FINSEC components is done using the data objects of the FINSTIX data model. This model integrates information relevant to the security and financial sectors. These objects are sequences of key-values passed as JSON strings.

Figure 2 depicts the implementation of the Predictive Analytics Service in the overall FINSEC RA, closing the adaptive loop of data collection (Monitor), Predictive

Analytics Service (Analyse), and Mitigation Service (Adapt), and Mitigation Enabler (Configure) through a feedback control loop. It works as follows. The probes of the P2P payment and blockchain nodes push collected data to the Data Collector using the FINSEC Data Collection API. The Data Collector (Monitor) stores the collected data in the Big Data Infrastructure using the FINSEC Data Access API. The Predictive Analytics Service gets data from the Big Data Infrastructure using the FINSEC Data Access API and analyses the data, predicts security attacks and generates x-attack objects in FINSTIX format, and stores this x-attack object to the Big Data Infrastructure. The Mitigation Service reads the x-attack object and suggests mitigation action through the Mitigation Enabler which in turn passes policy objects with the course of action to the P2P payment probe to re-configure itself to mitigate the predicted attack before it happens. This closes the end-to-end (e2e) loop (flow) of the FINSEC RA and it has been tested and validated successfully. Such preemptive inferring of future attacks [14] and increased (semi-)automation can enable the prevention, detection, response and mitigation of combined physical and cyber threats to critical financial infrastructure to be adapted to the predicted attacks.

4   Evaluated ML Algorithms and Results

To ensuring the exchange of monitored information, a dedicated and protected environment has been set up. It was built as a private and permissioned blockchain, composed of a subset of blockchain nodes and a P2P payment system. To identify abnormal behaviors on the P2P payment solution prior to attacks, end users' habits were analyzed. To assess the blockchain network behavior, the blockchain infrastructure was set up to monitor a set of parameters. In both cases, the analyzed parameters are characterized by typical values for normal situations. Consequently, deviations from these values can be interpreted as indicator of possible security threats/attacks. Datasets were produced separately for each of the trends, and contained roughly 20,000 points for each trend. Further details of the datasets are restricted by the EU regulations. The data were preprocessed in the temporal domain using a discrete wavelet transform.

4.1   Evaluated ML Algorithms

In this work, the prediction of attacks is a typical classification problem. To address the problem, we selected and evaluated a set of ML classification algorithms from the scikit-learn, xgboost.python libraries, and the Keras API as follows.

- A number of ensemble methods [15],
- Support Vector Machines [16],

- Decision Tree Classifier [17],
- K-Nearest Neighbors algorithm (KNN) [18],
- Stochastic Gradient Descent algorithm (SGD) [19], and
- Multilayer Perceptron classifier [20].

Ensemble methods represent a family of ML algorithms that build ML models by aggregating the prediction models of several base estimators, thus improving robustness of a single estimator. Two distinctive families of ensemble methods exist, averaging methods and boosting methods. Averaging methods randomly select several estimators and average their prediction results. This technique reduces the variance. The aim of the boosting technique is to reduce bias. Several weak classifiers are combined to produce a strong one. The following algorithms of this family have been used for the evaluation: Random Forest Classifier, AdaBoost Classifier, Gradient Boosting Classifier, Extreme Gradient Boosting.

SVM is an ML method that builds hyperplanes in the feature space to classify the points. The objective is to define the planes that maximise the distances between the classes. The main idea of the K-Nearest Neighbors algorithm is to find a predefined number of labeled points closest in distance to the new point, and use them to predict the label of the new point. Stochastic Gradient Descent algorithm is an iterative algorithm that performs optimization using gradient descent. It can also be used to optimize neural networks, often for back propagation.

A Multilayer Perceptron classifier was trained and tested with the Keras API [20]. The neural network consisted of three layers (one hidden layers); we used Rectified Linear Unit as the activation functions for the first two layers and Sigmoid function for the final output layer; the loss function was binary cross-entropy and the optimizer algorithm "adam". The models were trained for 10 epochs and tested using a fivefold cross validation. To train an Extreme Gradient Boosting model, the xgboost.python library was utilized. For the rest of the algorithms, we used the scikit-learn library. For the Random Forest, AdaBoost, Gradient Boosting, and Extreme Gradient Boosting Classifiers, we used the following set of estimators: {100, 200, 300}.

4.2 Evaluation Metrics

To select the models that are optimal or suboptimal for applying for specific scenarios, datasets and/or selection of features, the evaluation is conducted using reputable metrics. The optimal choice of these metrics is important for the evaluation process to be able to interpret correctly the model results. We have used the evaluation metrics that include: accuracy, precision, and recall, which are well established classification metrics and are available in the scikit-learn. Accuracy (a) defines the ratio of correctly predicted points to the total number of points. Precision (p) is the ratio of correctly predicted positive values to the total number of values predicted as positive while recall (r) is the ratio of correctly predicted positive values to all positive values.

Combined together these metrics can give us a balanced view about how the different algorithms behaved for specific scenarios, datasets, and selection of features.

To better interpret the evaluation results, a composite metric m has been defined as a weighted average of the described above elementary metrics. The composite metric is depicted in Eq. 1. In this work, all elementary metrics are considered equally significant, and all weights are set to $1/3$.

$$w_a a + w_p p + w_r r = m \qquad (1)$$

4.3 Results

The results of the evaluation of ML algorithms for the elementary metrics and the weighted average are depicted in Figs. 3 and 4, respectively. In general, all evaluated ensemble methods showed close to 100% results for all three metrics. More specifically, the weighted average of the Random Forest and the AdaBoost algorithms with 200 and 300 estimators was 100%. For 100 estimators, the Random Forest algorithm performed slightly better than the AdaBoost algorithm. The SVM, KNN, and SGD algorithms were considerably less efficient.

Based on the evaluation results of the algorithms, the models for the PAS were built using the Random Forest and AdaBoost algorithms with 200 and 300 estimators. These model are deployed in the PAS and are used for the analysis of continuously arriving data from the FINSEC probes and prediction of the attacks.

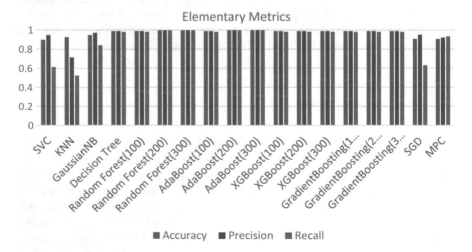

Fig. 3 The evaluation results of the algorithms using the selected elementary metrics. The x-axis shows the ML algorithms under evaluation. The y-axis shows the results of the classification evaluation metrics

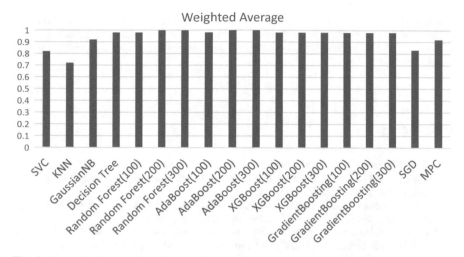

Fig. 4 The evaluation results of the algorithms using the composite metric. The x-axis shows the ML algorithms under evaluation. The y-axis shows the results of the weighted average metric

5 Conclusion and Future Work

In this paper, we developed the predictive analytics service (PAS) and its workflow. We have selected a set of the ML algorithms and the metrics for evaluating these algorithms. The results of this evaluation show that the ensemble methods achieve favourable results for the investigated datasets. These were close to 100% for all three metrics. However, we realize that more elaborate datasets are needed. Thus, further data collection, feature engineering and method investigations are required and planned for the next stage. Publicly available datasets will be considered. More importantly, considering a conglomeration of metrics rather than one metric alone gave us an experience of how a model selection process can be designed in the next step. The algorithms have been tested and successfully validated using the metrics. The ML models have been trained using these algorithms and deployed in the PAS that can be used by the financial institutions.

In our future work, a deeper insight in how the performance metrics are used is needed. To define how the models are selected for specific scenarios, datasets and/or selection of features, the study will be performed using reputable metrics [21–23]. Since the prediction is done as a part of the service, which is supposed to be light, the efficiency, scalability, stability and sensitivity of the ML and DL models are important requirements and will be carefully investigated. Further, the automation of the model selection and training process will be done for retraining the models online with newly available data.

Since the data labeling process is expensive and not always possible, we plan to apply unsupervised and semi-supervised learning. Using small sets of labeled data with unlabeled data collected from the probes for retraining the models will improve prediction results.

Acknowledgements This work has been carried out in the scope of the FINSEC project (contract number 786727), co-funded by the European Commission in the scope of its H2020 program. The authors gratefully acknowledge the contributions of the funding agency and of all the project partners.

References

1. AKAMAI Homepage (2020). https://www.akamai.com/uk/en/multimedia/documents/state-of-the-internet/soti-security-financial-services-hostile-takeover-attempts-report-2020.pdf. Last Accessed 28 Sep 2020
2. ESRB Homepage (2020). https://www.esrb.europa.eu/pub/pdf/reports/esrb.report200219_sys temiccyberrisk~101a09685e.en.pdf. Last Accessed 28 Sep 2020
3. Norges Bank Homepage (2020). https://static.norges-bank.no/contentassets/8c65f4c19bcb49b e9e49985629b41968/financial_infrastructure2019.pdf?v=05/23/2019160249&ft=.pdf. Last Accessed 28 Sep 2020
4. Xin Y, Kong L, Liu Z, Chen Y, Li Y, Zhu H, Gao M, Haixia (2018) Machine learning and deep learning methods for cybersecurity. IEEE Access 6:35365–35381
5. Berman DS, Buczak AL, Chavis JS, Corbett CL (2019) In: A survey of deep learning methods for cyber security, information 10:122
6. Wickramasinghe C, Marino D, Amarasinghe K, Manic M (2018) Generalization of deep learning for cyber-physical system security: a survey. In: Proceedings 44th annual conference of the IEEE industrial electronics society, IECON 2018, Washington DC, USA, Oct. 21–23. PDF https://doi.org/10.1109/iecon.2018.8591773
7. Dey S (2018) A proof of work: securing majority-attack in blockchain using machine learning and algorithmic game theory. I.J. Wireless Microwave Technol 5:1–9
8. Dinh TN, Thai MT (2019) AI and blockchain: a disruptive integration. 51(9):48–53, IEEE
9. Scicchitano F, Liguori A, Guarascio M, Ritacco E, Manco G (2020) A deep learning approach for detecting security attacks on blockchain. CEUR-WS.ORG/Vol-2597/paper-19.pdf
10. Podgorelec B, Turkanovic M, Karakatic S (2020) A machine learning-based method for automated blockchain transaction signing including personalized anomaly detection. Sensors 20:147
11. Lara-Rubio J, Villarejo-Ramos AF, Liébana-Cabanillas F (2020) Explanatory and predictive model of the adoption of P2P payment systems. Behaviour Information Technology
12. Soni S, Bhushan B (2019) A comprehensive survey on blockchain: working, security analysis, privacy threats and potential applications. In: 2019 2nd International conference on intelligent computing, instrumentation and control technologies (ICICICT), Kannur, Kerala, India, pp 922–926
13. GitHub Homepage, Structured Threat Information eXpression (STIX) 2. https://oasis-open.git hub.io/cti-documentation/
14. Husák M, Komárková J, Bou-Harb E, Čeleda P (2019) Survey of attack projection, prediction, and forecasting in cyber security. IEEE Commun Surveys Tutorials 21(1):640–660. First quarter. https://doi.org/10.1109/comst.2018.2871866
15. Rokach L (2005) Ensemble methods for classifiers. In: Maimon O, Rokach L (eds) Data mining and knowledge discovery handbook. Springer, Boston, MA
16. Hearst MA (1998) Support vector machines. IEEE Intell Syst 13(4):18–28
17. Breiman L, Friedman JH, Olshen RA, Stone CJ (1984) Classification and regression trees. Wadsworth and Brooks/Cole Advanced Books and Software, Monterey, CA. ISBN 978-0-412-04841-8
18. Cover TM, Hart PE (1987) Nearest neighbor pattern classification. IEEE Trans Info Theory 13(1):21–27

19. Robbins H, Monro S (1951) A stochastic approximation method the annals of mathematical statistics. 22(3):400–407
20. Keras Homepage. https://keras.io/getting_started/
21. Mishra A (2018) Metrics to evaluate your machine learning algorithm
22. Brownlee J (2020) Tour of evaluation metrics for imbalanced classification
23. Hossin M, Sulaiman MN (2015) A review on evaluation metrics for data classification evaluations. Int J Data Mining Knowledge Manage Process

Deterring SLCAs by Establishment of ACSC Based on Internet Peace Principles

Young Yung Shin

Abstract Currently, State-Led CyberAttacks (SLCAs), directly and indirectly supported by various states, have increased continuously and the damage has been significant. Events such as the WannaCry incident and the hacking of India's nuclear facilities can have devastating and fatal consequences. However, deterrence technology against cyberattacks remain insufficient, and cooperation among countries is also inadequate. Prevention of SLCAs requires not only technical methods but legal and policy solutions. Considering that Asian countries particularly are in the early stages of ICT applications in the manufacturing industry as part of the Fourth Industrial Revolution era, Asian industries overall and the digital economy are exposed to the possibility of sophisticated cyberattacks and exploitation of weak cybersecurity framework. The study proposes that Asian countries establish what is termed the Asia Cybersecurity Convention (ACSC) to deter SLCAs, which consists of government representatives and private technical, legal, institutional experts and organizations. The study also suggests the Internet Peace Principles (IPPs) as a cybersecurity norm for the ACSC, by which cyber capabilities should not be used to attack other countries or as a means by which to launch malicious detour attacks.

Keywords SLCAs · WannaCry · ACSC · Fourth industrial revolution · IPPs

1 Introduction

As the use of the Internet increases, information about almost all areas of life, such as politics, the economy, society, culture, and the arts, are shared in cyberspace, and connections and interdependencies are deepening. However, the increased number of users as well as various levels of users, such as non-state actors and the entities known as Anonymous and LulzSec, as well as vulnerabilities related to cyberspace caused by transnationality issues, anonymity, openness, and easy access and technical

Y. Y. Shin (✉)
171 Dongdaejeon-ro (155-3 Jayang-dong), Dong-gu, Daejeon 34606, Republic of Korea
e-mail: inewhero@wsu.ac.kr

© The Author(s), under exclusive license to Springer Nature Singapore Pte Ltd. 2021   83
H. Kim et al. (eds.), *Information Science and Applications*, Lecture Notes
in Electrical Engineering 739, https://doi.org/10.1007/978-981-33-6385-4_8

susceptibility of connections during communication with other devices endanger the safe use of the Internet and computers.

The most serious threats are SLCAs using advanced offensive capabilities to attack a target according to a certain purpose. Targets include public institutions, government agencies, the critical infrastructure of societies, and private companies, and attacks cause social instability and threaten national security. The purpose of this study is to prevent SLCAs in advance and minimize the damage even after attacks by establishing an international cybersecurity organization.

This research raises the following study questions. What are the characteristics of cyberattacks and what are typical examples? What stops each country from punishing those who commit the crime of a SLCA? What types of security measures should each country take?

To answer these questions, the study investigates and compares seven typical SLCAs and three PLCAs according to their characteristics, means and methods, and results, among other factors. Next, the reasons for the lack of punishment of cyberspace criminals are found to be anonymity, transnationality, and extraterritoriality, among others. Accordingly, the study tries to find new ways to regulate illegal activities in the cyberspace preserving the advantages of the Internet.

As we enter the initial stage of the Fourth Industrial Revolution, the Asian region is exposed to many external attacks due to the vulnerabilities that arise during the process of applying ICT to various areas of the manufacturing industry and society. Other causes are the wide gap in ICT development between countries, the inherent weaknesses of modern technology itself. Thus, the study suggests the Asia Cyber Security Convention (ACSC), a regional organization for cybersecurity cooperation that strengthens state responsibility and enhances cooperation among states in cyberspace to punish the attackers by legitimizing a regional engagement. Then ACSC expands or joins international cybersecurity organizations.

The ACSC is planned based on the existing ASEAN Plus Three Cooperation (Korea, China, and Japan), and is open to the United States, India, Russia, European countries, and other countries who want to participate. Also, as an international cybersecurity consultative body, it can work closely with the UN GGE (Group of Governmental Experts) and OEWG (Open-Ended Working Group), which are currently active internationally. In addition, this study suggests Internet Peace Principles [1, 2] to support the ACSC, with the claim of not using the Internet as a means of attacking other countries or not using it to engage in malicious detour attacks.

The scope of the research includes the UN Charter and Draft Articles on the Responsibility of States for Internationally Wrongful Acts (Responsibility of States) that govern relationships between Member countries; the Council of Europe Convention on Cybercrime (CECC, Budapest Convention) [3], which addresses computer crimes; and the Tallinn Manual 2.0. International Law Applicable to Cyber Operations (Tallinn Manual), which provides guidelines for each country when using the Internet. Also included are the contents of agreements of ongoing UN GGE and UN OEWG discussions. Regarding the research method, the study includes a literature review of established works, including various published white papers, research reports, results from the investigative authority of each country involved such as Ministry of Science and ICT, Korean Public Prosecutors' Office, Korea Police Agency, U.S. Federal Bureau of Investigation, and testimony from North Korean defectors [4]. Also, the study utilizes a comparative analysis of several SLCAs and PLCAs as well as a correlation analysis to examine the relevance of causes, means and methods, and results of cyberattacks.

2 Characteristics and Cases of SLCAs

2.1 The Characteristics of SLCAs

A cyberattack is defined as any attack that illegally invades, disturbs, paralyzes, destroys, or steals or damages national information and communication networks by electronic means such as hacking, computer viruses, logic bombs, mail bombs, and service interruptions.

SLCAs have several characteristics that differ from those of ordinary cyberattacks, so called Private-Led CyberAttacks (PLCA). First, SLCAs have various purposes according to their motivation by supporting individuals, groups, or themselves, such as threatening national security, neutralizing a target government's functions or private companies, extorting money from banking systems such as the Society for Worldwide Inter-bank Financial Telecommunications (SWIFT), intervening in other country's election through the manipulation of public opinion, and targeting organization involved in coronavirus vaccine development and research. Like this, SLCAs such as cyber espionage and hacktivism increase by 42% in 2019 [5].

For example, in the beginning, North Korea hackers concentrated on attacks on key facilities, such as government offices and broadcasting stations, but recently, they have focused on foreign currency earnings by attacking crypto-currency exchanges and banks to overcome economic difficulties due to sanctions from the United Nations.

Second, SLCAs use their own Internet Protocols (IPs), specific attack methods, and malware. For example, during Russia's cyberattacks against Georgia and Estonia, attackers used Russian IPs. Also, North Korea attackers used a North Korean IP by which the source IP could easily be traced, or they used a dedicated IP leased from China by the (North) Korea Post and Telecommunications Corporation (KPTC) in China. In addition, attackers reused malware. This behavior differs from an ordinary hackers' practice of not taking such risks to conceal their identity.

The behavior of attacking countries allows us to guess that they regard IP exposure and the reuse of attack methods as irrelevant with regard to origin identification; they also ignore criticism from the international community, or make the mistake that the attacked countries will not be able to counterattack because they are misinformed.

Third, cyber warriors are dispatched to foreign countries to disguise their identity. For example, North Korea manages foreign residents who live seemingly normal lives in China or Southeast Asia, and when ordered, they focus on attacking certain targets until they succeed, typically by spear-phishing and watering hole methods.

Fourth, because the attacks are performed in the virtual space, they do not cause death or injuries directly to the target of the attack, making it possible to circumvent the scope of the ban on "hostilities" and "use of force" agreed upon by UN Member countries.

2.2   Analysis of Seven Typical SLCA and Three PLCA Cases

The study selects seven typical SLCA cases considering different motivations, such as disrupting nuclear facilities, means and methods such as DDoS attacks and the use of USB, targets such as private companies, and results such as worldwide infections by malware from many SLCAs. These are listed below.

(S1)   The WannaCry ransomware attack affected nearly 230,000 computers globally in 2017 [6]

(S2)   Hacking against the Korea Hydro & Nuclear Power Co., Ltd. (KHNP) in South Korea in 2014 [7]

(S3)   Attack on Sony Pictures Entertainment (Sony) in 2014, which destroyed and leaked information as the first SLCA that targeted a private corporation [8]

(S4)   The Stuxnet attack on Iran's nuclear facility in 2010 [9]

(S5)   The July 4 and 7 consecutive cyberattacks against the USA and South Korea in 2009 [10]

(S6)   DDoS attack against Georgia along with a conventional military operation in 2008 [11]

(S7)   DDoS attack against Estonia in 2007, representing the first SLCA [12].

To compare, the study chooses three typical PLCAs that took place recently as follows.

(P1)   Hacking against Hyundai Capital in 2011, which leaked information 133 of 420,000 customers exploiting loose management of Intrusion Prevention System 134 in Korea [13]

(P2)   Ransomware attack against NAYANA Web hosting company in 135 2017, which infected client sites with malware in Korea [14]

(P3)   Hacking against G company in 2000, which stole business secrets in Korea [15].

Targets of SLCAs It is the target of the attack that suitably represents the characteristics of a SLCA. South Korea is the most frequently attacked by SLCAs and the USA, Iran, Georgia, and Estonia follow. The primary attack targets in these countries are critical infrastructures, such as energy systems (S2, S4), government organizations (S5, S6, and S7), financial institutions, and press (S5). Attackers also target defense agencies and weapon systems(S4), the general public (S1, S5, P1, P2, P3), and private corporations (S3, P1, P2, P3). In addition, SLCAs steal intellectual property, confidential research results, individual information, draw money from SWIFT networks and ATMs unlawfully, and steal money by means of ransomware (S1, P2).

Origin Countries There are several ways to identify the attackers, such as by the origin of the IP address and corresponding country, and the attack means and methods. However, in the current Internet environment, attackers can easily spoof their origins by compromising other computers and can use a detour method through a virtual private network (VPN) via third countries. Therefore, circumstantial evidence such as the geopolitical situation, the nature of the target, and the attack timing, among other factors, should be considered when attempting to determine the origins of the attackers [16].

Attack Means and Methods Attackers decide upon the attack means and methods depending on the purposes and vulnerability of the targets, implying that the means and methods are the second important aspects by which to recognize the attackers. Among many means and methods of cyberattacks, hacking, the spreading of malicious code, distributed denial of service (DDoS) attacks and a combination of them are the most frequently used in cyberattacks. Advanced persistent threat (APT) methods in S2, S3, and S4, social engineering, phishing emails in S2, P2, and P3, DDoS attacks in S5, S6, and S7, and malware injections in S1, S2, S3, S4, S5 and P2 are the most popularly used SLCA strategies.

Attack Routes Attack routes are another indication of the source of the attack. SLCAs passed through many ISPs and VPNs of numerous countries from origin to destination countries including the United States and China. Thus, as described later, implementing the four establishment principles and ten implementation principles for ACSC is very important to prevent direct and detoured attacks. SLCAs detour through many more countries to conceal the attackers' identity than do ordinary PLCAs.

Consequences This depends on the purposes, means, and methods of the attack. Above mentioned S1 crippled computers in at least 150 countries and cost $40 billion economic damage, S4 destroyed about 1,000 centrifuges at Natanz, which violated an international norm by infiltration into a prohibited area such as nuclear facility, S2 revealed stolen nuclear power plant drawings and asked money to threat social stability and national security, S5, S6, and S7 infiltrated illegally another countries governments offices and critical facilities, S3 stole confidential document and leaked personal information, and S5 targeted 14 the US and 22 Korean government institutions, news media, and financial websites to damage 1,300 units of PCs and infected 11 million units of PC [17]. In addition, SLCAs leaked individual information, stole money from cryptocurrency exchanges, infiltrated and paralyzed banks and broadcasters during March 20 in South Korea, and interfered in the 2016 U.S. presidential election. As such, SLCAs caused economic damage, harm society, and national security threats in a wide range of area, so there is a claim that SLCAs violated international norms such as UN charter, Responsibility of States, etc. On the other hand, it can be noted that the purpose of PLCAs is focused on the extortion and stealing information for economic benefits using malware and hacking.

3 Deterring SLCAs by the Establishment of ACSC Based on IPPs

3.1 International Efforts for Cybersecurity

The main discussions of the international community to develop international cyber-security standards include the Group of Governmental Experts(GGE), which is an UN-mandated working group in the field of information security; the Open-Ended Working Group on developments (OEWG) in the Field of ICTs, which is another UN-mandated working groups; CECC; Shanghai Cooperation Organization (SCO), made up of central Asian countries; G20; APEC; ASEAN Regional Forum (ARF); Tallinn Manual, which is a guide by which to resolve cyber disputes created by reinterpretations of how and when to apply current international norms, such as those by the United Nations Charter, the Responsibility of States norm, the Geneva Convention tenets, and the Hague Convention doctrine.

3.2 Establishment of ACSC Based on IPPs

Vulnerable Cybersecurity Environment in Asia The Asian region, especially the Southeast Asian region, is dynamically developing based on ICT, but the focus is on economic development, with neglectful investments in cybersecurity. Thus, the number of various PLCAs and SLCAs are increasing. In the Southeast Asian market, the cryptocurrency market is growing, allowing many to avoid surveillance and tracking and simplifying monetization and money laundering. However, due to the absence of a proper security system to support this and a lack of regulations in place, SLCAs can exploit these vulnerabilities at any time. The median time for cyber-attackers to stay in the Asia–Pacific region was 498 days, five times longer than the 101-day global median time [18]. Moreover, Asian countries' IDIs (Global ICT Development Indexes) differ by country and are relatively good compared to those of other regions as of 2017.

In Southeast Asia, as the Internet penetration rate increases, cybercrime is becoming more popular. Most cyberattacks are in the form of DDoS, ransomware, and email attachments containing malicious software. Because middle and small businesses are not focused on cybersecurity given the focus mainly on products or User Interface, cyberattacks continue, not only hacking, but also involving even home appliances connected to the Internet, and there are cyber threats through entities known as crawlers and bots. In this way, numerous vulnerabilities, with social engineering, data breaches, system outages, and denial of service attacks, third-party threats, are exposed during the move to the digital economy.

Against these vulnerable cybersecurity situations, ASEAN Member States have made lots of efforts to prevent SLCAs and PLCAs. However, as shown in the Table 1,

Table. 1 ASEAN member states' cybersecurity efforts

	ITU GCI (2018)* [19]	Cybersecurity investment [20]	Regulatory framework** [21]	Information security [22]	Cybersecurity policy and legislation [23]
Singapore	High	Fortified	Advanced	High	Established
Malaysia	High	Tipping point	Advanced	High	Established
Thailand	High	Nascent	Established	High	Established
Philippines	Medium	Nascent	Established	Low	Established
Vietnam	High	Nascent	Nascent	Low	Established
Indonesia	High	Nascent	Nascent	High	Established
Myanmar	Low	None	Absent	Low	N/A
Cambodia	Low	None	Absent	Low	N/A
Brunei	Medium	None	Absent	Low	N/A
Laos	Low	None	Absent	Low	N/A

*Cybersecurity Legal, Technical. Organizational, Capacity building, Cooperation
**National strategy development, Registration, Governance and operational entities, Sector-specific focus, International cooperation, Awareness building, Capacity building

there are great variations between ASEAN Member States' cybersecurity preparedness such as ITU Global Cybersecurity Index, Cybersecurity investment, Regulatory framework, Information security, and Cybersecurity policy and regulation. Also, they lack a regional cybersecurity coordination platform that harmonizes Member States' efforts and shares security information.

Four Establishment Principles for ACSC New international cybersecurity norms should be able to prevent SLCAs that exploit the characteristics and vulnerabilities of cyberspace by allowing the tracking and punishing of attackers. To this end, the study proposes to create ACSC with Four Establishment Principles based on the analysis of security vulnerability of ASEAN region and then suggests implementing Ten IPPs to support ACSC and prevent SLCAs as reflected in Fig. 1. The principles of establishment show the philosophical background of the ACSC creation and four principles are as follows.

First, Application of Current Cyber Security Norm to Cyberspace Many countries have agreed on current international norms that apply to cyberspace. Accordingly, new international cybersecurity norms must initially rely on the current international norms such as the UN Charter. It is necessary to establish a foundation based on the formulated rules of accountable state conduct on the internet to legalize this international commitment.

Second, Maintenance of advantages of Cyberspace The characteristics of cyberspace, such as anonymity, accessibility, openness, and transnationality, for instance, should not be restricted.

Third, Adoption of International Agreements It is necessary to reflect international agreements, such as the GGE and OEWG.

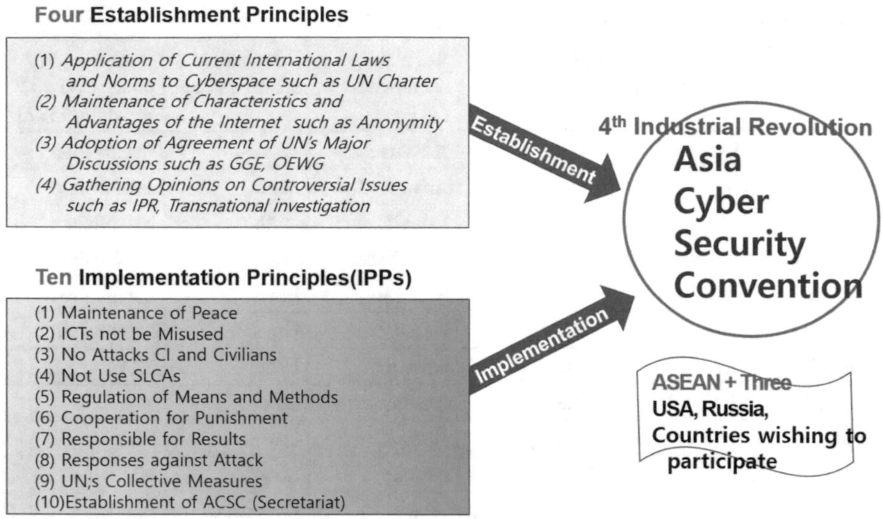

Four Establishment Principles

(1) *Application of Current International Laws
    and Norms to Cyberspace such as UN Charter*
(2) *Maintenance of Characteristics and
    Advantages of the Internet such as Anonymity*
(3) *Adoption of Agreement of UN's Major
    Discussions such as GGE, OEWG*
(4) *Gathering Opinions on Controversial Issues
    such as IPR, Transnational investigation*

Ten Implementation Principles(IPPs)

(1) Maintenance of Peace
(2) ICTs not be Misused
(3) No Attacks CI and Civilians
(4) Not Use SLCAs
(5) Regulation of Means and Methods
(6) Cooperation for Punishment
(7) Responsible for Results
(8) Responses against Attack
(9) UN;s Collective Measures
(10)Establishment of ACSC (Secretariat)

Establishment

Implementation

4th Industrial Revolution
**Asia
Cyber
Security
Convention**

ASEAN + Three
USA, Russia,
Countries wishing to
participate

Fig. 1 Four establishment principles and ten implementation principles for ACSC

Fourth, Reflection of Improvement of Current Cyber Norm It is necessary to secure the participation and support of more member countries by determining the reasons why countries are reluctant to sign current international cybersecurity treaties, such as the Budapest Convention.

Ten Implementation Principles (IPPs) for ACSC

In addition, the study suggests adoption of Ten IPPs for the successful establishment and development of ACSC, which the Internet should not be used for attacking other countries or as a means of detouring attack. Considering ASEAN's cybersecurity vulnerabilities, digital divide between countries, and legal and institutional differences between countries, the study proposes the phased development of ACSC. To begin with, this study suggests that among the 10 principles of the IPPs, principles that are urgent for crime prevention should be addressed in the first stage. This study then suggests that principles that require a lot of time to agree and are hard to agree will be executed later as illustrated in Fig. 2.

First, Reinforcement of State Responsibility Each country should manage its personnel or cyberinfrastructure so as not to harm other countries, according to due diligence in cyberspace. Also, As provided by Tallinn Manual "A State bears international responsibility for a cyber-related act that is attributable to the State and that constitutes a breach of an international legal obligation" in Rule 14. As a concrete way of attribution, a state can check the SCLAs by allowing the ISP to inspect the suspicious packets traffic at the country gates when it has or receives threat information or finds a flow of suspected information. Another possible measure is that when users enter the Internet, they must validate their real ID at least one time. After that, users utilize a pseudonym or nickname and adequate message encryption could be used. There is no need to examine users' log records for privacy if there is no cyberattacks or cybercrime.

Second, Transit State's Responsibility States should not knowingly allow their cyber infrastructure to be used for attacking other countries. If cyberattacks on other countries are caused using manpower, equipment, and facilities under the jurisdiction of a specific country, the country is obliged to cooperate actively during the investigation. IP is regarded as potential evidence that can confirm the criminal, and if the specific IP used for the crime denies the crime or the attack, there is a need for the owner of the IP to be liable given the 'proof of responsibility'.

Third, Protection of Critical Infrastructure and Civilians States should prohibit preemptive cyberattacks against critical infrastructure, including nuclear power plants and innocent civilians and their facilities in cyberspace [2].

Fourth, Cooperation for Information, Punishment, and Identification States should cooperate to exchange information, including the identification of real names upon the issuance of a valid search warrant, in order to assist each other and to prosecute terrorists and the criminal use of the Internet [2].

Fifth, State Response against Cyberattacks An attacked state is entitled to request compensation and take legitimate self-defense and countermeasures. For this Principle, states and the international community should establish a global standard to respond to the cyberattacks. When private or public organizations are attacked by SLCAs or PLCAs, or even by hired contractors to conduct cyber operations, if two conditions are met: attributability and violation of international obligation, the attacked countries, as a guardian of the target corporations, are eligible to respond proportionately to the origin of the attack [24].

Sixth, ACSC Operation under the UN The Asia Cyber Security Convention is expected to run its organization based on the authority of the United Nations, as it is necessary to handle state-related tasks such as the management of member states and the punishment and return of a country to a normal member after the punishment. The ACSC's responsibility is to undertake "investigations and prosecution, with jurisdiction over cyberattacks, provision of international cooperation, and the enactment of related procedures" (similar to the CECC's Articles 14–21) and is to operate organizations such as a board of governors, general conferences, and a secretariat of the ACSC.

Besides, *four additional principles of IPPs* can be implemented in the second stage as seen in Fig. 2.

Seventh, Conditions of Legitimate Cyberattacks: States should prohibit the use of SLCAs except for in legitimate self-defense and for countermeasure purposes.

Eighth, Regulation of Means and Methods: States should regulate the unrestricted utilization of the methods and means of cyber warfare just like with conventional weapons.

Ninth, Offending state's Responsibility The attacking state should take the responsibility for the results of its cyber offenses. For this Principle, states and the regional community need to cooperate to come up with a legal and institutional system, including a substantial law and procedural law in cyberspace: when and how states' activities become an internationally wrongful act of a State in cyberspace, when and

Staged Implementation of IPPs

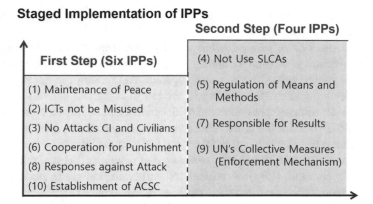

Fig. 2 Ten internet peace principles for ACSC

how NSAs' activities come to an illegal action, and how damage can be accurately calculated [2]. Cybercrime is attributed to a country when non-state actors carry out cybercrime under the state's effective or overall control.

Tenth, Enforcement Mechanism: The UN may take useful collective actions for the restraint and exclusion of cybersecurity threats and for the suppression of cyberattacks and may consider cyber blockades.

4   Concluding Remarks and Further Works

As Asian society becomes more hyper-connected, a hyper-intelligence society develops based on 5G, AI, IoT, big data, and cloud computing, among other technologies. However, there remains no international cybersecurity agreement or regional cooperation between countries. As hostile states conceive cyberterrorism plans more systematically, national critical infrastructure becomes more vulnerable to catastrophic disasters.

The purpose of the ACSC is to apply the current international norms to cyberspace, to maintain the anonymity and transboundary advantages of cyberspace, to respect ongoing cybersecurity discussions such as those by the UN GGE and UN OEWG, and to improve the constraints of current international cybersecurity norms such as the CECC. Furthermore, it emphasizes national responsibility to punish criminals of SLCAs under the jurisdiction of the United Nations and supplements the IPPs to support the ACSC.

The Asian region has various characteristics at the ICT level and has ICT-related laws and systems at the economic level. Hence, this study suggests the promotion of the ACSC gradually while respecting ASEAN way, which seeks to manage the diverse needs of its members, either outside or within the region.

The first step is to establish the ACSC as a loose, non-mandatory but mutually beneficial association through information exchanges such as those related to cyber threats and vulnerability diagnoses, framework construction, and risk-sharing for cyber accident types to respond to actual cybercrimes and terrorism threats effectively. The second step is to strengthen cooperation and achieve practical results by implementing additional four principles of IPPs. Achieving these goals needs a whole project of procedures, technologies, state law, and global partnership and execution with encouragement from like-minded international organizations.

References

1. Lee JK (2015) Research framework for AIS grand vision of the bright ICT initiative. MIS quarterly 39.2:3-12
2. Shin YY, Jae KL, Myungchul K (2018) Preventing state-led cyberattacks using the bright internet and internet peace principles. J Assoc Info Syst 19.3:3
3. https://www.coe.int/en/web/conventions/fulllist/conventionstreaty/185/signatures?p_auth= Z9kSyEKq
4. Kim H-K (2020) revealed North Korea's Cyber Unit' Reality. https://www.rfa.org/korean/wee kly_program/rfa_interview/rfainterview-01262015095622.html
5. https://www.securitymagazine.com/articles/91889-concerned-about-nation-state-cyberatta cks-heres-how-to-protect-your-organization
6. Mohurle S, Manisha P (2017) A brief study of WannaCry threat: ransomware attack. Int J Adv Res Comput Sci 8.5
7. Lee K, Lim J (2016) The reality and response of cyber threats to critical infrastructure: a case study of the cyber-terror attack on the Korea Hydro and nuclear power Co., Ltd. KSII Trans Internet Info Syst 10.2 (2016).
8. Sanger DE, Martin F (2015) NSA breached North Korean networks before sony attack, officials say. New York Times 18:A1
9. Zetter K (2011) How digital detectives deciphered Stuxnet, the most menacing malware in history. Wired Magaz 11:1–8. https://www.wired.com/2011/07/how-digitaldetectives-deciph ered-stuxnet
10. Sang-Hun C, John M (2009) Cyberattacks jam government and commercial web sites in US and South Korea. New York Times 9
11. Kozlowski A (2014) Comparative analysis of cyberattacks on Estonia, Georgia and Kyrgyzstan. Europ Sci J 3:237–245. https://eujournal.org/index.php/esj/article/viewFile/2941/ 2770
12. Ruus K (2008) Cyber war I: Estonia attacked from Russia. European Affairs, 9(1–2). https://www.europeaninstitute.org/index.php/component/content/article?id=67:cyber-war-iestonia-attacked-from-russia
13. https://koreajoongangdaily.joins.com/2011/04/10/finance/Hyundai-Capital-admits-info-leak-was-a-lot-larger/2934644.html
14. https://www.bbc.com/news/technology-40340820
15. https://webcache.googleusercontent.com/search?q=cache:xa668XHjB6wJ:https://intellige nce.na.go.kr:444/intelligence/reference/reference01.do%3Fmode%3Ddownload%26arti cleNo%3D663003%26attachNo%3D453157+&cd=1&hl=ko&ct=clnk&gl=kr
16. Schmitt MN (2017) Tallinn manual 2.0 on the international law applicable to cyber operations. Cambridge University Press
17. Clark RA, Knake RK (2010) Cyber war
18. Marsh, McLennan (2019) Advancing cyber risk management, Fire Eye special report

19. https://www.itu.int/dms_pub/itu-d/opb/str/D-STR-GCI.01-2018-PDF-E.pdf
20. https://www.kearney.com/digital-transformation/gsli/2019-full-report
21. Overview of Cybersecurity status in ASEAN and the EU prepared by: Sociedade Portu-guesa de Inovação (SPI) (2020)
22. ITU Global Cyber security Index
23. Sunkpho J, Ramjan S, Ottamakorn C (2018) Cybersecurity policy in ASEAN countries. In: 17th Annual Security Conference
24. Messerschmidt JE (2013) Hackback: permitting retaliatory hacking by non-state actors as proportionate countermeasures to transboundary cyberharm. Colum J Transnat'l L 52:275

Research on Life Cycle Model of Data for Handling Big Data-Based Security Incident

MinSu Kim

Abstract Artifact analysis is used to investigate various types and characteristics of attacks to the system in case of security incident. Moreover, the accuracy of security incident analysis varies depending on artifact data. However, highly developed network technology and generation of a massive amount of artifacts in today's hyper-connected society make management difficult. The accuracy of analysis can increase if all artifacts regarding security incident are collected and saved, but this is associated with disadvantages of cost and efficiency in respect of managing artifacts. Therefore, this study aims to suggest the plan of effective handing of security incident in addition to the reduction of managerial load based on an "Artifact life cycle" model.

Keywords Artifact · Life cycle model · Security incident · Security event log · Standardizing log

1 Introduction

An immense volume of personal information leakage and security incident happens every year in the knowledge information society as a product of constant development of network technology and hyper-connected society. It is crucial to collect artifacts for the initial handling to analyze the route, size and period of intrusion against the attacks in case of security incident. In other words, the analysis of security incident depends on the storage of data by period regarding the collection of artifacts. Therefore, maximum artifacts are collected in storage capacity of server against security level of system and event log. However, currently, the load of management and cost rise to keep the minimized storage period of event logs as a huge amount of artifacts are collected in accordance with exponential increase of recent data transmission. Looking into personal information leakage of YAHOO in 2014 regarding security incident [1], it took no less than two years to recognize the leakage after the incident.

M. Kim (✉)
Department of Information Security Engineering, Joongbu University, Dongheon-ro,
Deogyang-gu, Goyang-si, Gyeonggi, South Korea
e-mail: fortcom@hanmail.net

Like this, until the leakage of information is acknowledged, it takes 146 days world-wide on average, 469 days in Middle East and Africa, 520 days in the Asia–Pacific region and 1035 days in South Korea [2–4]. Observing the average period of information leakage aknowldegement, it is assumed that it is difficult to collect enough event logs to analyze post-security incident. To solve this problem of log, managerial load must decrease through optional selection of important event log required for handling security incident by setting priority of logs, moreover, event logs need to be collected to respond to the incident. Therefore, this study aims to propose life cycle model of artifact suitable for effective handling of incident by diminishing managerial load regarding event log collection.

2 Research Method

2.1 Event Log

Log is the sole data allowing track security incident the same way the data records main operation of the system. In addition, log file can be used as legal evidence and monitoring data to prevent security incident [5–7].

Such event log analysis makes it possible to establish security response system together with regular diagnosis of weakness [8], development of the integrated platform that collects, saves, processes, searches, analyzes and visualizes has been carried out by connecting functional limitation that is challenging to do a preliminary diagnosis to Big Data technology [9].

The data of illegal intrusion is collected and analyzed by relying on log file in case of system error or security incident in the managerial respect [10].

2.2 Procedure of Handling Security Incident

2.2.1 Procedure of Handling Security Incident in South Korea

Korea Internet and Security Agency (KISA) suggests seven steps of handling security incident as the following figure shows. The handling steps are classified by steps of preparing for handling security incident prior to occurrence of incident, detection of abnormal signs, systemization of handling strategy by carrying out initial investigation and incident type when detecting incident and establishing security policy after writing an objective report on security incident through collecting and analyzing data for investigation of incident [11] (Fig. 1).

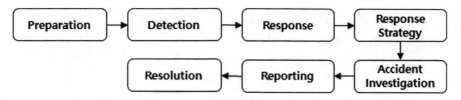

Fig. 1 Procedure of handling security incident (KISA)

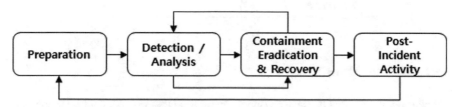

Fig. 2 Incident response life cycle (Detection and analysis)

2.2.2 Procedure of Handling Security Incident in Foreign Countries.

National Institute of Standard and Technology (NIST) suggests the handling model that classifies the procedure of security incident by four steps in SP 800–61 "Computer Security Incident Handling Guide" as seen in the Fig. 2 [12].

The model of handling security incident consists of preparation for handling, detection/analysis, removal and recovery of infection and post-activity, and has the cyclic structure of post-activity step, preparation for handling step, detection/analysis step and removal and recovery of infection step.

2.3 Algorithm

2.3.1 Selection Algorithm

Alg Selection algorithm receives arrangement A, first beginning index I, last index j and rank k as the input, while s begins from 1 and increases to rank k as a temporary rank.

tmin is the index of minimum value in the entire arrangement and amin is the value of tmin. It is the algorithm that searches rank k each time the performance is carried out and searches the minimum value in the arrangement except rank k repeatedly [13–15].

$$
\begin{aligned}
&\text{Alg } Selection(A, i, j, k) \\
&\quad for\ s\ from\ 1\ to\ k\ do \\
&\qquad t_{min} \leftarrow i + s - 1;\ a_{min} \leftarrow A(t_{min}) \\
&\qquad for\ t\ from\ i + s\ to\ j\ do \\
&\qquad\quad if\ A(t) < a_{min}\ then\ a_{min} \leftarrow A(t);\ t_{min} \leftarrow t \\
&\qquad A(t_{min}) \leftarrow A(i + s - 1);\ A(i + s - 1) \leftarrow a_{min} \\
&\quad return(A(i + k - 1))
\end{aligned}
$$

2.3.2 Priority Algorithm

Scheduling of priority Queue is an operation method that proceeds with Queue with higher priority by applying weight according to priority of data. Therefore, it is capable of processing emergent data first. Figure shows scheduling of Queue with priority [16, 17] (Fig. 3).

Queueing delay time of highest priority can be shown according to average number N_H of data remained in Queue and arrival rate of data λ like (1). Here, N_H can be shown with (2).

$$
w_i = \frac{\overline{N_H}}{\lambda} \tag{1}
$$

$$
\overline{N_H} = \frac{P_t + P_{t+1} + \geq \ldots + P_{t-n+1}}{T_i} \tag{2}
$$

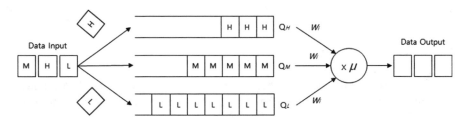

Fig. 3 Priority algorithm

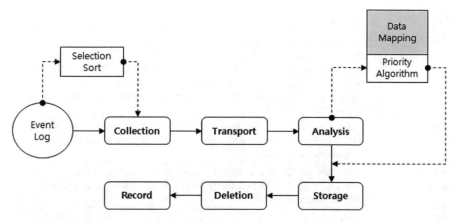

Fig. 4 Life cycle of artifact

2.4 Test and Result Analysis

The origin of artifact collector for handling the existing security incident refers to the time immediately after the incident. Therefore, it is handled by collecting pre-incident logs by adopting the "Preparation" step (Fig. 4).

2.4.1 Collection Step

It is necessary to set the policy and define which artifacts should be collected in the Collection step. There are artifacts that are required for handling security incident in general such as auth (secure), audit, access, error and history as a setting collection policy, however, the setting that is suitable for the server to handle the incident in a more precise way [18–22] (Table 1).

2.4.2 Transmission Step

One of methods that handle removal of artifacts by attackers after security incident in the Transmission step is to protect data by real time transmission of artifacts to other storage servers. Moreover, the transmitted artifacts are required to guarantee the integrity of data by hashing for each step.

2.4.3 Analysis Step

Field is needed to be divided first in the Analysis step to implement parsing as the step selecting the fields is required to handle security incident and proceeds with

Table 1 Linux system log

Linux log	Contents
Secure	They are logs related to authentication of Linux. It leaves main logs related to security and contains the log related to user authentication. It collects logs as it is capable of checking non-approved authentication log. Log-in at an abnormal time slot, repeated log-in failure and abnormal su command must be checked in an intensive/thorough manner
Audit	It is a log that provides audit function. ISO/IEC 27,001 expresses the importance of audit logging. It collects the logs of users' activities, exceptional cases and information security events
Apache access	It records and collects all requests that server processes as a server access log. There are injection attacks that are frequently made such as query union and selected as the attack to web
Mysql transaction	If the transaction log of mysql is activated, record of sql remains as seen in the figure above. In particular, in case of web server, it collects the logs as there are many cases where the attacker performs an injection attack or a web shell upload attack
Messages/ Syslog	It records general logs of system. It collects logs generated in system due to its capability
Cron	It collects schedule logs in system due to its capability
.bash_history	It collects command history of the account due to its capability. It is able to analyze what activities the account has done based on the data

standardization. The following is the example that divides the fields of Linux system log [23–26] (Table 2).

Artifacts have the different field values each other and parsed artifact can be processed by standardization. In particular, timeline is important to interpret the flow regarding handling security incident and it is required to process it through standardization as time format left by each artifact is not regular. For example, time stamp can be processed by match function in Logstatsh (Table 3).

The meaning can be deduced from the standardized artifact, however, the meaning that is suspected as an attack from web log, SQL transaction log and authentication log can be extracted in case of a web server.

In addition, in case of using the engine such as Elasitc search or Splunk, visualization can be implemented through web and it would be more helpful to respond to security incident.

2.4.4 Storage Step

Cloud can be used and an in-house storage server can be established and operated in the Storage step. In case of an operating storage server, the external access to the storage server must be disabled by the separation of network from the server.

Table 2 Field division of log

Web Access: Apache Access		Snort			
Remote Host IP	10.10.10.1	SID	1:1,000,007:0		
Remote Logname	–	Rule msg	KGU_Telnet_Test		
		Priority	0		
		Date	05/24–14:45:47.547117		
User ID	–	Src IP:Port	10.10.20.150:63,062		
Date	08/May/2018:21:46:20 + 0900	Dst IP:Port	10.10.20.151:23		
Client Request	"GET /wizboard/skin/sandle_secret/images//dotline.gif HTTP/1.1"	Protocol: TTL	128	Type of Service	0 × 0
Status Code	304	ID	20,448	Seq	0 × B6F84B75
		IpLen	20	Ack	0 × 0
Size	-	DgmLen	52	Win	0 × FAF0
		Flag	DF	TcpLen	32
Task Schedule: Cron		Certification: Secure, Auth, Audit			
Date	May 8 21:38:12	Date	May 8 21:46:53		
Hostname	Victim	Hostname	Victim		

(continued)

Table 2 (continued)

Process Name [PID]	anacron[2850]	Process Name [PID]	Passwd
		Message	pam_unix(passwd:chauthtok): password changed for root
Message	Job 'cron.daily' terminated	Port	
		Etc	
DataBase: Mysql transaction		System: Messages, Syslog	
Date	27	Date	May 8 21:38:06
Source	Query	Hostname	Victim
Message	SELECT * FROM wizTable_root_board02 ORDER BY UID DESC LIMIT 0, 5	Message	syslogd 1.4.1: restart

Table 3 Example of standardizing log (Apache Access)

```
date {
match = > ["timestamp", "dd/MMM/yyyy:HH:mm:ss Z"]
target = > "log_date"
timezone = > "Asia/Seoul"
}
```

2.4.5 Deletion Step

Deletion of unnecessary fields and normal artifacts is processed and the work of selecting the fields and artifacts only required for handling security incident is carried out.

It is needed to find out when (Time stamp), who (ID), where (IP address), how the intrusion (port) is penetrated and which program (PID) is executed with regard to handling of security incident. It is easier so that the analysts can read it in a better way by sorting the various fields of artifacts, and the capacity of storage ca reduce as well (Table 4).

Table 4 Definition of log

Item	Content
Apache access	Remote log name and User ID are deleted in web access log. IP must not be deleted to find out from where it accesses as well as date and time, request as activity, Status code as how the access has been done
Secure, Auth, audit	There is no field to delete in the log related to security authentication. Authentication log related to security has the important information related to security incident such as log-in at an abnormal time slot and repeated log-in failure. Therefore, the incident can be responded more promptly if normal logs are excluded. Both logs of the users with the authority of root and exceptional users must not be deleted
Cron	There is no field value to be deleted in log of work schedule. Normal scheduling logs, which the manager registers in advance are not required to be checked in point of handling view
Messages, syslog	Abnormal re-boot and error message related to system can be checked in this log. As it has almost all records about system, it is capable of storing a huge amount If normal I/O of process can be checked, it is capable of filtering normal log
Snort	Log detected by the rule that has been preliminarily set can be checked. It is important to find out when, where and by which rule it has been detected. Therefore, filtering can be proceeded based on Snort Rule msg, date, Src IP and Dst IP

2.4.6 Record Step

The logs selected only from the steps above are saved among the logs entirely saved through batch script in the Record step. The selected artifacts can be periodically managed by registering bat script in work scheduler. The core artifacts are requiere to be compressed as they grow older by registering and sorting the script by period.

3 Test and Result Analysis

Existing log analysis programs focus on service management other than security. The life cycle of data suggested in this study insists that the artifacts before occurrence of security incident should be possessed as much as possible to respond to promptly security incident and fast handling is required through standardization and visualization in case of incident as the research to respond to the incident in point of security view (Table 5).

As the comparative analysis results of each model related to response to security incident by classifying them into sustainability, speed, fragmentation and degree of response, the sustainability of data had no difference for each model as it follows institutional standards regarding the period of possessing the artifacts. Nontheless, this life cycle model had advantages in terms of the extension of period caused by capacity or selective storage of logs.

Moreover, speed of data analysis increases based on fragmentation and standardization of the logs through parsing and the degree of entire response can deduce the discriminated results according to the selection and priority regarding life cycle model.

Table 5 Comparative analysis

Item	Sustainability	Speed	Fragmentation	Degree of response (general)
KISA model of analyzing security incident	○	△	△	△
NIST model of handling security incident	○	△	△	△
ICS Model	○	△	△	△
Life cycle model	○	○	○	○

(○:High, △:Normal, × :Low)

# 4	Conclusions

In case of security incident, the analysis is carried out based on artifact and different speed of responding to security incident according to the manifested holding capacity of the data and period.

Nevertheless, this paper suggests the efficient data management model through management period and field selection under the concept of "Artifact life cycle" since demanding cost and management to possess all artifacts is burdensome.

In addition to that, it is difficult to respond to targeted security incident that recently happened due to the difference of holding amount and period of artifacts as it has the type of Advanced Persistent Threats.

Therefore, it checked the degree of responding to entire response of each model by classifying them into sustainability, speed and fragmentation against the steps suggested from existing models regarding response to security incident and verified the discriminated models of handling security incident by Life Cycle Model. Hereafter, the model that raises accuracy will be proposed by developing and applying multifarious algorithms regarding selection and priority against event log.

Acknowledgements This paper was supported by Joongbu University Research & Development Fund, in 2020

References

1. https://www.ddaily.co.kr/news/article/?no=147633
2. https://byline.network/2016/09/1-338/
3. https://www.yna.co.kr/view/AKR20160908161100017
4. https://www.postek.co.kr/bbs/board.php?bo_table=news&wr_id=21
5. Chae H, Lee S (2014) Security policy proposals through PC security solution log analysis. J Korea Inst Info Secur Cryptol 24(5)
6. Lee S, Cho H, Kim D, Pang S (2015) Design of camel-mahout model for aggregating/Anlayzer of security Log/Eventsm. Korea Inst Commun Sci 781–782
7. Deswarte Y, Powell D (2016) Internet security: an instrusion-tolerance approach. Proc IEEE 432–411
8. Kim H (2003) Need for log analysis. Information Security 21C Contribution, pp 1
9. Han K, Jeong H, Lee D, Chae M, Yoon C, Noh K (2014) A Study on implementation model for security log analysis system using big data platform. J Digital Convergence 12(8)
10. Lim S, Lee D, Kim J (2015) Methodology of log analysis for intrusion prevention based on LINUX. J Convergence Secur 15(3)
11. Korea Internet and Security Agency (2010) Procedure of handling security incident
12. Cichonski P, Milla T, Grance T, Karen S (2012) In: Special publication 800–61 revision 2 computer security incident handling guide. Computer Security Division Information Technology Laboratory National Institute of Standards and Technology 9
13. Triguero I, Derrac J, Garcia S, Herrea F (2012) A taxomomy and experimental study on prototype generation for nearest neighbor classification. IEEE Trans Syst Man Cybernet Part C (Application and Reviews) 42(1):86–100 I
14. Baek B, Hwang D (2019) Hyper-rectangle based prototype selection lgorithm preserving class regions. Korean Inst Info Sci Eng 829–831

15. Arturo Olvera-Lopez J, Ariel Carrasco-Ochoa J, Francisco Martinez Trinidad J, Kittler J (2010) A review of instance selection methods. Artif Intell Rev 34(2):133–143
16. Ryu M, Lee S, Song M, Kim J, Won K, Cho K (2011) Self-sustainable system-based emergent data ransmission algorithm using priority queue in wireless sensor networks. Korea Inst Info Telecommun Facil Eng 25–27
17. Yoo B, Hwang S, Kim Y, Kim W (2018) Real time priority dangerous objects detecting algorithm of artificial intelligence and internet of things. Korea Info Sci Soc 2345–2347
18. Hwang Y, Kim K, Kwon O, Moon I, Shin G, Ham J, Park J (2017) Analyzing box-office hit factors using big data: focusing on korean films for the last 5 years. J Info Commun Convergence Eng 15(4):217–226
19. Park J, Kim S (2010) The design for security system of linux operating system. Korea Info Electron Commun Technol 35–42
20. Cozzi E, Graziano M, Fratantonio, Balzarotti D (2018) Understanding linux malware. In: IEEE Symposium on security and privacy, pp 161–175
21. Smalley S, Fraser T (2005) A security policy configuration for the security-enhanced linux. NAI Labs Technical Report
22. Hwang J-h, Lee T-J (2019) Study of static analysis and ensemble-based linux malware classification. Korea Inst Info Secur Cryptol 29(6):1327–1337
23. Singh S (2018) Forensic and automatic speaker recognition system. Int J Electri Comput Eng 8(5):2804–2811
24. Baek N, Shin J, Chang J, Chang J (2019) Spark-based network log analysis aystem for detecting network attack pattern using snort. Korea Contents Soc 48–59
25. Sim H, Bae J, Park H (2016) Generate extended snort rules by edit distance. Korean Inst Info Sci Eng 1821–1823
26. Kim M, Jo H, Park H (2014) Analysis of detection rules and development classification model for optimization on duplicated detection rules on snort. Korean Inst Info Sci Eng 1657–1659

Feature Selection Based on a Shallow Convolutional Neural Network and Saliency Maps on Metagenomic Data

Toan Bao Tran, Nhi Yen Kim Phan, and Hai Thanh Nguyen

Abstract In recent years, personalized medicine has been discovered by scientists to improve existing curative methods. These studies are mainly performed on metagenomic datasets which is the large dataset related to many human diseases, especially genetic data. The development of machine learning models and related algorithms has enabled us to speed up computation and improve disease diagnosis accuracy. However, due to the large dataset and the rather complicated processing of the data, we encountered certain difficulties. Therefore, we propose an approach to the task of selecting features based on the explanatory model. This approach is made up of proposing a small set of features from the original, implemented with Explanations with Saliency Maps. The results exhibit better performances comparing to random feature selection. Explanations generated by Saliency Maps have provided a promising method in selecting features and are expected to apply in practical cases.

Keywords Feature selection · Explanation · Metagenomic data · Disease prediction · Saliency maps · Machine learning

1 Introduction

Personalized medicine or precision medicine, is a model that divides patients into different groups to decide, intervening with medical products that are suitable for each patient based on specific information. Patient's score and genetic information. Personalized medicine has applied intensive studies and complex techniques based

N. Y. K. Phan · H. T. Nguyen (✉)
College of Information and Communication Technology, Can Tho University, Can Tho 900100, Vietnam
e-mail: nthai.cit@ctu.edu.vn

T. B. Tran
Center of Software Engineering, Duy Tan University, Da Nang 550000, Vietnam
e-mail: tranbaotoan@dtu.edu.vn

Institute of Research and Development, Duy Tan University, Da Nang 550000, Vietnam

© The Author(s), under exclusive license to Springer Nature Singapore Pte Ltd. 2021   107
H. Kim et al. (eds.), *Information Science and Applications*, Lecture Notes
in Electrical Engineering 739, https://doi.org/10.1007/978-981-33-6385-4_10

on the individual patient's genetic map to achieve high results in disease diagnostic performance, consistent with the characteristics. Genetics at the molecular level of each patient.

The genome plays an important role in the human body, it contains all the information needed to build and maintain the body. Many studies have shown that the genome directly affects diseases related to genetic factors. Therefore, the application of modern techniques to research on genomes will be an important step in the field of medicine, helping to effectively diagnose and treat diseases for each patient, and limit side effects. Compared with traditional therapeutic methods, applying a single treatment regimen for a particular disease in all patients. The regimen chosen is the best one based on the doctor's diagnosis, drug side effects, and patient reactions. With each treatment, if the chosen regimen is not appropriate, the doctor will adjust it until the desired results are achieved. From the above, we can see that individualized medical methods will bring better diagnosis and treatment results with the motto: the right person, the right medicine, the right time, the right dose. Physicians can easily devise an individualized treatment regimen for each patient, based on the medical pre-treatment, response level, and genetic characteristics of each individual.

Metagenomics is the study of genetic material, recovered directly from the samples, and requires no culture. Currently, metagenomics is being interested in and researched by the scientific community and has acquired a certain number of achievements, contributing to the field of medicine. Studies of microbial populations have shown their effects on human health, the microbiota in the human body gives us a lot of information regarding the diseases they are facing. However, because the association of the bacteria is still individual and a large part of the bacteria remains undetected. Furthermore, the metagenomic dataset is so complex that it is still a great challenge to learn, study, and detect the remaining bacteria.

2 Related Work

Many studies have shown that the microbiota is closely related to personalized medicine and plays an important role in human health [1–4]. Investigation of 16rRNA sequences from undetermined microorganisms uncovered new microbial communities [5–7]. Metagenomics has also provided a lot of information about the evolutionary history of the species. In a recent study by Udugama et al. [8], experts extracted the first genome of SARS-CoV-2 based on RNA sequencing. Researchers have explored the diversity of microbial communities and their contributions to the natural environment introduced in [9, 10].

The authors in [11] presented an approach using machine learning techniques to differentiate between viral and bacterial and find out the new viruses based on the VirFinder tool. The viral sequences detection is performed in aquatic metagenomes and using the ecosystem-focused models targeted to marine meta-genomes. The study also stated the limitation of retrieving the low abundance of viral sequences and proposed the potential biases. The study [12] using several machine learning techniques namely Random Forests and Gradient Boosting Trees for extracting the features on the Amazon Fine Food Reviews dataset. The authors take advantage of selected features for several classification tasks and gained promising results.

The authors in [13] investigated the performance of deep learning techniques for discovering viruses based on metagenomics data. The viral RefSeq dataset is used in this study and the proposed method gained better performance in comparison with the state-of-the-art methods. Furthermore, the authors stated that the accuracy of under-represented viral recognition can be raised by using an abundance of viral sequences. The proposed method also applied to human gut metagenomic data and recognized vast viral sequences of colorectal carcinoma. A deep learning framework named Seeker is presented in the study [13] to identify the phages in sequence datasets and clean the difference of phage sequences. The authors stated the existing approaches based on the similarity of bacteriophage sequences and presented the performance of unknown phages detection can be applied to identifying the undiscovered phages from investigated phage families.

The advantages of the virMine [14] are to identify the viral and bacterial communities or the viral genomes from representative of viral. The authors stated that this method can be used to seek information easily. In comparison with the other framework, the proposed method is not affected by the insufficient representation of viral variety in public data repositories. The virMine removed the non-viral sequences and applied an iterative approach for detecting the viral genomes. As a consequence, the bacteriophages, related viruses, eukaryotic viruses, and novel species can be detected easier. The proposed tool also investigated the performance of three different habitats of microbial.

In this study, we investigate the performance of six metagenomic datasets by using the Convolutional Neural Network model combining with explainability models for feature extraction. Our contributions include:

- We use the Convolutional Neural Network model to classify the IBD disease dataset and extracted the most important features with the learned model and on the original datasets.
- We take advantage of Saliency maps [15] to discover the most important features on one dataset and use them for predicting the rest.
- We compared the performance of the proposed method on the considered datasets and comparing the performance of the proposed features selection approach with features are selected randomly. Furthermore, performance is investigated in several tasks.

In the rest of this study, we present the information about the considered datasets in Sect. 3. The learning model and feature selection approach are described in Sect. 4. Our experimental results are presented in Accuracy, Area Under the Curve (AUC), and Matthews correlation coefficient (MCC) in Sect. 5. Section 6 conducts some closing remarks for the work.

3 Dataset Benchmark

We looked at the Crohn's disease dataset, an intestinal disease. Patients who are infected will develop gastrointestinal inflammation, which can lead to fatigue, weight loss, malnutrition, and sometimes life-threatening complications. This dataset is Sokol's lab dataset described in [16] Based on the phenotype, the dataset consists of two subsets Ulcerative colitis (UC), Crohn's disease (CD). For Crohn's disease, the dataset is further subdivided into Crohn's disease (iCD). We looked at two conditions: remission (r) and flare (f) (worsening of symptoms, worsening of patient condition). Each dataset we examine will contain four key pieces of information: the number of features, healthy samples, patients, and the number of samples (Table 1).

Six datasets reviewed for inflammatory bowel disease (IBD) totaled 228 healthy people and 326 patients. Specifically, the CDf dataset contains 98 samples with 60 patients and 38 healthy people. CDr includes 38 healthy individuals and 77 patients. Review the iCDf dataset with a total of 247 features and a total of 82 samples including 44 patients and 38 controls. For iCDr dataset, the number of positive cases is 59 out of 97. UCf dataset consists of 41 patients and 38 healthy individuals out of 79 samples. UCr disease dataset with a total of 82 samples and 44 cases positive for the disease. The CDf dataset dimension is 98×259 and CDr is 115×237. The datasets iCDf and iCDr have the corresponding size are 82×247 and 97×257. UCf has size 79×250 and UCr has size 82×237.

This feature shows the abundance of species, it indicates the proportion of bacteria in each sample in the human gut environment. Total abundance of all features in one sample is sum up to 1 which is calculated using the following formula (Eq. 1):

$$\sum_{i=1}^{k} f_i = 1 \qquad (1)$$

With:

- k is the number of features for a sample.
- f_i is the value of the i-th feature.

Table 1 Information on six datasets considered Inflammatory Bowel Disease (IBD)

Dataset	Healthy	Patients	Samples	Features
CDf	38	60	98	259
CDr	38	77	115	237
iCDf	38	44	82	247
iCDr	38	59	97	257
UCf	38	41	79	250
UCr	38	44	82	237

4 Feature Selection Based on Important Features Extracted from Saliency Maps

4.1 Learning Model

We considered implementing a shallow Convolutional Neural Network which contains two Convolutional layers, each layer includes 64 filters of 3 × 3, followed by a Max-Pooling of 2 × 2 (stride 2), a dropout rate of 0.1 and a Fully Connected layer with 64 neurons. We also implemented the CNN with Adam optimizer [17], the default learning rate is 0.001, and the network uses a batch of size 16.

We investigated the performance by computing the Accuracy, AUC, and MCC on 10-folds cross-validation. We carried out the features selection by Saliency map and the best model. The range value of both ACC and AUC is from 0 to 1 which higher is better. Whereas MCC is from −1 to +1, +1 represents for a good classifier, 0 means the model makes predictions randomly, and −1 denotes the disagreement between observations and predictions.

4.2 Feature Selection Approach with Saliency Maps

Saliency maps present the important regions that the model focus on and contributes the most to the output. In other words, a heatmap contains the highlighted pixels that affect the decision of the model is created by Saliency maps. Figure 1 visualizes the

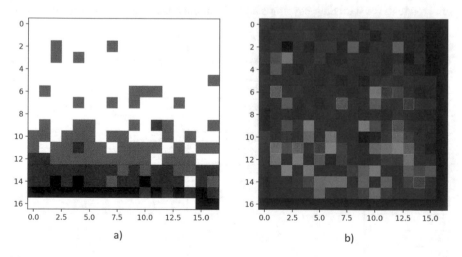

Fig. 1 **a** Sample in the IBD dataset; **b** Heatmap includes highlight pixels marked by Saliency maps

important pixels within an image by calculating the gradient of a class output via back-propagation by a sample in the iCDr dataset.

As mentioned above, we used six datasets to investigate the performance of the features selection method in this study. We extracted the most important features on a specific dataset and applied them for the rest classification tasks. We considered using 10 the most important features on each dataset from the original features. For specific, we extracted the most important features on the first dataset and predict the rest by using the features of the first dataset. We restated the features selection progress on six datasets.

5   Experimental Results

The performance of the proposed method is investigated on 10-fold cross-validation. We also compared the performance of selected features by Saliency maps and randomly selected features by computing the average Accuracy, AUC, and MCC.

We presented the comparison between our selected features by the Saliency method with randomly selected features based on the CDr dataset. Tables 2, 3, 4, 5, 6 and 7 contain 2 main rows, the Accuracy, AUC, and MCC with selected features by the Saliency maps is presented in the first row. The second and last row exhibits the performance of randomly selected features.

Table 2 presents the performance of using features on the CDr dataset with Saliency maps and randomly selected features. As observed, CDf dataset has 0.6545 of Accuracy and 0.2042 of MCC when using features on Saliency higher than Random. All 4 datasets iCDr, iCDf, UCr, UCf achieved outstanding performance using selected features with Saliency maps and all three measures gave better results when

Table 2 Performance of feature selection method by saliency maps and randomly selected features based on CDr dataset

Dataset	Features	Accuracy	AUC	MCC
CDf	Saliency	**0.6545**	0.6111	**0.2042**
iCDr		**0.68**	**0.6167**	**0.2377**
iCDf		**0.6361**	**0.6721**	**0.2535**
UCr		**0.6361**	**0.6783**	**0.2535**
UCf		**0.6071**	**0.59**	**0.2111**
CDf	Random	0.6444	**0.6527**	0.14
iCDr		0.5	0.5417	−0.2722
iCDf		0.5556	0.45	0
UCr		**0.5847**	0.5763	0.1415
UCf		**0.5446**	0.55	0.0649

Table 3 Performance of feature selection method by saliency maps and randomly selected features based on CDf dataset

Dataset	Features	Accuracy	AUC	MCC
CDr	Saliency	**0.6264**	**0.4844**	**0.0615**
iCDr		**0.6028**	**0.5333**	**0.0743**
iCDf		**0.5890**	0.5	0
UCr		0.5361	**0.4891**	**0.0059**
UCf		0.5304	**0.5467**	**0.0486**
CDr	Random	0.6014	0.455	−0.0066
iCDr		0.5556	0.5	0
iCDf		0.5486	0.4856	0.0466
UCr		0.5361	0.4560	−0.0511
UCf		**0.5339**	0.5035	0.0258

Table 4 Performance of feature selection method by saliency maps and randomly selected features based on iCDr dataset

Dataset	Features	Accuracy	AUC	MCC
CDr	Saliency	**0.7778**	**0.9**	**0.5976**
CDf		**0.8889**	**0.8**	**0.7906**
iCDf		**0.6667**	0.6	**0.3162**
UCr		**0.6667**	0.6	**0.3163**
UCf		**0.625**	**0.875**	**0.2582**
CDr	Random	0.6444	0.6375	0.1176
CDf		0.7111	0.7229	0.3803
iCDf		0.5722	**0.6404**	0.1161
UCr		0.5736	**0.6641**	0.1047
UCf		0.5589	0.5847	0.0971

Table 5 Performance of feature selection method by saliency maps and randomly selected features based on iCDf dataset

Dataset	Features	Accuracy	AUC	MCC
CDr	Saliency	**0.75**	**0.7333**	**0.4667**
CDf		**0.625**	**0.75**	**0.2581**
iCDr		**0.75**	**0.7333**	**0.4667**
UCr		**0.875**	**0.9375**	**0.7746**
UCf		**0.75**	**0.8125**	**0.5**
CDr	Random	0.6428	0.6967	−0.1293
CDf		0.5392	0.5546	0.0459
iCDr		0.6446	0.695	0.1151
UCr		0.575	0.7124	0.2211
UCf		0.6714	0.7187	0.3999

Table 6 Performance of feature selection method by saliency maps and randomly selected features based on UCr dataset

Dataset	Features	Accuracy	AUC	MCC
CDr	Saliency	**0.6607**	**0.58**	**0.1087**
CDf		**0.5553**	**0.5733**	**0.1021**
iCDr		0.625	0.6	0
iCDf		**0.75**	**0.8125**	**0.5773**
UCf		**0.5464**	**0.6416**	**0.1441**
CDr	Random	0.6482	0.4766	0.0466
CDf		0.5107	0.5270	0.0345
iCDr		**0.6642**	**0.675**	**0.1395**
iCDf		0.4982	0.4520	0.0258
UCf		0.4732	0.3791	−0.0377

Table 7 Performance of feature selection method by saliency maps and randomly selected features based on UCf dataset

Dataset	Features	Accuracy	AUC	MCC
CDr	Saliency	**0.6964**	**0.62**	**0.2232**
CDf		**0.5804**	**0.6317**	**0.1083**
iCDr		**0.6250**	**0.7333**	0
iCDf		**0.5000**	**0.5000**	0
UCr		**0.5571**	**0.6541**	**0.1734**
CDr	Random	0.6392	0.6117	0.0920
CDf		0.5518	0.5550	**0.0897**
iCDr		**0.6375**	0.6020	**0.0788**
iCDf		0.4839	**0.6375**	**0.02**
UCr		0.4839	0.6291	0.0200

randomly selected features. Specifically, iCDr disease has specific values of Accuracy, AUC, and MCC, respectively 0.68, 0.6167, and 0.2377.

Considering Table 3, based on the CDf dataset, we compare the performance of geographic feature selection according to Saliency maps with randomly selected features on 5 datasets: CDr, iCDr, iCDf, UCr, UCf. In general, the results with the method we propose are better than Saliency maps. 3 datasets CDr, iCDr, and iCDf with Accuracy, AUC, and MCC values when using Saliency maps are all higher than random for selected features. The UCr dataset has Accuracy value when applying two equal methods and this value is 0.5361 for Saliency maps and Random for 0.6014, besides, the AUC value of this disease's MCC when using Saliency is both higher than the method of the remaining methods, namely 0.4891 of AUC and 0.0059 MCC. The UCf set has a higher AUC and MCC value than Random and that is 0.5467 and 0.0486.

We applied the features selection method on the iCDr dataset and presented the performance details in Table 4. The Accuracy, AUC, MNCC on the CDr dataset obtained the highest at 0.8889, 0.8, and 0.7906 respectively. The performance on CDf is also close with 0.778 of Accuracy, 0.9 of AUC, and 0.5976 of MCC. In general, the selected features by Saliency maps on the iCDr dataset revealed better performance in comparison with the random method but the AUC of the random method on the iCDf and UCr datasets are quite higher.

The performance of extracted features by Saliency maps from the iCDf dataset outperform the performance of randomly selected features on five comparative datasets. More specifically, the Accuracy, AUC, and MCC on CDr, CDf, iCDr, UCr, UCf of our features selection method are higher than randomly selected features. The UCr dataset reveals the exceptional performance with Accuracy of 0.875, 0.99375 of AUC, and 0.7746 of MCC. The UCf is also close with 0.75 of Accuracy, 0.8125 of AUC, and 0.5 of MCC. The further details are presented in Table 5. Furthermore, Table 6 exhibits the performance of features from the UCr dataset by the proposed method and random method. As observed, the performance on CDr, CDf, iCDf, and UCf dataset by our method outperform the random method. Besides, the performance of iCDr dataset by the random method is quite better. On iCDf dataset, the Accuracy obtained 0.75, 0.8125, and 0.5773 of AUC and MCC respectively. The results reveal the selected features by our method gained a good performance on iCDf dataset. We also presented the performance of the features selection method on UCf dataset in Table 7. The features selection by Saliency maps outperforms on CDr, CDf, and UCr dataset. Furthermore, the Accuracy and AUC of the iCDr dataset by our method is better whereas the MCC is not good enough in comparison with the random method. The extracted features by Saliency maps from the UCf dataset reach the highest performance on the CDr dataset with 0.6964 of Accuracy, 0.62 of AUC, and 0.2232 of MCC.

6 Conclusion

In this study, we recommend based on an explanation model with Saliency maps explanations. We select features from one dataset and will use the features for classification on other datasets to evaluate the effectiveness of the selected features. From the above experimental results, the feature selection method when using Convolutional Neural Network Models has a higher performance than the set of attributes from the original feature. Research has shown potential for use with datasets other than IBD datasets. Moreover, we can determine the most important bacterial or viral in the metagenomic datasets by the proposed method. Further studies can work on various data types to examine and select an optimal set of features for classification tasks.

References

1. Behrouzi A et al (2019) The significance of microbiome in personalized medicine. Clin Trans Med 8(1):16. https://doi.org/10.1186/s40169-019-0232-y
2. Kashyap PC et al (2017) Microbiome at the frontier of personalized medicine. Mayo Clin Proc 92(12):1855–1864. https://doi.org/10.1016/j.mayocp.2017.10.004
3. Gilbert JA, Quinn RA, Debelius J et al (2016) Microbiome-wide association studies link dynamic microbial consortia to disease. Nature 535(7610):94–103. https://doi.org/10.1038/nature18850
4. Petrosino JF (2018) The microbiome in precision medicine: the way forward. Genome Med 10:12. https://doi.org/10.1186/s13073-018-0525-6
5. Handelsman J (2004) Metagenomics: application of genomics to uncultured microorganisms. Microbiol Mol Biol Rev 68(4):669–685. https://doi.org/10.1128/MMBR.68.4.669-685.2004
6. Ma B, France M, Ravel J (2020) Meta-pangenome: at the crossroad of pangenomics and metagenomics. https://doi.org/10.1007/978-3-030-38281-0_9
7. Jang SJ, Ho PT, Jun SY, Kim D, Won YJ (2020) Dataset supporting description of the new mussel species of genus Gigantidas (Bivalvia: Mytilidae) and metagenomic data of bacterial community in the host mussel gill tissue. Data Brief 30:105651. Published 2020 Apr 29. https://doi.org/10.1016/j.dib.2020.105651.
8. Udugama B, Kadhiresan P, Kozlowski HN, Malekjahani A, Osborne M, Li VYC, Chen H, Mubareka S, Gubbay JB, Chan WCW (2020) Diagnosing COVID-19: The Disease and Tools for Detection. ACS Nano 14(4):3822–3835
9. Do TH, Nguyen TT, Nguyen TN, Le QG, Nguyen C, Kimura K, Truong NH (2014) Mining biomass-degrading genes through Illumina-based de novo sequencing and metagenomic analysis of free-living bacteria in the gut of the lower termite Coptotermes gestroi harvested in Vietnam. J Biosci Bioeng 118(6):665–671. https://doi.org/10.1016/j.jbiosc.2014.05.010
10. Chroneos ZC (2010) Metagenomics: theory, methods, and applications. Human Genomics 4(4):282–283. https://doi.org/10.1186/1479-7364-4-4-282
11. Ponsero Alise J, Hurwitz Bonnie L (2019) The promises and pitfalls of machine learning for detecting viruses in aquatic metagenomes. Front Microbiol 10
12. Tran PQ, Trieu NT, Dao NV, Nguyen HT, Huynh HX (2020) Effective opinion words extraction for food reviews classification. Int J Adv Comput Sci Appl (IJACSA) 11(7) https://doi.org/10.14569/IJACSA.2020.0110755
13. Auslander N et al (2020) Seeker: alignment-free identification of bacteriophage genomes by deep learning. bioRxiv. https://doi.org/10.1101/2020.04.04.025783
14. Garretto A, Hatzopoulos T, Putonti C (2019) virMine: automated detection of viral sequences from complex metagenomic samples. PeerJ. https://doi.org/10.7717/peerj.6695
15. Simonyan K et al (2014) Deep Inside Convolutional Networks: Visualising Image Classification Models and Saliency Maps. CoRR abs/1312.6034
16. Le Chatelier E, Nielsen T, Qin J et al (2013) Richness of human gut microbiome correlates with metabolic markers. Nature 500:541–546. https://doi.org/10.1038/nature12506
17. Kingma DP, Ba JL (2014) Adam: a method for stochastic optimization. arXiv:1412.6980v9

A Word + Character Embedding Based Relation Extraction Frame for Domain Ontology of Natural Resources and Environment

Ngoc-Vu Nguyen, Mai-Vu Tran, Hai-Chau Nguyen, and Quang-Thuy Ha

Abstract Building domain ontology is a challenging problem, and there are many different approaches for domain ontology construction. However, most of these approaches are still mainly using manual methods [1]. Ontology enrichment is a fairly standard approach in domain ontology construction, in which semi-automated methods and automated methods of ontology learning from a derived ontology. Relation extraction is one of the ways for ontology enrichment. Relation extraction techniques include law-based techniques, machine learning-based techniques with three typical methods: supervised learning, semi-supervised learning, and unsupervised learning. This paper proposes a word + character embedding-based relation extraction frame for the Vietnamese domain ontology of natural resources and environment. The model's effect was demonstrated by experiments in the domain of natural resources and the environment and achieving promising results.

Keywords Relation extraction · Vietnamese relation extraction · Domain ontology · Natural resources and environment

N.-V. Nguyen (✉)
Department of Information Technology, Ministry of Natural Resources and Environment (MONRE), 10 Ton That Thuyet, Cau Giay, Hanoi, Vietnam
e-mail: nnvu@monre.gov.vn

M.-V. Tran · H.-C. Nguyen · Q.-T. Ha
Vietnam National University, Hanoi (VNU), University of Engineering and Technology (UET), 144, Xuan Thuy, Cau Giay, Hanoi, Vietnam
e-mail: vutm@vnu.edu.vn

H.-C. Nguyen
e-mail: chaunh@vnu.edu.vn

Q.-T. Ha
e-mail: thuyhq@vnu.edu.vn

© The Author(s), under exclusive license to Springer Nature Singapore Pte Ltd. 2021   117
H. Kim et al. (eds.), *Information Science and Applications*, Lecture Notes in Electrical Engineering 739, https://doi.org/10.1007/978-981-33-6385-4_11

1   Introduction

Semantic relations represent relationships between concepts in the form of a hierarchical structure through associations. The classification of semantic relations is diverse, depending on the semantic features and the purpose and object. There are currently two semantic relation classification systems used quite commonly in semantic relations extraction problems, namely WordNet and Girju's [2].

Semantic relation extraction is an essential task in natural language processing (NLP) between two entities in a text. It is an intermediate step in various NLP applications, such as question-answering (Q&A), automatic knowledge base completion, information extraction, etc. [3]. The relation extraction task identifies the semantic relations between two entities, and with input is a given sentence and a predefined relation type. There are proposed techniques for semantic relations extraction. The most common and direct approach is to treat this problem as a classification problem. There have been several studies on the classification approach based on datasets and various deep learning and machine learning techniques [4–8], but these approaches all need the labeled datasets with good quality and large enough for training the classification model. Building these training datasets (for Vietnamese with fewer language resources) is often expensive and more expensive with the domain of natural resources and environment. One solution to this problem is a weakly supervised machine learning with much less training data [9]. The most notable weak monitoring machine learning method for relational extraction is bootstrapping. This method starts with a small set of the relation instance seeds and repeatedly learns to extract the patterns, and this method has been widely used [10, 11].

This paper presents a word + character embedding-based relation extraction frame for the Vietnamese domain ontology of natural resources and environment. Our work's main contributions are proposing a model to detect semantic relationships between concepts from a documentary repository on the domain of natural resources and environment. The relational sampling-based model uses the improved Snowball bootstrapping method when combining additional similar metrics based on different Vietnamese word + character embeddings.

The remainder of this paper is organized as follows. The next section presents the proposed word + character embedding based relation extraction frame for domain ontology. Experiments are described and discussed in Sect. 3, related studies are explored in Sect. 4, and the last section presents the conclusion of the paper.

2 The Word + Character Embedding Based Relation Extraction Frame for Domain Ontology of Natural Resources and Environment

2.1 Problem Formulation

The problem of semantic relations extraction for the domain of natural resources and environment is stated as follows:

Input: Concept/entity pairs from the Vietnamese domain ontology of natural resources and environment, unstructured document sets from the repository on the domain of natural resources and environment.

Output: We are discovering new concept/entity pairs with semantic relations.

Snowball's idea is quite simple, starting with a set of entity pairs (seeds) that are related according to the target relationship types. For example, if the target relationship is HeadquarteredIn, we could use seeds like (Microsoft, Redmond), (Google, Mountain View) and (Facebook, Palo Alto).

In a sizeable linguistic set, we need to look for the above seeds' simultaneous occurrence at a close distance. It is assumed that if two entities are related according to the type of target relationship appearing in the immediate vicinity, the context in which they appear is most likely the candidate relationship pattern. For example, we can find parts of sentences like "Google's headquarters in Mountain View" and "Redmond-based Microsoft" and extract patterns like "ORG's headquarters in LOC" and "LOC based ORG." We find additional pairs of entities (ORG, LOC) in the corpus that contain the HeadquarteredIn relation for these patterns. We add new entity pairs to the original seeds and repeat the process. More samples and pairs of entities will be added to the result until a specific condition is met.

2.2 The Proposed Word + Character Embedding Based Relation Extraction Frame for Domain Ontology

We propose a word + character embedding based relation extraction frame for Vietnamese domain ontology of natural resources and environment include six phases (Fig. 1):

(i)   Pre-train word + character embedding and data pre-processing.
(ii) Find sentences that contain seed tuples.
(iii) Tag entities.
(iv) Pattern clustering.
(v)   Find sentences that contain patterns.
(vi) Generate new seed tuples.

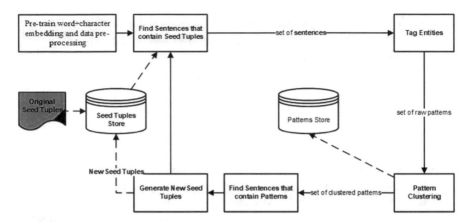

Fig. 1 The word + character embedding based relation extraction frame for domain ontology of natural re sources and environment

2.3 Pre-train Word + Character Embedding and Data Pre-processing

The resource repository of the domain of natural resources and environment in Vietnamese collected from 38,272 documents (614,395 pages) includes legal documents (19,666 documents), scientific articles (14,535 papers), scientific reports (2637 documents), technical standards (1434 documents) in PDF and Word format.

Data pre-processing using Vietnamese natural language processing technicals, specifically including:

Step 1: **Convert to text**

We used ABBYY software to convert from PDF files to text files automatically. Data after converting to text has a capacity of about 860 Mb.

Step 2: **Split sentences, word segmentation, POS tagging, NER and dependency parsing**

We used VnCoreNLP[1] (a Vietnamese natural language processing) toolkit for splitting sentences, word segmentation, POS tagging. The result of this step is a text file containing the data set that will be input to pre-train word + character embedding model, a total of 2,421,138 sentences.

Use the DeepLML-NER model proposed in [12] for Vietnamese named entity recognition (NER) with input is the set of 2.421,138 preprocessed sentences in step 2. The result is a text file in which all entities labeled CoNLL[2] format.

Step 3: **Pre-train word + character embedding for the domain of natural resources and environment**

---

[1]https://github.com/vncorenlp/VnCoreNLP.

[2]http://www.clips.uantwerpen.be/conll2002/ner/.

We used the Word2vec model to pre-train Vietnamese word + character embedding for natural resources and environment. Using Gensim[3] API for training the model with parameters dimensionality of the embedding vectors (size) is 300, and window size is 5. The word + character embedding for the domain of natural resources and environment (**nreEBD**) after pre-training published at the address: https://zenodo.org/record/3976712#.XzDKACgzaUk.

Step 4: **Retain sentences with more than two concepts/entities as input to the next steps.**

The data input of phase 2.3 is a set of sentences that contain more than two concepts/entities. So, in this step, we have to remove sentences that only contain one or no contain concept/entity. The result of this step is a text file containing a total of 1,980,356 sentences. We added NER tags for each NER label in the sentence with each sentence in the text file (**NREsentences**).

For example:

<ORG > Uỷ_ban_nhân_dân tỉnh Quảng_Nam </ORG > và < LOC > thành_phố Đà_Nẵng </LOC > phải báo_cáo < ORG > Bộ Tài_nguyên và Môi_trường </ORG > (< ORG > People's Committee of Quang Nam Province </ORG > and < LOC > Danang city </LOC > must report to < ORG > Ministry of Natural Resources and Environment </ORG >).

2.4 Find Sentences that Contain Seed Tuples

Initially, the seed tuples store contains a set of original seed tuples that built from the relationships contained in the Vietnamese domain ontology of natural resources and environment[4]. With each seed tuples from the seed tuples store, looking for the set of sentences containing seed tuples from the resource repository.

This step's output is a set of sentences containing seed tuples and at this step, using the Vietnamese word + character embedding to measure the semantic similarity to identify the sentences in the document containing concepts/entities.

2.5 Tag Entities

Based on the expanded set of entities, finding and tagging labels for entities in the sentence set are obtained in phase 2.4. After the entities are labeled, define the left, right, and middle components for the entities contained in the seed tuples set based on the resulting set. The above 5-component combination is called a formalized pattern $P = \langle L, C_1, T, C_2, R \rangle$ where:

---

[3]https://radimrehurek.com/gensim/index.html.

[4]https://zenodo.org/record/3836597#.XyjL9CgzaUl.

- L: left element (words, phrases before concept/entity number 1).
- C_1: concept/entity number 1.
- T: middle component (words, phrases between concept/entity number 1 and concept/entity number 2, usually represent the relationship of two concepts/entities).
- C_2: concept/entity number 2.
- R: right ingredient (words, phrases after concept/entity number 2).

For example with the sentence: *khi* **mực nước sông** *dâng cao* **ngập lụt** *thành phố* (*when* **the river level** *rises* **floods** *the city*), where:

- L: *khi (when)*
- C_1: **mực nước sông (the river level)**.
- T: *dâng cao (rises)*.
- C_2: **ngập lụt (floods)**.
- R: *thành phố (the city)*.

2.6 Pattern Clustering

After generating a large number of patterns, performing pattern grouping based on the Single-pass clustering algorithm with a predefined threshold. The similarity of the patterns is calculated based on the following formula:

$$sim(P_n, P_m) = \alpha * \cos(L_n, L_m) + \beta * \cos(T_n, T_m) + \gamma * \cos(R_n, R_m)$$

where:

- $sim(P_n, P_m)$: similarity between two patterns n and m.
- $\cos(L_n, L_m)$: cosine similarity of two left components.
- $\cos(T_n, T_m)$: cosine similarity of two middle components.
- $\cos(R_n, R_m)$: cosine similarity of two right components
- α, β, γ are parameters and $\alpha + \beta + \gamma = 1$

The grouped clusters will generate general patterns to serve the extraction. For example: with four patterns identified from phase 2.5

- *khi* **mực nước sông** *dâng cao* **ngập lụt** *thành phố* (*when* **the river level** *rises* **floods** *the city*)
- *khi* **triều cường** *dâng cao* **ngập lụt** *thành phố* (*when* **the tides** *rises* **floods** *the city*)
- *liên tục* **mưa lớn** *gây ra* **ngập lụt** *nhiều tuyến phố* (*continuous* **heavy rain** *caused* **flooding** *of many streets*)
- *liên tục* **mưa lớn** *gây ra* **ngập cục bộ** *nhiều tuyến phố* (*continuous* **heavy rain** *caused* **local flooding** *of many streets*)

Two general patterns were generated, example:

- <khi, C_1, *dâng cao*, C_2, thành phố > (< when, C_1, *rises*, C_2, the city >)
- <liên tục, C_1, *gây ra*, C_2, nhiều tuyến phố > (< continuous, C_1, *caused*, C_2, of many streets >)

2.7 Find Sentences that Contain Patterns

The identification and grouping of patterns in phase 2.6 is an essential step in determining important general patterns. To increase the semantics of similar measures, we propose applying more similar vocabulary level measures based on word embedding models. The formula for the similarity of the two patterns as follows:

$$\text{sim}(P_n, P_m) = \alpha_1 * \cos_{\text{tf--idf}}(L_n, L_m) + \alpha_2 * \cos_{\text{embed}}(L_n, L_m)$$
$$+ \beta_1 * \cos_{tf-idf}(T_n, T_m) + \beta_2 * \cos_{embed}(T_n, T_m)$$
$$+ \gamma_1 * \cos_{tf-idf}(R_n, R_m) + \gamma_2 * \cos_{embed}(R_n, R_m)$$

where:

- $\text{sim}(P_n, P_m)$: similarity between 2 patterns n and m
- $\cos_{tf-idf}$: cosine similarity of the two components is based on the tf-idf weight.
- $\cos_{embed}$: cosine similarity of the two components based on word embedding weights.
- $\alpha_1, \alpha_2, \beta_1, \beta_2, \gamma_1, \gamma_2$ are the parameters and $\alpha_1 + \alpha_2 + \beta_1 + \beta_2 + \gamma_1 + \gamma_2 = 1$

In phases 2.4, 2.5, and 2.6, the cosine similarity between phrases is calculated base on a vector representing that phrase. The vector representation for a phrase (concept/entity) as below example:

$$\mathbf{V}_{\text{tài nguyên môi trường}} = \mathbf{W}_{\text{tài}} + \mathbf{W}_{\text{nguyên}} + \mathbf{W}_{\text{mòi}} + \mathbf{W}_{\text{trường}}$$
$$(\mathbf{V}_{\text{natural resources and environment}} = \mathbf{W}_{\text{natural}} + \mathbf{W}_{\text{resources}} + \mathbf{W}_{\text{and}} + \mathbf{W}_{\text{environment}})$$

where $\mathbf{V}$ is the vector representing the phrase, $\mathbf{W}$ is the feature vector of a token.

The feature vector of the token i is W_i which is the concatenation of feature vectors of token i, including:

$$W_i = W_i^w \oplus W_i^c \oplus W_i^f$$

where:

- W_i^w is word + character embedding-based vector (using nreEBD word + character embedding pre-trained in phase 2.1).
- W_i^c is character-level feature vector (trained by using the proposed Deep-LML model in [12]).

- W_i^f is prefix feature vector (trained by using the proposed Deep-LML model in [12]).

2.8 Generate New Seed Tuples

Based on the general patterns (P), search in the corpus for pairs of $\langle A', B' \rangle$ concepts/entities (new seed tuples) satisfying the pattern. After finding new patterns, the general patterns are evaluated according to their reliability.

New seed tuples $T = \langle A', B' \rangle$ will fall into one of the following cases:

- Positive: If $\langle A', B' \rangle$ was on the seed tuples list.
- Negative: If $\langle A', B' \rangle$ only one of two (A' or B') appears in the seed tuples list.
- Unknown: If $\langle A', B' \rangle$, neither A' or B' will appear in the seed tuples list. The Unknown set is considered the set of new seed tuples for the next loop.

Snowball will calculate the accuracy of each pattern based on its Positive and Negative numbers and select the top N patterns with the highest scores. The reliability of the pattern is calculated by the formula:

$$belief(P) = \frac{positive}{positive + negative}$$

Find new seed tuples (T) for the next loop:

For each pattern in the top N list selected will be the pairs in the new seed tuples set, further added to the new loop.

Similar to the pattern, these pairs are estimated as follows:

$$conf(T) = 1 - \prod_{i=0}^{|p|}(1 - belief(P))$$

The system will select M pairs with the best evaluation and these M pairs will be used as seed tuples for the next sampling. The system will continue to go back to step 2.3. The above process continues to repeat until the system does not find a new pair or repeats it by the number of times that we specified.

3 Experimental Results

3.1 Data

Training data is a set of 1,980,356 sentences (**NREsentences**) pre-processed in phase 2.3, each sentence containing more than two concepts/entities and name entity's labels tagged.

Vietnamese pre-train word + character embedding is **nreEBD** (Step 2, phase 2.3) with dimensionality of the embedding vectors (size) is 300, and window size is 5.

Character-level feature vectors and prefix feature vectors for word in the dictionary trained by Deep-LML model [12] on training data is a set of 2,421,138 Vietnamese sentences for the domain of natural resources and environment.

Seed tuples set: 114 relationships from the domain ontology.

Evaluation: the evaluation of new relationship pairs is done manually.

3.2 Training

The experiment is applied on different word embedding in Vietnamese and on the parameters $\alpha_1, \alpha_2, \beta_1, \beta_2, \gamma_1, \gamma_2$ and different threshold R to choose the optimal result. Do the test with the following scenarios:

Scenario 1 (NRE$_{NoFull}$): Using only tf-idf weighted features, not using word features (embedding words), character level features (embedding characters), and prefix features.

Scenario 2 (NRE$_{Word}$): Using the tf-idf weighting feature, word feature (embedding word: word2vecVN) but not using: character level feature (embedding character) and prefix feature.

Scenario 3 (NRE$_{Word+Char}$): Using the tf-idf weighting feature, word feature (embedding word: word2vecVN), character level feature (embedding character) but not using the prefix feature.

Scenario 4 (NRE$_{Full}$): Full utilization of features based on tf-idf weights, word features (embedding of words), character level features (embedding of characters), and prefix features. With this scenario, divided into 3 scenarios: **NRE$_{Full+word2vecVN}$** (the set of embedding words is word2vecVN[5]), **NRE$_{Full+PhoBERT}$** (the set of embedding words using is PhoBERT[6]) and **NRE$_{Full+nreEBD}$** (the set of embedding words using is nreEBD[7]).

---

[5]https://github.com/sonvx/word2vecVN.

[6]https://github.com/VinAIResearch/PhoBERT.

[7]https://zenodo.org/record/3976712#.XzDKACgzaUk.

3.3   Evaluation and Analysis

The results of the experiment were run a grid-search on the parameters $\alpha_1, \alpha_2, \beta_1, \beta_2, \gamma_1, \gamma_2$ and R, the optimal parameters set is selected: $\alpha_1 = 0.1; \alpha_2 = 0.1; \beta_1 = 0.3; \beta_2 = 0.3; \gamma_1 = 0.1; \gamma_2 = 0.1; R = 0.85$

Scenario	Total of concepts	Total of concepts (exactly)	F1 (%)
NRE_{NoFull}	2479	1735	70
NRE_{Word}	2545	1835	72.10
$NRE_{Word+Char}$	2584	1892	73.21
$NRE_{Full+word2vecVN}$	2610	1954	74.86
$NRE_{Full+PhoBERT}$	2572	1926	74.88
$NRE_{Full+nreEBD}$	2558	1952	76.30

4   Related Work

Zhang et al. [13] proposed a framework for constructing semantic bootstrapping models that incorporate trigger words as semantic constraints to guide the iterations. In this work, a rule for defining the trigger word, the method for sample representation, the method for similar measurement and the evaluation method are provided. Additionally, a bottom-up kernel method is defined to define a pattern from a new sentence with a relationship orientation or not. We concentrate data representations in relation extraction based on a Feature-inferring architecture of Word + character.

Jagan et al. [14] proposed a bootstrapping-based approach to extract the relationship between two different concepts in the domain of travel and news of Tamil language based on using of Universal Networking Language (UNL) to present sentence with features before entering the bootstrapping loop. In our frame, because the data domain fixed then the pre-processing step and Feature-inferring architecture of Word + character has performed only one time.

5   Conclusions and Future Work

This paper proposed a word embedding based relation extraction frame for the Vietnamese domain ontology of natural resources and environment. Experimental plans to investigate the effects of improved similarity measure participants conducted: (i) not using semantic features of words, character-level features, and prefix features; (ii) using semantic features (embedding words) but not using character-level features, and prefix features; (iii) using a combination of semantic features of words, character-level features, and prefix features. Experimental results have shown that the accuracy increases according to the above test scenarios, and the use of the pre-train word embeddings for the domain of natural resources and environment (nreEBD) gives the highest results.

References

1. Sammut C, Webb GI (eds) Ontology learning. In: Encyclopedia of machine learning and data mining, Boston, MA, Springer, US, pp 937–938
2. Girju, C.R.: Text Mining for Semantic Relations. PhD. Thesis. The University of Texas at Dallas (2002)
3. Chan YS, Roth D (2010) Exploiting background knowledge for relation extraction. COLING 152–160
4. Chan YS, Roth D (2011) Exploiting syntactico-semantic structures for relation extraction. ACL 551–560
5. Jiang J, Zhai CX (2007) A systematic exploration of the feature space for relation extraction. HLT-NAACL 113–120
6. Kambhatla N (2004) Combining lexical, syntactic, and semantic features with maximum entropy models for information extraction. ACL (Poster and Demonstration)
7. Zhou G, Su J, Zhang J, Zhang M (2005) Exploring various knowledge in relation extraction. ACL 427–434
8. Brin S (1998) Extracting patterns and relations from the World Wide Web. WebDB 172–183
9. Mintz M, Bills S, Snow R, Jurafsky D (2009) Distant supervision for relation extraction without labeled data. ACL/IJCNLP 1003–1011
10. Nguyen N-V, Nguyen T-L, Nguyen Thi C-V, Tran M-V, Nguyen T-T, Ha Q-T (2019) Improving named entity recognition in vietnamese texts by a character-level deep lifelong learning model. Vietnam J Comput Sci 6(4):471–487
11. Li Qing, Li Lili, Wang Weinan, Li Qi, Zhong Jiang (2020) A comprehensive exploration of semantic relation extraction via pre-trained CNNs. Knowl Based Syst 194:105488
12. Meng Qu, Ren Xiang (2018) Yu Zhang. Weakly-supervised relation extraction by pattern-enhanced embedding learning. WWW, Jiawei Han, pp 1257–1266
13. Zhang Chunyun, Weiran Xu, Ma Zhanyu, Gao Sheng, Li Qun, Guo Jun (2015) Construction of semantic bootstrapping models for relation extraction. Knowl Based Syst 83:128–137
14. Jagan B, Parthasarathi R, Geetha TV (2019) Bootstrapping of semantic relation extraction for a morphologically rich language: semi-supervised learning of semantic relations. Int J Semantic Web Inf Syst 15(1):119–149

A Prototype Implementation of NNEF Execution Framework with CUDA Acceleration

Nakhoon Baek⦿

Abstract Recently, we have many research works on the neural networks and their related issues. For exchangeability of neural network frameworks, the Neural Network Exchange Format (NNEF) specification is now widely used. Due to very large size of these neural networks, their accelerations are actively explored, and can be achieved through parallel processing techniques. In this paper, we present a prototype implementation of NNEF execution system with parallel-processing accelerations based on CUDA (compute unified device architecture). We will tune the prototype acceleration to achieve more remark-able speed ups.

Keywords Neural network · NNEF · Parallel processing · CUDA · Acceleration

1 Introduction

In the practical field of deep learning methods, we have several widely used neural network frameworks: *TensorFlow* [1], *Caffe* [2], *Keras* [3], and others.

To exchange the neural network structures and their related data sets, we need standard protocols and/or file formats. The *Neural Network Exchange Format* (NNEF) [4] is exactly the de facto standard file format, for the neural network frameworks, and managed by the *Khronos Group* [5], an industrial standard organization.

Although the NNEF specification provides the file format only, the file format can be parsed and even executed to simulate the specified neural network. Recently, the Khronos NNEF official *GitHub* site presented an NNEF interpreter, implemented in C++ templates [6]. To accelerate the NNEF execution, *OpenCL* and *OpenMP* have been used in the previous works, [7] and [8].

In this paper, we present an NNEF interpreter, accelerated with the parallel execution features of *CUDA* (compute unified device architecture) [9], as shown in Fig. 1. Though our system is based on the original Khronos NNEF GitHub implementation [6], our modification shows remarkable speed-ups.

N. Baek (✉)
School of Computer Science and Engineering, Kyungpook National University, Daegu 41566, Republic of Korea
e-mail: nbaek@knu.ac.kr

© The Author(s), under exclusive license to Springer Nature Singapore Pte Ltd. 2021   129
H. Kim et al. (eds.), *Information Science and Applications*, Lecture Notes in Electrical Engineering 739, https://doi.org/10.1007/978-981-33-6385-4_12

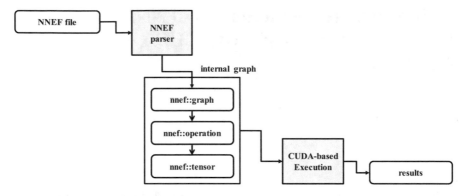

Fig. 1 The overall flow of our NNEF execution system

2 Design and Implementation

Based on the original Khronos NNEF interpreter [6], we used its NNEF parser routines, as is. Thus, after parsing an NNEF file successfully, we have an NNEF graph structure with the following three major components:

- **A single NNEF graph**: it contains the whole NNEF graph representation. Internally, an NNEF graph consists of NNEF operations and NNEF data nodes.
- **A sequence of NNEF operations**: this sequence represents all the operations in the NNEF graph. Each NNEF operation node has its corresponding NNEF data nodes.
- **A set of NNEF data nodes**: the set contains all the data nodes used by the NNEF operations.

Our implementation of the NNEF execution process is actually a graph traversal algorithm, which retrieves all the NNEF operations, with their corresponding NNEF data nodes. Our current implementation performs each NNEF operation, during the graph traversal process.

For the acceleration of this execution process, we used CUDA kernel programs, which are carefully designed for a set of NNEF operations or a single NNEF operation, according to the operation complexities. Our current implementation can provide most NNEF operations, and some not-supported operations will be covered soon.

The key idea with our implementation can be summarized in a step-by-step manner, as follows:

- **step 1. memory coalescing**—In the original NNEF parser implementation, they allocate all the NNEF data memory to its own local space. In contrast, we change the design and also the implementation to coalesce all the NNEF memory to a single and large memory area.

- **step 2. memory transfer to the CUDA area**—As an efficient CUDA implementation, we perform as many operations as possible, in the CUDA memory area. For this purpose, we copy the initial NNEF memory area to its corresponding CUDA memory area.
- **step 3. CUDA kernel executions**—For each NNEF operation, we implement its corresponding CUDA kernel, and execute it to perform the NNEF operation.
- **step 4. memory transfer back to the main memory**—At the end of all the NNEF operations, the CUDA memory area contains the final result of the NNEF execution. We copy back the CUDA memory area to the main memory, for further processing, as same as typical CUDA applications.

After these implementations, we compared the new CUDA-based execution to the original C++-template-based official implementation [6]. As an example, we used the *MNIST hand-writing training* case [10, 11], with the NNEF file shown in Fig. 2. After processing this NNEF file, with proper data sets from the MNIST case, we get the sample results, as shown in Fig. 3. Including this one, our test cases show that our implementation works well, and show some remarkable speed ups, in comparison to the original C ++ template implementation.

```
version 1.0;
graph mnist_query( input ) -> ( output ) {
  input = external<scalar>( shape=[784,1] );
  wih = variable<scalar>( shape=[200,784], label="mnist-wih" );
  who = variable<scalar>( shape=[10,200], label="mnist-who" );
  hidden = sigmoid( matmul( wih, input ) );
  output = sigmoid(matmul( who, hidden ) );
}
```

Fig. 2 An example NNEF file

0.04075291
0.00826926
0.05436546
0.02137949
0.01371213
0.03757233
0.00205216
0.90605090
0.02076551
0.03548543

Fig. 3 An example result from our NNEF execution

3 Conclusion

In this work, we presented an NNEF interpreter system, with CUDA-based accelerations. Since CUDA can provide massively parallel execution with GPUs, our final result shows remarkable speed ups. In contrast, our CUDA-based implementation may have some drawbacks, originated from the CUDA library itself. As an example, our CUDA-based implementation needs GPU support, for its acceleration. Thus, in some cases, where GPUs are not available, our work shows only limited performance. It can be solved through using OpenCL or other CPU-based parallel execution libraries, as another way of accelerating with CPUs. In the near future, we will release the full support framework for NNEF and its related specifications.

Acknowledgements This work has supported by Basic Science Research Program through the National Research Foundation of Korea (NRF) funded by the Ministry of Education (Grand No.NRF-2019R1I1A3A01061310).

References

1. Abadi M et al (2015) TensorFlow: large-scale machine learning on heterogeneous systems. white paper available from tensorflow.org
2. Jia Y et al (2014) Caffe: convolutional architecture for fast feature embedding. In: Proceedings 22nd ACM international conference on multimedia (MM'14)
3. Keras Homepage. http://www.keras.io. Last Accessed 30 May 2020
4. The Khronos NNEF Working Group (2019) Neural network exchange format. Version 1.0.2. Khronos Group
5. Khronos Group. http://www.khronos.org/. Last Accessed 30 May 2020
6. KhronosGroup/NNEF-Tools. https://github.com/KhronosGroup/NNEF-Tools. Last Accessed 30 May 2020
7. Yu M, Chen T, Lee J (2020) Accelerating NNEF framework on OpenCL devices using clDNN. IWOCL
8. Baek N, Park S-J (2020) An OpenMP-based parallel execution of neural networks specified in NNEF, ICA3PP 2020
9. CUDA Zone. http://developer.nvidia.com/cuda-zone. Last Accessed 09 Jun 2020
10. Rashid T (2016) In: Make your own neural network, 1st edn. CreateSpace Independent Publishing Platform
11. MNIST data set. https://github.com/makeyourownneuralnetwork/makeyourownneuralnetwork/mnist_dataset/. Last Accessed 09 Jun 2020

Condition Monitoring for Induction Motor Overload Using Sound

Nguyen Cong Phuong

Abstract Induction motors play an important role in electromechanical energy conversion and therefore are widely used not only in industry but also in home appliances. Overloading is among those which can shorten the operating life of electric motors. This research aims to use a microphone to distinguish between full load operation and overload operation of induction motors. Three acoustic features and six classification models are evaluated to establish the overload detection system based on sound analysis. Results show that this is a promising way to monitor induction motor overload.

Keywords Induction motor · Overload · Sound analysis · Machine learning

1 Introduction

Induction motors are used in various sectors of life. They offer many advantages, such as high reliability and low pollution. They are usually employed continuously for long duration, so condition monitoring and fault diagnosis for induction motors is recently almost compulsory in complex systems, because any interruption can cause huge losses.

According to [1, pp. 59], faults of induction motors can be divided into four classes: bearing (e.g. wear out of bearings), stator (insulation damages, for instance), rotor (broken rotor bars or cracked rotor end-rings), and other faults (eccentricity, for example).

Because induction motor is an electromechanical conversion device, so in order to detect its faults, one can use electrical and mechanical signals. Besides, thermal and chemical signals are also clues leading to some certain faults.

Electrical signals consist of current, voltage, magnetic flux, etc. Motor Current Signature Analysis (MCSA) [2] is probably the most popular approach. This is to detect faults such as eccentricity, broken rotor bars or cracked rotor end-rings,

N. C. Phuong (✉)
Hanoi University of Science and Technology, Hanoi, Vietnam
e-mail: phuong.nguyencong@hust.edu.vn

© The Author(s), under exclusive license to Springer Nature Singapore Pte Ltd. 2021   133
H. Kim et al. (eds.), *Information Science and Applications*, Lecture Notes
in Electrical Engineering 739, https://doi.org/10.1007/978-981-33-6385-4_13

opening or shorting of stator phase winding, bent shaft, bearing and gearbox failures. Voltage signals are employed to detect supply voltage unbalance [3] or stator winding inter-turn faults [4]. Magnetic flux can be used for fault detection in rotor cage [5] or eccentricity [6].

Vibration, noise, and torque are the most widely used mechanical signals for condition monitoring. Vibration analysis are employed in [7] for bearing fault diagnosis and in [8] for detection of eccentricity. Noise monitoring is another approach. It can be applied to detect eccentricity [9] and local defects of gearboxes [10], or even to predict resident lifetime [11]. For torque monitoring, one can employ it for gearbox fault detection [12].

Thermal monitoring is another good way for induction motors, for example, to identify turn-to-turn faults and bearing faults [13].

In chemical monitoring, insulation degradation can be detected chemically using coolant gas analysis [14]. Oil analysis is also a chemical tool to detect wear debris because faults such as misalignment or overload may lead to wearing [15].

In operating induction motors, overload is an abnormal condition. When a motor is overloaded, it can draw more current, causing excessive temperatures. A too high temperature may burn motors. Besides, overload can result in tooth breakage, or wear in roller bearings and gears. In order to detect these consequences, we can use approaches mentioned above.

We try to detect this phenomenon itself, not its consequences. In our research, the 100% overload (200% of full load) is distinguished from the full load operation by sound analysis. Our approach has three advantages. Firstly, sounds of overload appear earlier than its consequences (excessive temperatures, tooth breakage, etc.), so detecting an abnormal sound can prevent these consequences. Secondly, a sound sensor and its installation and maintenance are inexpensive. And finally, it does not need to stop the motor during the detection.

The organization of the paper is as follows. Section 2 informs the studies recently presented in the literature that refer to motor fault detection using audio signal. Section 3 is about the corpus used in this research. Section 4 describes in detail the proposed method, experiments, and results. Section 5 concludes the paper and presents future developments.

2 Related Works

Among mechanical signals, sounds of electric motors are reported to be capable of detecting some mechanical and electrical problems. Compared to other signals, sounds can be acquired inexpensively and easily using microphone(s). Authors in [16] employ a microphone and a high-resolution spectral analysis based on the MUSIC algorithm for detection of three faults: unbalance, bearing, and broken rotor bars. [17] measures acoustic signals by the acoustic camera (microphone arrays), the acoustic spectrum is then used to detect static eccentricity and soft foot. A method for detecting abnormal sounds of the motor for condition monitoring and fault diagnosis is proposed in [18]. Akcay and Germen [19] presents an approach to detect bearing and broken rotor bars faults. The acoustic signals are collected by five microphones positioned around the motor. Then the presence of faults is determined by calculating the power spectral density. In [20], acoustic signals (recorded with a digital voice recorder) are combined with the Bayes classifier and k-Nearest Neighbor (kNN) classifier to detect faulty rotor bars and shorted rotor coils. The approach in [19] is improved in [21] by using the Self-Organizing Maps method for the separation of healthy and faulty motors. The diagnosis of electric motor in [22] is done with acoustic noise and convolutional neural network. This approach is to detect tooth damage in the gearbox.

To our best knowledge, no study of using sounds to detect induction motor overload has been published so far.

3 The Corpus

The object of our study is a 4 kW three-phase induction motor. Sounds of full load and overload operations are recorded by a microphone. This microphone is connected directly to the audio input of a laptop (see Fig. 1), so the recorder is actually the laptop's sound card. The distance between the motor and the microphone is about 20 cm. soundtracks are acquired with the following parameters: sampling frequency is 44.1 kHz, bit resolution is 16, and mono channel. Totally, 137 s of full load signals and 112 s of overload signals are collected in the laboratory of Hanoi Electromechanical Manufacturing Company. This corpus is used in the next step of our research. Spectrograms of a full load sound and an overload one are in Fig. 2. Frequency of these signals ranges approximately from 10 Hz to about 8 kHz. It can be seen that the differences of those spectrograms are visible. When we listen to them, we notice that overload sounds are a little more jarring than full load sounds. But we can observe also the discontinuity of fundamental frequency, so if only F0 is employed to classify signals, the accuracy can be low.

    Three-phase induction motor   Microphone   Laptop

Fig. 1 The sound recording/monitoring system

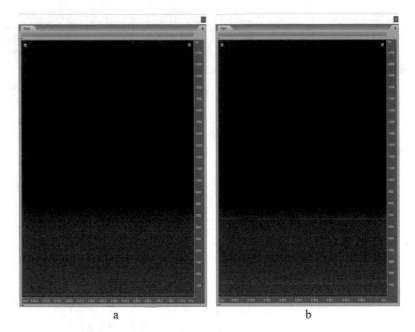

Fig. 2 Spectrograms of full load **a** and overload **b** sounds

4 The Experiments

In order to classify signals into full load sounds and overload sounds, a two-class classifier should be used. It includes a set of discriminant features and a classification model. Our discriminant features consist of F0, Mel-Frequency Cepstral Coefficients (MFCC, 12 coefficients [23]), and Band Energy Ratio (BER, 4 bands, [24]). These very popular features in audio signal processing form a starting feature set of 17 elements. Then the Principal Component Analysis (PCA [25]) is applied to reduce the dimension of our feature vector without losing too much information. Essentially, the PCA extracts the important information from the original feature set to rebuild them as a set of new orthogonal features (principal components), and hence to gain a better representation of classed by reducing the number of features. In order to find the appropriate number of new features, we test six classification models: artificial neural network (ANN [26]), decision tree (DT [27]), fuzzy inference system (FIS [28]), Gaussian mixture model (GMM [29]), kNN [30], and support vector machine (SVM [31]).

In the first phase of our experiments, for each classification model, the number of features varies from 1 to 10, and a tenfold cross validation is applied to find the best number of features for that model. The missed detection (MD) and the false alarm (FA) are employed in model evaluation. The tradeoffs between them (and therefore the corresponding numbers of features) can be found by the Receiver Operating Characteristics (ROC) curves in Figs. 3 and 4. Results of this step are in Table 1.

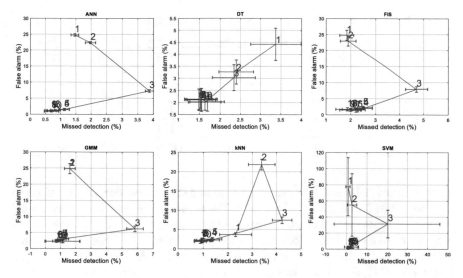

Fig. 3 ROC curves and deviations (full scale). 1, 2, 3, …: number of features. Horizontal bars show MD deviations, vertical bars are for FA deviations. When the number of features is small: curves of ANN, FIS, GMM, and SVM have a similar trend; MD and FA of DT are correlated

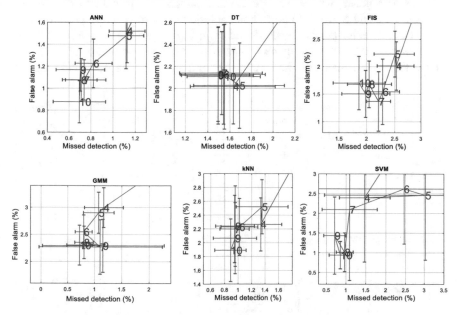

Fig. 4 ROC curves and deviations (zoom in). 4, 5, 6, …: number of features

Table 1 Performances of models with number of features varying from 1 to 10. σ: deviation

Model	Nb. of features	MD (%)	σ_{MD} (%)	FA (%)	σ_{FA} (%)
ANN, 20 hidden neurons	10	0.69	0.14	0.88	0.20
DT	4	1.63	0.29	2.02	0.42
FIS	9	1.98	0.37	1.51	0.42
GMM, 2 mixtures	6	0.80	0.16	2.55	0.31
kNN, 7 neighbors	4	1.34	0.31	2.26	0.39
SVM	10	0.85	0.29	0.94	0.35

These selected MDs and FAs are rather low, and there is no significant difference between the MD and the FA of a model. These parameters are stable because the corresponding deviations are rather small. Moreover, it should be noted that except FIS, for the five remaining models, MDs are smaller than FAs. This is very notable because normally, MD is more critical and important than FA in a detection problem like this. For DT, GMM, and kNN, when the number of features is kept increasing, their performances turn out to be worse. Their best numbers of features are only 4, 6, and 4, respectively.

In the next phase, we increase the number of features from 11 to 17, focusing on ANN, FIS, and SVM. The best numbers (and the corresponding performances) within that range for these models (found from Fig. 5) are listed in Table 2. These performances do not change much compared to the previous phase. MDs are still smaller than FAs. Finally, the best candidate for our system is still ANN because of

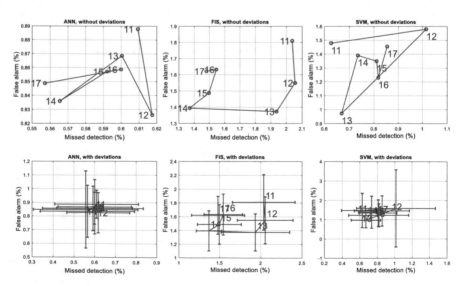

Fig. 5 ROC curves and deviations (number of features varies from 11 to 17). The first row is MD versus FA. The second row is MD versus FA with corresponding deviations

Table 2 Performances of models with number of features varying from 11 to 17

Model	Nb. of features	MD (%)	σ_{MD} (%)	FA (%)	σ_{FA} (%)
ANN, 25 hidden neurons	12	0.62	0.15	0.83	0.15
FIS	14	1.37	0.16	1.39	0.29
SVM	13	0.67	0.17	0.97	0.39

Table 3 Importance of features (without PCA), tested with ANN

Feature	Nb. of hidden neurons	MD (%)	σ_{MD} (%)	FA (%)	σ_{FA} (%)
Full set	40	0.60	0.21	0.83	0.22
F0	5	1.61	0.35	24.78	1.57
MFCC	20	0.61	0.19	0.81	0.24
BER	20	9.79	0.94	16.42	0.84

its performance. In addition, its computation speed is higher than those of the two other models.

To clarify the importance of these features, we evaluate each one separately using ANN. The obtained results (summarized in Table 3) demonstrate that BER is not very discriminative. For F0, it returns a low MD but a too high FA. On the contrary, those of MFCC are even better than first rows of Table 1 and Table 2. It should be noted that its deviations are a little higher and less balanced than those obtained with PCA. Because MFCC has 12 coefficients, the performance of MFCC in Table 3 can explain why the best number of features for ANN in Table 2 is 12. If the full set is employed, the number of hidden neurons is doubled compared to that of MFCC. Our experiments show that PCA can firstly reduce the number of inputs of a classifier (and therefore its complexity) and secondly can improve its operation's stability.

Finally, we choose F0, MFCC, BER, and the usage of PCA for our application, because this combination provides small and balanced deviations and less complicated model. It is difficult to compare the accuracy of our approach to the other ones, because we cannot find any published reports of accuracy of induction motor overload detection so far.

In order to detect overloads, the system in Fig. 1 is implemented. Sounds of the induction motor is acquired by a microphone. For each sound signal, a Hanning window is applied to frames of 1024 samples, the overlap is 512 samples, then F0, MFCC, and BER are computed from each frame and fed to PCA to obtain a set of 12 new features. They are then fed to a function fitting neural network with 12 inputs, one output, a hidden layer with 25 neurons, and trained with Levenberg–Marquardt backpropagation. One second of sound takes our algorithm (installed in a laptop with Intel Core i5 and 8 GB of RAM) about 0.25 s to process. The output of this network will tell that the recorded signal came from a full load or an overload. By detecting overload condition, this system can come to the root of some problems, such as tooth breakage, wear in roller bearings and gears, excessive temperatures, or even burning of motors.

5 Conclusions

This paper presents a method for monitoring induction motor overload using audio signals and artificial neural network. Audio signals are recorded by a microphone placed near an induction motor to monitor its overload. A feature set (including F0, MFCC, and BER) and six classification models are evaluated. The different importance of these features are also tested. Experiments prove that ANN fits our approach. Our proposed system can be an online monitoring method because it does not need to stop the motor. This system requires a microphone, so it is inexpensive. It is also flexible, meaning that if we can collect sounds of other faults, we can upgrade it by retraining it. Filtering techniques should be applied if this method is moved to industrial environment to reduce noises. Future developments can be related to other levels of overload (e.g. 10%, 20%, etc.) and other faults, such as eccentricity, bearing, rotor bars, etc.

References

1. Malik H, Iqbal A, Yadav AK (2020) In: Soft computing in condition monitoring and diagnostics of electrical and mechanical systems. Springer
2. Miljković D (2020) Brief review of motor current signature analysis. https://www.researchg ate.net/publication/304094187_Brief_Review_of_Motor_Current_Signature_Analysis. Last Accessed 07 Aug 2020
3. El Menzhi L, Saad A (2009) Induction motor fault diagnosis using voltage park components of an auxiliary winding-voltage unbalance. In: ICEMS 2009
4. Urresty JC, Riba JR, Delgado M, Romeral L (2012) Detection of demagnetization faults in surface-mounted permanent magnet synchronous motors by means of the zero-sequence voltage component. IEEE Trans Energy Convers 27(1):42–51
5. Cruz SMA, Cardoso AJM (2006) Diagnosis of rotor faults in closed-loop induction motor drives. In: Proceedings IEEE industry application society annual meeting, Tampa, FL, United States, CD-Rom, pp 1–8
6. Cruz SMA, Stefani A, Filippetti F, Cardoso AJM (2008) Diagnosis of rotor faults in traction drives for railway applications. In: 18th International Conference on Electrical Machines
7. Choi SD, Akin B, Rahimian MM, Toliyat HA (2012) Performance-oriented electric motors diagnotics in modern energy conversion systems. IEEE Trans Indus Electron 59(2):1266–1277
8. Dorrell DG, Thomson WT, Roach S (1997) Analysis of airgap flux, current, and vibration signals as a function of the combination of static and dynamic airgap eccentricity in 3-phase induction motors. IEEE Trans Indus Appl 33(1):24–34
9. Ellison AJ, Yang SJ (1971) Effects of rotor eccentricity on acoustic noise from induction machines. In: Proceedings of the institution of electrical engineers. vol 118. No. 1. IET Digital Library
10. Baydar N, Ball A (2001) A comparative study of acoustic and vibration signals in detection of gear failures using Wigner–Ville distribution. Mechanical Syst Signal Process 15(6):1091–1107
11. Scanlon P, Kavanagh DF, Boland FM (2013) Residual life prediction of rotating machines using acoustic noise signals. IEEE Trans Instrum Measurement 62(1):95–108
12. Henao H, Kia SH, Capolino GA (2011) Torsional-vibration assessment and gear-fault diagnosis in railway traction system. IEEE Trans Indus Electron 58(5):1707–1717
13. Milanfar P, Lang JH (1996) Monitoring the thermal condition of permanent-magnet synchronous motors. IEEE Trans Aerospace Electron Syst 32(4):1421–1429

14. Tavner PJ (2008) Review of condition monitoring of rotating electrical machines. Electric Power Appl IET 2(4):215–247
15. Laghari MS, Ahmed F, Aziz J (2010) Wear particle shape and edge detail analysis. In: Computer and automation engineering (ICCAE)
16. Garcia-Perez A, Romero-Troncoso RJ, Cabal-Yepez E, Osornio-Rios RA, Lucio-Martinez JA (2011) Application of high-resolution spectral analysis for identifying faults in induction motors by means of sound. J Vib Control 18(2011):1585–1594
17. Orman M, Pinto CT (2013) Acoustic analysis of electric motors in noisy industrial environment. In: 12th IMEKO TC10 workshop on technical diagnostics
18. Ono Y, Onishi Y, Koshinaka T, Takata S, Hoshuyama O (2013) Anomaly detection of motors with feature emphasis using only normal sounds. In: Proceedings of the 2013 IEEE international conference on acoustics speech and signal processing, pp 2800–2804
19. Akcay H, Germen E (2013) Identification of acoustic spectra for fault detection in induction motors. In: 2013 Africon, IEEE, pp 1–5
20. Glowacz A (2014) Diagnostics of DC and induction motors based on the analysis of acoustic signals. Meas Sci Rev 14:257–262
21. Germen E, Başaran M, Fidan M (2014) Sound based induction motor fault diagnosis using Kohonen self-organizing map. Mech Syst Signal Process 46:45–58
22. Choi DJ, Han JH, Park SU, Hong SK (2019) Diagnosis of electric motor using acoustic noise based on CNN. In: 22nd international conference on electrical machines and systems
23. Carey MJ, Parris ES, Lloyd–Thomas H (1999) A comparison of features for speech, music discrimination. In: ICASSP'99, pp 149–152
24. McCowan I, Gatica-Perez D, Bengio S, Lathoud G, Barnard M, Zhang D (2005) Automatic analysis of multimodal group actions in meetings. IEEE Trans Pattern Analy Machine Intell 27(3):305–317
25. Jolliffe I (1986) Principal component analysis. Springer, New York
26. Bishop CM (1995) In: Neural networks for pattern recognition, Oxford University Press
27. Breiman L, Friedman J, Olshen R, Stone C (1984) Classification and regression trees, Belmont, USA
28. Jang J-SR (1993) ANFIS: adaptive-network-based fuzzy inference systems. IEEE Trans Syst Man Cybernet 23(3):665–685
29. McLachlan G, Peel D (2000) Finite mixture models. Wiley Inc., Hoboken, NJ
30. Scheirer E, Slaney M (1997) Construction and evaluation of a robust multifeature music/speech discriminator. In: ICASSP' 97, vol 2. pp 1331–1334
31. Vapnik VN (1998) Statistical learning theory. Wiley, N.Y.

Analysing the Adversarial Landscape of Binary Stochastic Networks

Yi Xiang Marcus Tan, Yuval Elovici, and Alexander Binder

Abstract We investigate the robustness of stochastic ANNs to adversarial attacks. We perform experiments on three known datasets. Our experiments reveal similar susceptibility of stochastic ANNs compared to conventional ANNs when confronted with simple iterative gradient-based attacks in the white-box settings. We observe, however, that in black-box settings, SANNs are more robust than conventional ANNs against boundary and surrogate attacks. Consequently, we propose improved attacks against stochastic ANNs. In the first step, we show that using stochastic networks as surrogates outperforms deterministic ones, when performing surrogate-based black-box attacks. In order to further boost adversarial success rates, we propose in a second step the novel Variance Mimicking (VM) surrogate training, and validate its improved performance.

Keywords Adversarial machine learning · Stochastic neural network · Binary neural network · Black-box attack

1 Introduction

Artificial Neural Networks (ANNs) have been empirically successful in solving a plethora of tasks. Different variants of ANN have been used in image recognition [14] in all kind of forms, natural language processing, detecting anomalous behaviours in cyber-physical systems [10], or simply playing a game of Go [28].

Y. X. M. Tan (✉) · Y. Elovici · A. Binder
ST Engineering-SUTD Cyber Security Laboratory, Singapore, Singapore
e-mail: marcus_tan@mymail.sutd.edu.sg

Y. X. M. Tan · A. Binder
Information Systems Technology and Design Pillar, Singapore University of Technology and Design, Singapore, Singapore

Y. Elovici
Department of Software and Information Systems Engineering, Ben-Gurion University of the Negev, Beer-Sheva, Israel

© The Author(s), under exclusive license to Springer Nature Singapore Pte Ltd. 2021   143
H. Kim et al. (eds.), *Information Science and Applications*, Lecture Notes in Electrical Engineering 739, https://doi.org/10.1007/978-981-33-6385-4_14

Szegedy et al. [29] first showcased the vulnerability of ANNs to adversarial attacks, a phenomenon that involves the creation of perturbed samples from their original counterparts, imperceptible upon visual inspection, which are misclassified by ANNs. Since then, many researchers introduced other adversarial attack methods against such ANN models, whether under a white-box [5, 11, 16, 21, 22] or a black-box [3, 23] scenarios. This raises questions about the reliability of ANNs, which can be a cause for concern especially when used in cyber-security or mission critical contexts [27, 30].

Stochastic ANNs (SANNs) have also been used to perform image classification tasks. In this work, we focused on two sub-categories of such stochastic ANNs, one involving making both its hidden weights and activations are in a binary state [12], while the other only requiring its hidden activations to be binary [2, 25, 32]. They are known as Binarized Neural Networks (BNNs) and Binary Stochastic Networks (BSNs) respectively. These variants of networks use Bernoulli distributions in order to binarize its features.

References [9, 13] explored adversarial attacks against Binary Neural Networks (BNNs). To the best of our knowledge, we did not find prior work examining the adversarial robustness of networks employing the use of Binary Stochastic Networks (BSN). The authors in Galloway et al. [9] performed two white-box attacks and a black-box attack (Fast Gradient Sign Method (FGSM) [11], CWL2 and surrogate attacks proposed by Papernot et al. [23]) and showed that stochasticity in binary models does improve the robustness against attacks. However, a decision-based black-box attack (i.e. Boundary Attack) is also a highly viable attack vector an adversary could employ due to its generalisable nature to be applied to any network, which they did not consider. Furthermore, the assumption of surrogate-based attacks assumes that a surrogate could accurately learn the decision boundary of the target model for greater attack success. Galloway et al. [9] used a shallow MLP model as surrogates against a more complex Convolutional Neural Network (CNN) model with binarization as a target. Though [23] mentioned that the choice of the surrogate architecture does not highly affect the success of the attack, the surrogate should at least match the input-output relation of the target model (e.g. CNN for images). Also, Table 1 of [23] shows that using convolutional filters in surrogates to attack a CNN model yields better transferability. We postulate that [9] does not reflect the true susceptibility of BNNs against such attack vectors. We address these issues in our work.

Chen et al. [6] discusses the limitations of using randomness and discretization to provide for adversarial robustness. However, contrary to [6], we analyse networks that uses a Bernoulli distribution to binarize intermediate features rather than examining only the input image. References [8, 18, 19] studies the impact of noise on adversarial robustness by injecting noise to the input or intermediate feature layers, which falls under a different scope to our work. As such, we study the robustness of a relatively unexplored variant of model which does binarization through stochasticity. Our contributions are as follows:

1. We analyse to what extent binarized SANNs are susceptible to adversarial attacks by adopting a variety of differing attacks in our study. We also analyse the differing robustness levels between our stochastic network variants through the lens of loss surfaces.
2. For white box settings, we propose, inspired by Athalye et al. [1], an alternative objective function for the state-of-the-art CWL2 white-box attack and we analyse the robustness of the different network variants to samples generated via such attacks. Our focus lies on black-box attacks, however.
3. We propose a stronger black-box surrogate-based attack. This is based on using SANNs as surrogates, and as a novel contribution, Variance Mimicking. This is followed by a more in-depth analysis of our VM method.

The remaining of our paper is organised as such: Sect. 2 gives a reminder of the attacks used here. In Sect. 3, we show empirical results of white and black box attacks against our studied networks. This is followed by a discussion of our novel variance mimicking in Sect. 4. After which, in Sect. 5, we provide deeper analysis with regards to stochastic ANNs and our proposed method and conclude our work in Sect. 6.

2 Background on Attack Algorithms Used

To attack the model in a black-box setting, we used a decision-based method known as *Boundary Attack* [3]. This approach initialises itself by generating a starting sample that is labelled as adversarial to the victim classifier. Then, random walks are taken by this sample along the decision boundary that separates the correct and incorrect classification regions. These random walks will only be considered valid if it fulfils two constraints, (i) the resultant sample remains adversarial and (ii) the distance between the resultant sample and the target is reduced. Essentially, this approach performs rejection sampling such that it finds smaller valid adversarial perturbations across the iterations.

We used the *Basic Iterative Method* (BIM) [16] as one of the means to perform white-box attacks. This method is basically an iterative form of the FGSM:

$$x_{t+1} = x_t + \alpha * sign(\nabla J(F(x_t), y; \theta)) \tag{1}$$

where ∇J represents the gradients of the loss calculated with respect to the input space x_t and its original label y, t represents the iterations.

The second white box attack used is the CWL2 attack. It is based on solving the following objective function:

$$min_\delta ||\delta||_2 + c \cdot f(x + \delta) \tag{2}$$

$$such\ that\ f(x') = max(max_{i \neq t}(Z(x')_i) - Z(x')_t, -\kappa) \tag{3}$$

where $||\delta||_2$ minimises the L_2 norm of the perturbation while $f(x + \delta)$ ensures misclassification after perturbation. c is a constant. $Z(\cdot)_i$ refers to the logits of class index i and t refers to the target class. κ is the confidence value. This attack method is considered as close to state-of-the-art and can bypass several detection mechanisms [4].

3 Experiments and Results

Attacks: Here we focus on three prototypical attacks, a direct gradient-based one, a gradient attack with a distance regularizer, and a gradient-free boundary attack. We take these attacks as representatives for their classes. While there are many other adversarial attacks, analysing most of them would be beyond the length of a conference paper and is expected to result in correlations according to their classes. The extensions which we propose and evaluate further below, are also applicable to other attacks as well, which fall in these classes.

Networks: We explored five different variants of neural networks: ResNet18, two BSN architectures and two BNN architectures. The BSN architectures used are a 4-layered (BSN-4) Multilayer Perceptron, and a modified LeNet [17] (BSN-L). For the BNNs, we explored both deterministic (BNN-D) and stochastic binarization (BNN-S) strategies. The baseline image classification accuracies are summarised in Table 1. Though not state-of-the-art, it is not the focus of this work since we are interested in adversarial robustness.

Datasets: We used three datasets for our experiments, namely MNIST [17], CIFAR-10 [15] and, Patch Camelyon [31] which we refer to as PCam. The library we used in our experiments was PyTorch [24] for constructing our image classifiers. For attacks, we used the Foolbox [26] library at version 1.8.0.

Table 1 Baseline image classification performance for all models

	MNIST	CIFAR-10	PCam
Resnet18	0.988	0.842	0.789
BSN-4	0.972	0.542	0.733
BSN-L	0.990	0.642	0.788
BNN-D	0.989	0.876	0.798
BNN-S	0.967	0.647	0.780

3.1 White-Box Attacks Against Stochastic Neural Networks

While the focus of our paper are black-box attacks, we show here for completeness results of white-box attacks against stochastic networks. We report the proportion of successful adversarial samples and term it as Adversarial Success Rate (ASR; in range [0,1]). Furthermore, we report the mean L_2 norms per pixel, of the differences between natural images and their adversarial counterparts. In our experiments, we sub-sampled 500 samples from the test set of the respective datasets during the evaluation of the BIM attack and 100 samples for the evaluation of the other attacks, unless stated otherwise. Note that we selected only samples that were *originally correctly classified*. We adopted an average inference policy for SANNs, taking the average prediction across 10 forwards passes per sample.

For the Basic Iterative Method (BIM) attack, we varied the attack strength (ϵ measured in L_∞ space) while keeping the epsilon, step sizes and iterations fixed at $32/255$, 0.05 and 100 respectively. Two notable observation can be made about BIM from Table 2. Firstly, when comparing networks, BIM has the highest ASR against BNN-D other than ResNet18. Due to binarization, it becomes easier to effect a greater change in activation values. Secondly, when comparing attacks for a given architecture, BIM yields the highest ASR on BNN-D and BNN-S, however this is achieved at the cost of higher L2-norms.

For the Carlini & Wagner L2 (CWL2) attack, we used the default attack parameters as specified in Foolbox. For stochastic ANNs, we disabled binary searching for the constant c and showing the results when $c = 10$ in Table 2. We postulate having 'c' dynamic in attacking such a model would reduce the efficacy of the

Table 2 Adversarial success rate/mean L_2 norms per pixel (upscaled by 1000 for illustration) between the original image and its perturbed adversarial image for the attacks across the different variants of models. For CWAug, K in Eq. 4 was set to 5

Dataset	Attack method	Resnet18	BSN-4	BSN-L	BNN-D	BNN-S
MNIST	BIM	1.000/2.17	0.956/2.29	0.831/2.63	1.000/1.56	0.595/2.27
	CWL2	0.970/0.91	0.849/2.47	0.777/2.59	0.980/0.00	0.300/1.48
	CWAug	1.000/5.85	0.866/2.46	0.785/2.61	1.000/0.29	0.293/1.76
	Boundary	1.000/1.40	0.044/5.40	0.003/5.06	0.980/0.00	0.044/0.60
CIFAR-10	BIM	1.000/0.93	0.988/1.03	0.986/1.10	1.000/0.96	0.998/1.04
	CWL2	1.000/0.08	0.797/0.09	0.789/0.09	1.000/0.04	0.660/0.09
	CWAug	1.000/0.11	0.855/0.19	0.851/0.21	1.000/0.06	0.693/0.11
	Boundary	1.000/0.14	0.466/2.96	0.420/1.86	0.944/1.18	0.445/2.36
PCam	BIM	1.000/0.98	0.955/0.94	0.915/1.03	0.974/1.03	0.974/0.99
	CWL2	1.000/0.09	0.476/0.84	0.477/1.06	0.920/0.10	0.322/0.53
	CWAug	1.000/0.14	0.547/0.89	0.463/1.08	0.930/0.14	0.396/0.53
	Boundary	0.730/0.09	0.005/1.31	0.015/2.34	0.190/2.00	0.008/2.41

attack method since the model is never consistent. Exemplified by the results from ResNet18 in Table 2, the CWL2 attack is an extremely powerful attack. However, it is less effective against SANNs compared to BIM, where BIM outperforms CWL2 in all SANN settings. This observation, together with the inspiration from a prior art in [1], prompted us to propose a modification of the objective function of this attack. **Augmented Carlini & Wagner L2 Attack (CWAug)** We utilised randomness in augmenting input samples in the attack procedure by modifying Eq. 3 to include an additional term that performs random augmentations on the input image, both rotations and translations before optimisation, gaining inspiration from the Expectation over Transformation (EoT) [1] approach. Our goal is to investigate the effects of modifying an already powerful attack with EoT on adversarial robustness, to observe if it will be a more powerful attack. Athalye et al. [1] used the PGD algorithm to optimise with the EoT approach while we use the CWL2 objective function. Equation 4 describes our modified attack, CWAug.

$$min_\delta ||\delta||_2 + c \cdot f(x + \delta) + \frac{1}{K} \sum_{i=1}^{K} f(R(x + \delta)) \tag{4}$$

where K is the number of iterations to perform random transformations, symbolised by $R(\cdot)$, on the input sample. The first term penalises against having high degrees of perturbations while the second term enforces misclassification. We refer readers to [5] for more details about the first two terms of Eq. 4. Our function $R(\cdot)$ in the third term involves first making random rotations followed by random translations. Putting $R(\cdot)$ into the optimisation objective makes the adversarial sample more robust against small changes, stabilising it against the effects of stochasticity in the SANNs. In this work, we defined the allowable range of rotation angles to 180 degrees clockwise and counterclockwise, sampled from a uniform distribution. Also, we select at random the translation direction and pixels (integer from 0 to 10) to be applied on the image. We adopt the same experimental settings as in CWL2 attacks. In total, out of 12 configurations of SANNs in Table 2, CWAug performs in 9 configurations better compared to CWL2 in terms of ASR.

3.2 Black-Box Attacks Against Neural Networks

Boundary Attack This attack turns out to be the poorest performing for the case of stochastic ANNs, which indicates that the attack method is very sensitive to stochasticity. Note that the attack was performed on an average of queries for each data point. For SANNs, its weights and activations vary due to stochasticity, resulting in slightly varied predictions for the same sample at different times. Having a stochastic decision boundary compromises the ability to obtain accurate feedback for the traversal of adversarial sample candidates, explaining the poor performance of this attack. Section 5.2 illustrates this point further.

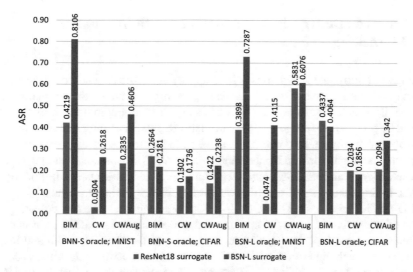

Fig. 1 Attack success comparison between ResNet18 (blue) versus BSN-L (orange) models as surrogates when targeting *BNN-S* and *BSN-L*. 100 samples were used. $c = 10$ for CW-based attacks and $\epsilon = 32/255$ for BIM

Surrogate-Based Black-Box Attacks We report the effectiveness of surrogate-based black-box attacks, as introduced by [23]. We considered two different model variants as our target classifier (i.e. oracle), namely BSN-L and BNN-S, and evaluated on both the MNIST and CIFAR-10 datasets, taking 20% of the test data to be used for training the surrogate. The remaining test data were used to evaluate the performance of the surrogate models. We also considered *SANNs (i.e. BSN-L) as surrogates*, which to the best of our knowledge, has not been explored before. Prior art frequently discussed surrogate-based attacks with deterministic models (e.g. ResNets) as surrogates [7, 9, 23] instead. As illustrated in Fig. 1, the CWAug attack achieves higher transferability as compared to the CWL2 attack, which is also consistent with our explanation provided in Sect. 3.1. However, the BIM attack is in general more efficient in attaining transferable adversarial samples than the other, more complex, attack methods.

We postulate that, using a deterministic surrogate against stochastic oracles, results in a reduced ASR due to the difficulty to account for the stochastic variance of the decision boundary. As querying the stochasticity of a target is trivial, the attacker can easily employ a stochastic surrogate. Making observations from Fig. 1 (orange bars), employing SANNs as surrogates increases the ASR against stochastic, with the exception of BIM on CIFAR-10. We postulate that using stochastic ANNs acts as a regularizer to prevent adversarial perturbations from becoming too small, similar to the concept for CWAug (see Sect. 3.1). This increases the chances of fooling ANNs whenever there is no highly accurate approximation of the decision boundaries by the surrogates.

4   Towards Better Black-Box Attacks with Variance Mimicking

The goal of surrogate-based attacks is to approximate the decision boundary of an oracle O. When faced against stochastic targets, approximating the prediction variance of the oracle is as important. This leads us to our proposed method of Variance Mimicking (VM), for usage against stochastic targets. Assuming the attacker uses a stochastic surrogate, he could inject noise in the surrogate training dataset before querying the oracle for hard labels (annotated by $\overline{O}(.)$). We assume that in the black-box setting, we can *obtain only hard labels* for a predicted class, but not the underlying logits. In this case, a small perturbation δ might be insufficient to induce label flips in the prediction (i.e. $\overline{O}(x) = \overline{O}(x + \delta)$ for some input x), thereby defeating the point of measuring variance. To mitigate that, the attacker calculates the variance σ_b^2 within a mini-batch of data d_b of size B, specific to the channel, width and height dimensions (each dimension has its own variance), and perturbs d_b with $\mathcal{N}(0, \sigma_b^2)$ for m separate times, yielding $m \cdot B$ samples. The attacker then trains the surrogate on this expanded mini-batch with their respective labels provided by the oracle, as usual like in [23]. Algorithm 1 describes our method procedure. Our modification replaces the training of the surrogate model in line 6 of Algorithm 1 in [23]. The other steps remain the same and m was defined as 10.

As shown in Fig. 2, the VM procedure (blue bars) outperforms the naive approach (green bars), in terms of ASR, except for two cases. While these results do not outperform white-box attacks (see Table 2), noting the limitations of using white-box attacks in realistic settings when the model is not accessible, this result shows

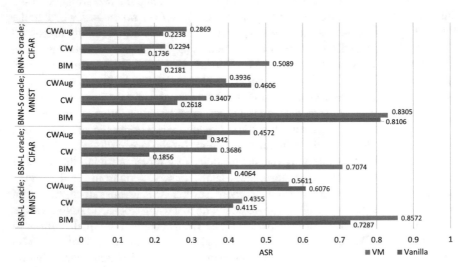

Fig. 2 Attack success comparison between Vanilla (green) versus VM (blue) surrogate training procedures when using *BSN-L* as surrogate. 100 samples were used. $c = 10$ for CW-based attacks and $\epsilon = 32/255$ for BIM

the moderate susceptibility of SANNs and the success of our proposed method. Observing our results across all black-box attacks against BNN-S and BSN-L models, *our VM approach yields the best ASR.*

Algorithm 1: Variance Mimicking - oracle O, surrogate model F, number of training epochs max_ρ, number of perturbations m and some surrogate training dataset D.

$\textbf{Input: } O, F, D, m, max_\rho$
for $\rho \leftarrow 0$ **to** max_ρ **do**
   for $b \in D$ **do**
      `// d_b is a data mini-batch`
      $F.\text{train\_step}(d_b, \overline{O}(d_b))$
      $\sigma_b^2 \leftarrow get\_variance(d_b)$
      for $i \leftarrow 0$ **to** m **do**
         $d_{perturb} \leftarrow d_b + \mathcal{N}(0, \sigma_b^2)$
         $F.\text{train\_step}(d_{perturb}, \overline{O}(d_{perturb}))$
      end
   end
end

---

5 Discussion

5.1 Analysis of the Effectiveness of VM

In order to show the effectiveness of mimicking, one can measure the degree of this approximation through computing L2 norm differences of logits. We collected these statistics between i) target and surrogate with VM and ii) target and surrogate without VM. We observe that L2 norm difference in (i) is consistently lower than in (ii), summarised in Table 3. It is clear that with VM, the L2 norm difference of logits is lower than without, which is more emphasised in CIFAR-10.

We also investigate if the improved performance is due to the increased data (due to line 7 of Algorithm 1) used for the case of training surrogates with VM through Table 4. Even after training with an equivalent batch size without VM on CIFAR-10, the baseline does not consistently perform better than VM, especially when the surrogate and oracle uses the same architecture.

Table 3 L2 norm differences of logits across (i) and (ii)

Dataset	VM(i)	no-VM(ii)
MNIST	5.901	6.425
CIFAR-10	4.539	12.457

Table 4 ASR of using BSN-L surrogate against oracles using the BIM attack on CIFAR-10. m is the number of perturbations; B is the batch size

Oracle	Batch size used	
	B	$m \cdot B$
BNN-S	0.2181	0.2131
BSN-L	0.4064	0.3528

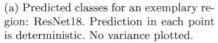

(a) Predicted classes for an exemplary region: ResNet18. Prediction in each point is deterministic. No variance plotted.

(b) Predicted classes and variances for an exemplary region: BSN-L. High variance appears as low gamma (fading colours).

Fig. 3 Predicted classes and variances around three data points (black dots)

5.2   Understanding the Impact of Stochasticity

We postulate that the difficulty of attacking SANNs in general is due to the variance in the definition of the decision boundary. Figure 3 shows for the BSN-L classifier the predicted labels and the variances, and for the Resnet18 only the predicted labels. This was computed on a section defined by a two-dimensional plane spanned by three exemplary data points. One can see two peculiarities: Firstly, the decision boundaries for BSN-L are much more spiky and noisy, as can be seen for the red labels in the lower right corner, and the purple in the upper right corner, with zones of high variance in white around them. Traversing along a decision boundary is more challenging for stochastic networks. This shows the difficulty for boundary attacks against stochastic models. Secondly, decision boundaries around the same three samples are substantially different for the two networks. The zones of high variance provide additional obstacles to correctly training surrogates. This explains the difficulties for transferability and surrogate attacks, and the reason why stochastic mixtures are able to provide the observed hardening.

5.3   Differing Robustness Behaviours Between BSN-L and BNN-S

As observed in Table 2, out of 9 different white-box attacks configurations, 7 of which BSN-L showcases higher susceptibility than BNN-S. We study this phenomenon

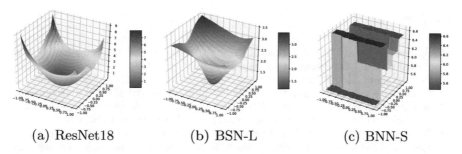

(a) ResNet18   (b) BSN-L   (c) BNN-S

Fig. 4 Loss surfaces of ResNet18, BSN-L and BNN-S when evaluated with CIFAR-10 training data. Colour spectrum scale in figures of ResNet18, BSN-L and BNN-S indicates 1 to 7, 1.5 to 3 and 5.6 to 6.6 respectively

through the lens of loss surfaces. We used a visualisation tool proposed by [20] to produce a 3D visualisation of the loss values within an exemplary region of BSN-L and BNN-S, illustrated by Fig. 4b, c respectively. Figure 4c shows the highly discontinuous nature of the loss surface of BNN-S. This could be attributed to the higher degree of quantization found in BNN-S as compared to BSN-L. As such, it becomes more difficult to craft adversarial samples using white-box attacks that employs gradients against BNN-S than BSN-L, where this difficulty is more emphasised for the MNIST dataset.

6   Conclusion

We investigated the adversarial robustness of two variants of binarized SANNs across MNIST, CIFAR-10 and PCam datasets. In white-box settings, they are similarly vulnerable to BIM attacks as deterministic ANNs. We can observe in black-box settings higher robustness for SANNs compared to CNNs. We show that the success rate for black-box attacks can be improved when using surrogate attacks with stochastic surrogates and the novel variance mimicking. The latter achieves a better approximation of the stochastic decision boundary in SANNs.

Acknowledgements This research is supported by both ST Engineering Electronics and National Research Foundation, Singapore, under its Corporate Laboratory @ University Scheme (Programme Title: STEE Infosec-SUTD Corporate Laboratory).

References

1. Athalye A, Engstrom L, Ilyas A, Kwok K (2018) Synthesizing robust adversarial examples. In: Dy J, Krause A (eds) Proceedings of the 35th international conference on machine learning. Proceedings of machine learning research, vol 80, pp 284–293. PMLR, Stockholmsmässan, Stockholm Sweden (10–15 Jul 2018). http://proceedings.mlr.press/v80/athalye18b.html
2. Bengio Y, Léonard N, Courville A (2013) Estimating or propagating gradients through stochastic neurons for conditional computation, pp 1–12. http://arxiv.org/abs/1308.3432
3. Brendel W, Rauber J, Bethge M (2018) Decision-based adversarial attacks: reliable attacks against black-box machine learning models. In: International conference on learning representations. https://arxiv.org/abs/1712.04248
4. Carlini N, Wagner D (2017) Adversarial examples are not easily detected: bypassing ten detection methods. http://arxiv.org/abs/1705.07263
5. Carlini N, Wagner D (2017) Towards evaluating the robustness of neural networks. In: Proceedings—IEEE symposium on security and privacy, pp 39–57. https://doi.org/10.1109/SP.2017.49
6. Chen J, Wu X, Rastogi V, Liang Y, Jha S (2019) Towards understanding limitations of pixel discretization against adversarial attacks. In: 2019 IEEE European symposium on security and privacy (EuroS&P). IEEE, New York, pp 480–495
7. Cheng S, Dong Y, Pang T, Su H, Zhu J (2019) Improving black-box adversarial attacks with a transfer-based prior. In: Advances in neural information processing systems, pp 10932–10942
8. Co KT, Muñoz-González L, de Maupeou S, Lupu EC (2019) Procedural noise adversarial examples for black-box attacks on deep convolutional networks. In: Proceedings of the 2019 ACM SIGSAC conference on computer and communications security, pp 275–289
9. Galloway A, Taylor GW, Moussa M (2017) Attacking binarized neural networks, pp 1–14. http://arxiv.org/abs/1711.00449
10. Goh J, Adepu S, Tan M, Lee ZS (2017) Anomaly detection in cyber physical systems using recurrent neural networks. In: 2017 IEEE 18th international symposium on high assurance systems engineering (HASE), pp 140–145 (Jan 2017). https://doi.org/10.1109/HASE.2017.36
11. Goodfellow IJ, Shlens J, Szegedy C (2014) Explaining and harnessing adversarial examples, pp 1–11. http://arxiv.org/abs/1412.6572
12. Hubara I, Courbariaux M, Soudry D, El-Yaniv R, Bengio Y (2016) Binarized neural networks. In: Advances in neural information processing systems (Nips), pp 4114–4122
13. Khalil EB, Gupta A, Dilkina B (2018) Combinatorial attacks on binarized neural networks, pp 1–12. http://arxiv.org/abs/1810.03538
14. Krizhevsky A, Sutskever I, Hinton GE (2012) Imagenet classification with deep convolutional neural networks. In: Advances in neural information processing systems, pp 1097–1105
15. Krizhevsky A et al (2009) Learning multiple layers of features from tiny images. Technical report, Citeseer
16. Kurakin A, Goodfellow I, Bengio S (2016) Adversarial examples in the physical world. arXiv preprint arXiv:1607.02533
17. LeCun Y, Bottou L, Bengio Y, Haffner P et al (1998) Gradient-based learning applied to document recognition. Proc IEEE 86(11):2278–2324
18. Lecuyer M, Atlidakis V, Geambasu R, Hsu D, Jana S (2019) Certified robustness to adversarial examples with differential privacy. In: 2019 IEEE symposium on security and privacy (SP). IEEE, New York, pp 656–672
19. Li B, Chen C, Wang W, Carin L (2019) Certified adversarial robustness with additive noise. In: Advances in neural information processing systems, pp 9464–9474
20. Li H, Xu Z, Taylor G, Studer C, Goldstein T (2018) Visualizing the loss landscape of neural nets. In: Advances in neural information processing systems, pp 6389–6399
21. Madry A, Makelov A, Schmidt L, Tsipras D, Vladu A (2017) Towards deep learning models resistant to adversarial attacks. arXiv preprint arXiv:1706.06083

22. Moosavi-Dezfooli SM, Fawzi A, Frossard P (2016) Deepfool: a simple and accurate method to fool deep neural networks. In: Proceedings of the IEEE conference on computer vision and pattern recognition, pp 2574–2582
23. Papernot N, McDaniel P, Goodfellow I, Jha S, Celik ZB, Swami A (2017) Practical black-box attacks against machine learning. In: Proceedings of the 2017 ACM on Asia conference on computer and communications security. ACM, New York, pp 506–519
24. Paszke A, Gross S, Chintala S, Chanan G, Yang E, DeVito Z, Lin Z, Desmaison A, Antiga L, Lerer A (2017) Automatic differentiation in pytorch
25. Raiko T, Berglund M, Alain G, Dinh L (2014) Techniques for learning binary stochastic feed-forward neural networks, pp 1–10. http://arxiv.org/abs/1406.2989
26. Rauber J, Brendel W, Bethge M (2017) Foolbox: a python toolbox to benchmark the robustness of machine learning models. arXiv preprint arXiv:1707.04131
27. Rosenberg I, Shabtai A, Elovici Y, Rokach L (2018) Low resource black-box end-to-end attack against state of the art API call based malware classifiers. http://arxiv.org/abs/1804.08778, https://doi.org/10.1145/TODO
28. Silver D, Huang A, Maddison CJ, Guez A, Sifre L, Van Den Driessche G, Schrittwieser J, Antonoglou I, Panneershelvam V, Lanctot M et al (2016) Mastering the game of go with deep neural networks and tree search. Nature 529(7587):484
29. Szegedy C, Zaremba W, Sutskever I, Bruna J, Erhan D, Goodfellow I, Fergus R (2013) Intriguing properties of neural networks. arXiv preprint arXiv:1312.6199
30. Tan YXM, Iacovazzi A, Homoliak I, Elovici Y, Binder A (2019) Adversarial attacks on remote user authentication using behavioural mouse dynamics. arXiv preprint arXiv:1905.11831
31. Veeling BS, Linmans J, Winkens J, Cohen T, Welling M (2018) Rotation equivariant CNNs for digital pathology (Jun 2018)
32. Yin M, Zhou M (2019) ARM: augment-reinforce-merge gradient for stochastic binary networks, pp 1–21. https://github.com/mingzhang-yin/ARM-gradient

Dynamic Pricing for Parking System Using Reinforcement Learning

Li Zhe Poh, Connie Tee, Thian Song Ong, and Michael Goh

Abstract The number of vehicles in urban cities has increased and raised attention towards the need for effective parking lot management in public areas such as hospital, shopping mall and office building. In this study, dynamic pricing is deployed with real time parking information to maximize the parking usage rate and alleviate traffic congestion. Dynamic pricing is a practice of varying the price of product of service reflected by the market conditions. This technique can be used to deal with vehicle flow around the parking area including peak and non-peak hour. During peak hours, the dynamic pricing mechanism will regulate the price of parking fee to a relatively high rate, and vice versa for non-peak hours. Reinforcement Learning (RL) is used in this paper to develop a dynamic pricing model for parking management. Dynamic pricing over time is divided into episodes and shuffled back and forth through an hourly increment. The parking usage rate and traffic congestion rate are regarded as the rewards for price regulation.

Keywords Parking management · Reinforcement learning · Dynamic pricing

1 Introduction

Huge increase in private vehicles is one of the pain points in many urban areas especially big cities like London, Hong Kong and Kuala Lumpur. The increase of vehicles brings negative impacts to the environment and also people around the cities. Increased number of vehicles leads to raising demand for parking lots. According to a survey published by New Strait Time (2017), the average time spent every day looking for parking areas in some major cities is about 25 min. The drivers keep circling around the destination and time is wasted to look for available parking lot. This leads to the increase of fuel combustion and carbon dioxide emission that destroys the atmosphere and causes greenhouse effect. The higher the time driver

L. Z. Poh (✉) · C. Tee · T. S. Ong · M. Goh
Faculty of Information Science and Technology, Multimedia University, Melaka, Malaysia
e-mail: lizhepoh@gmail.com

© The Author(s), under exclusive license to Springer Nature Singapore Pte Ltd. 2021   157
H. Kim et al. (eds.), *Information Science and Applications*, Lecture Notes
in Electrical Engineering 739, https://doi.org/10.1007/978-981-33-6385-4_15

spent on road, the higher the traffic congestion in that area. This creates a chain effect that further delays and causes aggravation for other drivers.

Recognizing the challenges associated with the increased number of vehicles, many researchers have tried to solve the traffic congestion problem and satisfy the huge parking lot demand. With technology such as sensor (loop or ultrasonic sensor) and ticket or e-payment systems, parking information such as availability of free parking space can be obtained real-time. This provides an opportunity to implement a dynamic pricing based smart parking solution. By using dynamic pricing, the parking operator can offer flexible price regulation based on different time periods to optimize their revenue and at the same time increase parking space usage rate.

In this paper, we present a dynamic pricing model that regulates the parking price based on vehicle volume and traffic congestion rate. We monitor dynamic pricing with different pricing schemes and observe how they are utilized in different settings to bring a higher reward to traffic congestion and revenue for the parking operators. The dynamic pricing model can forecast traffic congestion and vehicle volume and disperse vehicle flows that leads to traffic congestion. During peak hours at certain time of the day, the vehicle flows are dispersed to non-peak hours to increase the parking usage rate and reduce traffic congestion by suppressing driver's visit to the specific areas. This is achieved by price regulation through price reduction in terms of parking fee rebate.

The dynamic pricing model which provides traffic congestion forecast and parking lot availability enables time-saving in parking search. Drivers can go straight to the parking area with lower parking fee (lower parking usage). Within the pre-defined parking area, traffic congestion can be alleviated because less vehicles will cruise for parking lot around that area. This in turn reduces carbon dioxide emission and increases mobility efficiency.

The rest of this paper is organized as follows. Section 2 reviews related work such as dynamic pricing and smart parking system. Section 3 defined the proposed method. Section 4 concludes and discuss future improvements.

2 Literature Review

2.1 Dynamic Pricing

Dynamic pricing is a pricing strategy that has a huge influence in our society because e-commerce has become typical carrier of business model. Every trade-off can be made by the Internet and it has spared many physical expenses and gave rise to an easy entry to the market. Nowadays, many researchers focus on dynamic pricing in e-commerce due to the availability of big data that makes user behaviour visible.

Karpowicz and Szajowski [1] presented four pricing strategies for e-commerce and they are (1) time-based pricing strategy; (2) market segmentation and limited rations strategy; (3) dynamic marketing strategy; (4) the comprehensive use of above three types. On the other hand, Chen and Wang [2] introduced a dynamic pricing model for e-commerce based on data mining. The model was composed of three bottom–up layers: data layer, analytic layer and decision layer.

In a multi-agent environment, the optimal pricing policy of agent depends on the pricing policies of the other agents [3]. Han et al. [3] proposed a multi-agent reinforcement learning which integrates the observed objective actions as well as the subjective inferential intention of the opponents. Pan et al. [4] proposed a new continuous time model with price and time sensitive demand to consider dynamic pricing, order cancellation ration and different quality of service (QoS) levels in web network.

Chinthalapati et al. [5] proposed reinforcement learning (RL) as a tool to study price dynamics in an electronic retail market. The proposed method consisted of two competing sellers, price sensitive and lead time sensitive customer. They considered two representative cases: (1) no information case and (2) partial information case. Ujjwal et al. [6] proposed a bargaining agent which used genetic algorithm for implementing dynamic pricing on internet. Online bargaining is a "win–win" situation for both the seller and buyer because mutually agreed deal price is higher than the seller's reserved price but lower than the buyer's reserved price.

The authors in [7] proposed Pareto-efficient and subgame-perfect equilibrium and offered a bounded regret over an infinite horizon, where regret was defined as the expected cumulative profit loss as compared to the ideal scenario with a known demand model. They considered an oligopoly dynamic pricing under demand uncertainty but assumed that all sellers had the same marginal cost.

Another study in [8] investigated pricing strategy of firms in the context of uncertain demand. According to the study, reference prices and the price of competition were the two factors that affected demand dynamics. Simulations results demonstrated that firms could lower the volatility of their price path if they gathered and processed the information about consumers and competitors because this gave a greater control over uncertainty than ever before.

Wang [9] proposed a dynamic pricing model of the retailer in the case of supply exceeding demand vice versa. The study listed the equilibrium conditions for the optimal dynamic pricing strategies and derived the optimal dynamic pricing strategies.

2.2 Smart Parking Solutions

Parking solutions integrate interdisciplinary knowledge for successful large-scale deployment. Recent research like the 5G network also makes the Machine-to-Machine (M2M) communication in real time practicable. There are several review papers that provide a good viewpoint to the recent smart solutions using different technology. In [10], smart parking solutions are classified into three macro-theme according to the goals of the research fields: information collection, system deployment, and service dissemination.

The study in [11] proposed a smart parking system that was based on intelligent resource allocation, reservation and pricing. The system offered guaranteed parking reservation with the lowest possible cost and searching time for drivers and the highest revenue and resource utilization for parking manager. Mitsopoulou and Kalogeraki [12] proposed a crowdsourcing system known as ParkForU that aimed to find the available and most suitable parking options for users in a smart city. It was a parking-matching and price-regulator algorithm that allow user to specify their destination along with a set of preferences regarding price and distance (from the parking lot to their destination), as well as a maximum number of results they wish to receive.

In [13], Nugraha and Tanamas proposed a dynamic allocation method to reserve parking lot using Internet application. The method reduced the need for the driver to explore the entire parking spot by finding the vacant parking lot and performed the reservation for the driver. They used an event-driven schema allocation at the event of a vehicle arrived at the parking lots gate to maintain parking lot utilization level.

The authors in [14] proposed a ridesharing system with two matching algorithm that followed single and dual side search paradigm to offer more options on pick-up time and price, so that the travelers could choose the vehicle that matches their preferences best. The baseline algorithm of ridesharing matching algorithms inserted ridesharing request to a kinetic tree and it returned all possible pairs of pick-up time and price that did not dominate each other.

Jioudi et al. [15] proposed dynamic pricing policy that changes prices proportionally to the arrival rates on each parking and therefore reduces congestion and eliminates the driver's preference for some parking. They analyzed the parking time under near-real-life conditions using the Discrete Batch Arrival Process (D-BMAP). The drivers arriving according to D-BMAP were selected for service in random order (ROS) according to their parking time.

In [16], the authors aim to improve the rotation of attractive spot and set a usage-based parking assignment via appropriate incentive which was equivalent to using a strategic pricing to control parking dwell time and improve traffic conditions in the target area. Improving the utilization of the limited parking capacity in high demand areas by reducing the number of long-term parking spots combined with suitable measures allowing short-term parking are elemental.

The authors in [17] proposed two prediction models for qualitative and quantitative improvement of parking availability information. Quality issue refers to fixed interval to updated information and network delay between parking sensors and data server. Quantity issue refers to non-smart parking lots which are unmonitored. A future availability prediction model was introduced by learning the tendency of parking and changes of occupancy rate from historical data for quality improvement. They also proposed an availability prediction model for non-smart parking which were unequipped with sensors based on the assumption that prediction target's occupancy rate could be estimated with occupancy rate data of neighbor smart parking located in nearby for quantity improvement.

Recent research above focuses on reservation or price regulation under real time algorithm and feedback the result to user of smart parking system. For example, iParker [11] uses real time reservation with share time reservation to find the available and most suitable parking options for user while ParkForU [12] uses parking-matching and price-regulator algorithm, parking operator was notified after driver's selection and adjust their price dynamically to attract driver. Reservation-based parking system [13] reserves vacant parking slot by calculating driver's ETA and dynamically reallocate parking slot when certain car arrived without reservation. Jioudi et al. [15] uses dynamic pricing policy that changes prices proportionally to the arrival rates. Jioudi et al. [16] uses area zonificaion to encourage short and mid-term parking dwells by allowing them to park in the most attractive zones permit to increase short-term user. Kim and Koshizuka [17] use two prediction models for qualitative and quantitative improvement of parking availability information. In this paper, we planned to use dynamic pricing and reinforcement learning to forecast future arrival rate and traffic volume to define a satisfactory discount scheme to maximize parking occupancy usage and alleviate traffic congestion. With the proposed system, the simulation can be designed to different scenario with number of parking operator, its pricing policy and the traffic volume in the parking area. This had make the proposed method become reliable.

3 Proposed Method

This study proposes a Reinforcement Learning (RL) model for training the dynamic pricing mechanism. The proposed model performs as an environment in the parking industry market. In this environment, a parking operator acts as the player with the other opponents (the other parking operator). The reward is the number of vehicles that park in the player's parking area and the revenue obtained from parking.

Unlike recent smart parking research that focus on reservation for driver and price regulation by parking provider, we focus on smoothening the regression distribution model of the historical parking occupancy rate by increasing the occupancy rate at non-peak hours. According to [16], the driver behaviour can be assumed and captured using multinomial log-linear regression, also known as multinomial Logit model.

The sales process can be modelled as a Stackelberg game between the retailer and consumers. It is a two-stage sequential game where the retailer moves first by setting the prices and then customers follow by making their purchase decisions [9]. In this paper, the parking operator moves first by setting the prices and this is followed by the driver who makes his/her purchase decision whether to park the vehicle. The concept of sub-game perfect equilibrium is also practiced because the strategy of parking operator and driver constitute the best response to optimize their expectation. There is a Pareto efficiency between the parking operators because no individual or preference criterion can be better off without making at least one individual or preference criterion worse off in this parking environment with parking demand (Fig. 1).

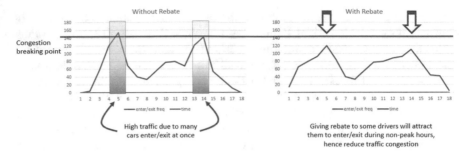

Fig. 1 Expected output of regression model after dynamic pricing

3.1 Architecture

Consider a district parking area containing multiple parking operators with pre-defined dynamic pricing policies, the RL model is constructed with entity and component to perform each single task to fulfil the notation and assumption in this paper. Below we describe the entities and modules in the proposed smart parking solution:

1. Parking Operator
    An entity who provides parking service. Parking operator is defined with its capacity, building type, price rate, and location. It performs as a player in this RL environment.
2. Pricing Engine
    Fetch the parking utility information including pricing policy used and current price rate and discount range. The engine run and design new price rate for next state according the pricing policy predefined such as arrival-time-dependent pricing (ATP), usage-aware pricing (UAP) [16] and flat rate (a fixed price in a day).
3. Driver
    An entity who seeks a parking space. The drivers are smartphone users who use access the smart parking application to get discount information of that parking area. The driver is defined with his/her current location, destination and preferences between price and distance (described as price-focused, balanced and distance-focused) [12]. The driver preferences are denoted as decimal number such as 0.7 to price and 0.3 to distance means the driver make decision tend to lower price parking area.
4. Reward Module
    The module to calculate the reward for state transition considering historical data, predicted reward with and without dynamic pricing, and make comparison to provide the final reward for each state of episode. In this module, the reward is normalized with player capacity and discounted price to avoid bias.
5. Arrival Process
    Batch arrival prediction for interval of 1 h using Auto Regressive Integrated Moving Average (ARIMA). It learns the tendency of parking and changes of

occupancy rate from historical data then forecast future parking occupancy. This module will input the weekday and period section of episode and return forecasted occupancy.

3.2 Notation and Assumption

In this study, we assume a pricing scheme for both the player and opponents. The pricing scheme includes the price range of parking fee for the driver. Both player and opponents should determine the total volume of parking lot, and also the maximum and minimum parking fees at the start of the simulation. To model the relationship between vehicle flow and the traffic congestion, the following assumptions are made:

1. Episode is defined as weekly incremental, each episode consists of 7 days.
2. Problem formulated for parking operators: maximize parking usage, maximize driver satisfaction and maximize global revenue.
3. Parking operators with different pricing policies: Arrival-time-dependent pricing (ATP), Usage-aware pricing (UAP), progressive pricing, discount on flat rate, season pass, incentive pricing.
4. Pricing engine takes action three times per day, make three section such as morning, afternoon and night.
5. Estimate parking market needs by calculating history vehicle flow as supply and demand rate.

3.3 Reinforcement Learning Mechanism

The proposed model follows the general RL mechanism as an outer layer. The environment receives the environment variables such as vehicle flow and traffic congestion rate. The player learns to increase its reward by taking an action from the Q-table. An epsilon value is used to increase exploration to avoid sticking in local optimum.

In the reward function, the environment entity calculates the in and out flow of the player and opponents. With the calculated in and out flows, the number of drivers learn to recognize their parking area with their distance and price preference. At the same time, the player learns to alleviate traffic congestion by increasing their parking revenue. Meanwhile, the drivers also learn to choose the parking area. Algorithm 1 shows the procedure of the reward function.

Algorithm 1:
(1) **Input:** batch arrival rate of in and out flow of vehicle
(2) **Input:** player's current state (occupancy and discounted price)
(3) **Input:** opponents' current state (occupancy and discounted price)
(4) **Initialize** reward, in and out flow of vehicle to 0
(5) **Foreach** hour in period section
(6) Calculate in and out flow of this hour
(7) Calculate reward by referring vehicle in / out flow and discounted price
(8) **Update** occupancy of player and opponent

3.4   *Experiments and Results*

We assume there are one player and one opponent in the environment. The action space is set to 4 which means the parking fee will be discounted between 0 and 3. Each episode goes through seven days with three section. The morning section will start from 5 am to 12 pm, afternoon section from 12 to 5 pm and evening section from 5 pm to 4 am. In every section, the player will make an action to select its price regarding the action space and loop through the period's hours to get the arrival and exit rate from the Arrival Module. The in and out vehicle volume of player and its opponents will be counted based on driver's decision by Driver Module. The reward is calculated by vehicle in volume, player's revenue and traffic condition.

In this experiment, the driver's preference ratio are set to (price, distance). For price-focused driver, the preference is (0.7, 0.3), while it is (0.5, 0.5) for balanced driver and (0.3, 0.7) for distance-focused driver. The driver's parking fee is counted based on his/her arrival time by Pricing Engine. Each action, rebate will be applied to player parking fee which attracts driver to park at the player's parking area (Fig. 2).

From the simulation result, the reward reaches its optimal value at 20 thousand episodes. The episode reward is normalized using the discounted parking fee, which means less discount, the higher the reward return and profit of parking operator. With regulating the price of parking lot, decreasing the vehicle flow on the peak hour by increasing the rebate on non-peak hour which gently the daily parking occupancy and decrease traffic congestion on the peak hours. From the result, we also can see the proposed method decreases the traffic congestion and increases parking occupancy. However, it still needs more effort to relate the simulation to real environment such as classifying the drivers as long-term, medium-term or short-term visitors.

4   Conclusion

In this paper, a dynamic pricing model for parking industry is presented for traffic congestion forecast and parking price regulation. Firstly, the features of traffic

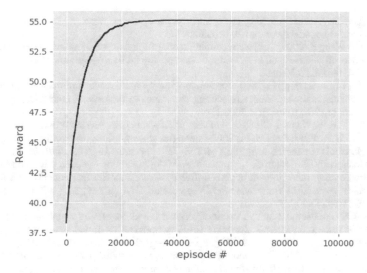

Fig. 2 Simulation result with 50 thousand episodes

congestion and parking revenue are represented. Then a dynamic pricing model for parking industry is proposed and trained by reinforcement learning. It is shown that the dynamic pricing is influential to the optimal solution. The new concepts introduced in this paper is the smart parking system with dynamic pricing to increase the revenue of parking operator by providing a discount scheme to maximize the parking occupancy usage and alleviate the traffic congestion by considering traffic volume as an objective. This proposed method is not handling real-time parking but a forecasting method on vehicle arrival rate and parking occupancy. With the outputted discount scheme to maximize revenue of parking operator and alleviate traffic congestion, the discount scheme can be promoted to frequent visit driver or scheduled visitor. However, there are still problems to be solved for the model, because currently the research focuses on predefined parking operator with pre-defined details of parking operator and its opponents such as location and occupancy. We planned to increase the reliability of the dynamic pricing algorithm to fit with multiple variety type of parking operator such as the building type, occupancy and pricing scheme.

Acknowledgements This work was supported by the TM and TM R&D [Project SAP ID: MMUE/180024]

References

1. "(PDF) Double optimal stopping times and dynamic pricing problem (2020) Description of the mathematical model ResearchGate. https://www.researchgate.net/publication/225715

898_Double_optimal_stopping_times_and_dynamic_pricing_problem_Description_of_the_ mathematical_model Accessed 20 Feb 2020

2. Chen Y, Wang F (2009) A dynamic pricing model for e-commerce based on data mining. In: 2009 Second international symposium on computational intelligence and design. https://doi. org/10.1109/ISCID.2009.99

3. Han W, Liu L, Zheng H (2008) Dynamic pricing by multiagent reinforcement learning. In: 2008 International symposium on electronic commerce and security, August 2008, pp 226–229. https://doi.org/10.1109/ISECS.2008.179

4. Pan W, Yue W, Wang S (2009) A dynamic pricing model of service provider with different QoS levels in web networks. In: 2009 International symposium on information engineering and electronic commerce, May 2009, pp 735–739. https://doi.org/10.1109/IEEC.2009.160

5. Chinthalapati VLR, Yadati N, Karumanchi R (2006) "Learning dynamic prices in MultiSeller electronic retail markets with price sensitive customers, stochastic demands, and inventory replenishments. IEEE Trans Syst Man Cybernet Part C (Appl Rev) 36(1):92–106. https://doi. org/10.1109/TSMCC.2005.860578

6. Ujjwal K, Aronson J (2007) Genetic algorithm based bargaining agent for implementing dynamic pricing on internet. In: 2007 IEEE symposium on foundations of computational intelligence, April 2007, pp 339–343. https://doi.org/10.1109/FOCI.2007.372189

7. Zhai Y, Zhao Q (2016) Oligopoly dynamic pricing: a repeated game with incomplete information. In: 2016 IEEE International conference on acoustics, speech and signal processing (ICASSP), Shanghai, March 2016, pp 4772–4775. https://doi.org/10.1109/ICASSP.2016.747 2583

8. Wu L-LB, Wu D (2016) Dynamic pricing and risk analytics under competition and stochastic reference price effects. IEEE Trans Ind Inf 12(3):1282–1293. https://doi.org/10.1109/TII.2015. 2507141

9. Wang Y (2016) Dynamic pricing considering strategic customers. In: 2016 International conference on logistics, informatics and service sciences (LISS), Sydney, Australia, July 2016, pp 1–5. https://doi.org/10.1109/LISS.2016.7854471

10. Lin T, Rivano H, Le Mouel F (2017) A survey of smart parking solutions. IEEE Trans Intell Transport Syst 18(12):3229–3253. https://doi.org/10.1109/TITS.2017.2685143

11. Kotb AO, Shen Y-C, Zhu X, Huang Y (2016) iParker—a new smart car-parking system based on dynamic resource allocation and pricing. IEEE Trans Intell Transport Syst 17(9):2637–2647. https://doi.org/10.1109/TITS.2016.2531636

12. Mitsopoulou E, Kalogeraki V (2018) ParkForU: a dynamic parking-matching and price-regulator crowdsourcing algorithm for mobile applications. In: 2018 IEEE International conference on pervasive computing and communications workshops (PerCom Workshops), Athens, March 2018, pp 603–608. https://doi.org/10.1109/PERCOMW.2018.8480321

13. Nugraha IGBB, Tanamas FR (2017) Off-street parking space allocation and reservation system using event-driven algorithm. In: 2017 6th International conference on electrical engineering and informatics (ICEEI), Langkawi, November 2017, pp 1–5. https://doi.org/10.1109/ICEEI. 2017.8312456.

14. Chen L, Zhong Q, Xiao X, Gao Y, Jin P, Jensen CS (2018) Price-and-time-aware dynamic ridesharing. In: 2018 IEEE 34th International conference on data engineering (ICDE), Paris, April 2018, pp 1061–1072. https://doi.org/10.1109/ICDE.2018.00099

15. Jioudi B, Sabir E, Moutaouakkil F, Medromi H (2019) Estimating parking time under batch arrival and dynamic pricing policy. In: 2019 IEEE 5th World forum on internet of things (WF-IoT), Limerick, Ireland, April 2019, pp 819–824. https://doi.org/10.1109/WF-IoT.2019.876 7179

16. Jioudi B, Sabir E, Moutaouakkil F, Medromi H (2019) Congestion awareness meets zone-based pricing policies for efficient urban parking. IEEE Access 7:161510–161523. https://doi.org/ 10.1109/ACCESS.2019.2951674

17. Kim K, Koshizuka N (2019) Data-driven parking decisions: proposal of parking availability prediction model. In: 2019 IEEE 16th international conference on smart cities: improving quality of life using ICT & IoT and AI (HONET-ICT), Charlotte, NC, USA, October 2019, pp 161–165. https://doi.org/10.1109/HONET.2019.8908028

Malware Detection by Merging 1D CNN and Bi-directional LSTM Utilizing Sequential Data

Seung-Pil W. Coleman and Young-Sup Hwang

Abstract Due to the popularity of the android platform, there is a growth in the number of devices and threats. For this reason, it is essential to build reliable tools that can detect malware android application packages (APK) on this platform. Creating effective models requires the use of rich features that are hard to generate. In this work, we extracted the Dalvik executable (.dex) byte-codes from APKs. Android application binaries are opcode sequences. Then, we trained one-dimensional convolutional Neural networks (CNN) using those sequential data. These one-dimensional CNNs detect local features and reduce the feature size. We went even farther to combine one-dimensional CNNs with a bi-directional long-short term memory network (LSTM) to detect malware. Experimental results show that our model, trained on a balanced number of samples, got an error rate of merely 5.4% on a dataset of 20,000.

Keywords Android malware detection · Data section · One dimensional convolutional neural network · Bi-directional LSTM · Sequential data

1 Introduction

The android platform is the most popular today, and it contains several hundred thousand applications in different markets. This has led to smartphones running on the Android operating system becoming a target for black hat hacker developers that have malicious intentions. Android is vulnerable compared to other platforms because it allows applications installation from multiple third-party markets. Recent studies have announced that mobile malware is finding new ways to hide, and the number of mobile malware seems to be increasing [1]. This is evident that there is a need to create a robust security solution.

S.-P. W. Coleman (✉) · Y.-S. Hwang
Department of Computer Science and Engineering, Sun Moon University, Asan, South Korea
e-mail: spil3141@naver.com

Y.-S. Hwang
e-mail: young@sunmoon.ac.kr

© The Author(s), under exclusive license to Springer Nature Singapore Pte Ltd. 2021   167
H. Kim et al. (eds.), *Information Science and Applications*, Lecture Notes
in Electrical Engineering 739, https://doi.org/10.1007/978-981-33-6385-4_16

The most popular methods used for android malware detection are static and dynamic analysis. Static analysis is a technique widely used by researchers and industries. It involves the APK file being scanned before they can be executed on an android system. In such a case, the file is disassembled by a disassembler to obtain information such as API calls, permission lists, among others, which can then be examined. On the other hand, dynamic analysis involves methods that can monitor the behavior of applications at run-time. Some examples of this method are implemented by tools like TaintDroid [2], DroidRanger [3], and DroidScope [4]. Even though these are effective methods they have limitations. For example, even though dynamic analysis is effective at identifying malware, there is a caveat of overhead. And as for static analysis, it is fast and efficient but can easily be dodged by malware writers who can trick the disassemblers into producing incorrect code. This is accomplished by inserting errors into the source which leads to the actual code execution path being hidden or obfuscated. In this work, we choose to use the static analysis method because this method is essentially helpful on low-power and memory-limited devices such as Android devices. High optimization for performance is essential on the android operating system.

Data used to train deep learning models can come in the form of spatial, temporal, and more. Spatial data refers to location-aware information, a common example of this is a digital image. Temporal data are time-series that are collected as time progresses. These two concepts have been researched and powerful analysis tools in machine learning and deep learning have been created.

In this paper, we held the assumption that the DEX file binary, the bytes of the Dex file, is in the form of a time-series data [4]. The android binary file can be seen as containing sequences of opcode. We targeted the data section of the android application package (APK file). Our work is among the first to utilize CNN and RNN architectures. Particularly, one-dimensional convolutional neural network and bi-directional long-short term memory RNN.

The rest of this paper is arranged in the following way: we provided insight into previous works relating to this domain in Sect. 2. Our technique methodology is explained in Sect. 3. Experiments and Results with related information are presented in Sect. 4. Finally, Sect. 5 concludes the paper.

2   Related Works

The number of researches relating to android malware detection has seen an increase since the discovery of deep learning as a possible alternative to older techniques. Deep learning applications relating to the areas of speech recognition, image classification, and natural language processing are among the several pioneers. Deep learning for android malware detection can be seen in [5] which was among the first to utilize deep learning for android malware detection. They extended the research work of [6] by employing long short-term memory (LSTM) on a large-scale. In [7], R2-D2 translates android apps into RGB (red, green, and blue) color code and transforms them into a fixed-sized encoded image for image classification.

In [8], they proposed an end-to-end solution using one-dimensional CNN, where both the spatial and temporal data features in bearing fault diagnosis are extracted and utilized. In this paper, we extended this approach in the field of android malware detection. Convolutional neural network (CNN) models, developed for image classification, can also work well on one-dimensional sequential data. In our case, this refers to the raw bytes of android applications, we extract features from sequential data and maps to a vector space. A one-dimensional CNN works the same way as two- or three-dimensional CNN. The difference is in the structure of the input data and the convolution kernel movement.

3 Methods

In this section, we describe the methodology of our research. The core of our approach can be divided into three parts. The first is the data section extraction, next is data preprocessing, and finally the model design Fig. 2.

3.1 Data Section Extraction

The goal of this step is to gather the raw bytes of the data section, embed it in a vector space, and group them to form the dataset. A DEX file contains many sub-sections shown in Fig. 1 among these are the header, string ids, data section, code item, etc. [11]. Our method extracts information on the data section binary from the header which contains information on every other section. This information includes the offset and size of the data section.

The tool used to disassemble the APKs was Androguard [9]. Android applications are developed in Java and compiled into optimized bytecodes for the Dalvik virtual machine. This bytecode can be directly accessed with the help of Androguard.

3.2 Data Preprocessing

In this step, the features were made suitable for the deep learning model. This involved padding the features to a fixed length and scaling it with normalization. Normalization is a feature scaling method to change the values of individual features to use a common scale (a range of 0 ~ 1) without distorting differences in the range of values or losing information.

The input shape of our model must be of dimension [timestep, feature]. However, the resultant dataset after the data extraction step from Sect. 3.1 contains elements of

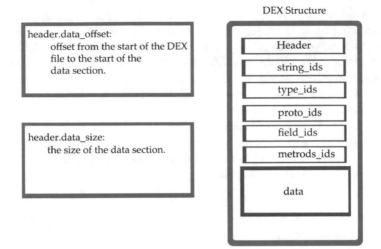

Fig. 1 Data section extraction

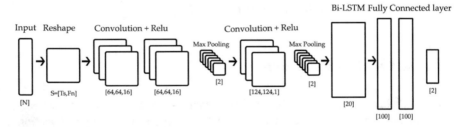

Fig. 2 Deep learning neural network architecture (one dimensional convolutional neural network and Bi-directional long-short term memory network)

various lengths. We reshape every sample in our dataset to achieve this fixed length. The process involved the use of a post-padding algorithm that appends the value of zero to the end of every sample less than a predefined threshold or cut if greater. Next, we used a window size to reshape the features into a sequential timestep. This new fixed input shape "S" and the window size can be defined as follows

$$S = [T_s, F_n]; \quad T_s = N/F_n; \quad F_n = 4 \tag{1}$$

where T_s is the timestep (window size), and F_n is the number of features per timestep for a given sample. For clarity, our final dataset has N equals ten million features, the new shape S becomes (2.5 million Timestep, 4 features). Before deciding on the number of features per timestep, other sizes were tested (4, 8, 16, etc.) but a timestep with 4 bytes gave the best performance.

3.3 Model Design

With the sample data padded and scaled, the next step is to decide on the malware detection architecture. Instead of using machine learning algorithms for pattern recognition and feature extraction, a deep learning solution was chosen. This work made use of the convolutional and recurrent deep learning architectures. As shown in Fig. 2, our model is composed of three one-dimensional CNN layers, a Bi-directional layer, and two dense layers (fully connected layers).

The one-dimensional CNN layers extract spatial and local temporal features from the sequences of normalized features. The pooling layers reduce the size of each feature map thus leading to a reduction in computational efficiency. The bi-directional LSTM layer extracts long-term temporal patterns that are analyzed by the fully connected layers. The fully connected layers then perform binary classi-fication. Binary classification is accomplished by a classifier that can distinguish between two classes or labels.

4 Experiments

4.1 Dataset

All the data are from the following sources: Google play (the period between October 2016 and February 2017), Amazon, APK pure, AMD, and Drebin [10]. The malware samples are from the Drebin and AMD archives while the benign samples were combined samples from Amazon, Google play, and APK pure. The dataset we used contains 20,000 applications comprising of malware as well as benign android pack-ages with a ratio of 1:1 (10,000 malicious applications and 10,000 benign ones). For the experiments, we used 19,000 samples for training, 500 for validation, and 500 for testing, summing up to a total of 20,000 (Table 1).

Table 1 Dataset sources

Source	Malware (%)	Benign (%)
Google play	0	20
Amazon	0	12.5
APK pure	0	17.5
AMD	35	0
Drebin	15	0

Table 2 Evaluation result from tested models

Models	Train accuracy	Test accuracy	Precision	Recall	F1-Score
3 × 1D Conv, 1 × Bi-LSTM	0.983	0.946	0.947	0.946	0.945
1 × 1D Conv, 1 × Bi-LSTM	0.91	0.902	0.9	0.9	0.9

Table 3 Experiment to determine F_n

Models	Epoch	Train accuracy	Test accuracy
(N/ F_n,16)	10	0.81	0.806
(N/ F_n,8)	10	0.84	0.832
(N/ F_n,4)	10	0.84	0.846

4.2 Experiment Result

The table below depicts the result of experiments undertaken to evaluate the performance of our deep learning model.

Table 2 shows the training accuracy, test accuracy, precision, recall, and F1-score of our final experiments. The table displays results from the two best models. The first is the 3 1D CNN layers architecture, and the other was a simpler model with a single convolutional layer. Our best model generalized on the test dataset with merely a 5.4% error rate. Also, this experiment proved that going deeper with convolutional layers yields better performance.

Table 3 shows the result of the experiment to determine the number of features per timestep. A simplified version of our dataset with 4, 8, and 16 feature channels enabled us to achieve our desire outcome. Each sample in this simplified dataset was made of merely 30,000 bytes. The model simplification involved reducing the layers and units per layer. Using 4 bytes per timestep gave the best testing model performance. Out of the result, we realized that larger feature channels did not correlate to better generalization.

5 Conclusion

In this work, we addressed the challenges of android malware detection through the introduction of a possible solution utilizing bi-directional LSTM and one-dimensional convolutional neural networks. Our method handles android binary files as sequences of opcode for malware detection.

In practice, this work demonstrates the process of analyzing both the temporal and spatial aspects of an android application for malware detection. To improve the achieved results, in future work, we plan to investigate methods that handle the large input size of our proposed model and the case of malware family detection.

Acknowledgements This research was supported by the National Research Foundation of Korea (NRF) and funded by the Ministry of Science and ICT (no. 2018R1A2B2004830).

References

1. McAfee Mobile Threat Report Q1 (2020)
2. Enck W, Gilbert P, gon Chun B, Cox LP, Jung J, McDaniel P, Sheth A (2010) Taintdroid: an information-flow tracking system for realtime privacy monitoring on smartphones. In: Proceedings of USENIX symposium on operating systems design and implementation (OSDI), pp 393–407
3. Zhou Y, Wang Z, Zhou W, Jiang X (2012) Hey, you, get off of my market: Detecting malicious apps in official and alternative android markets. In: Proceedings of network and distributed system security symposium (NDSS)
4. Yan L-K, Yin H (2012) Droidscope: seamlessly reconstructing os and dalvik semantic views for dynamic android malware analysis. In: Proceedings of USENIX security symposium
5. Bilar D (2007) Opcodes as predictor for malware. Int J Electron Secur Digit Forensics 1(2):156–168
6. Vinayakumar R et al (2018) Detecting android malware using long short-term memory (LSTM). J Intell Fuzzy Syst 34(3):1277–1288
7. Hsien-De Huang TT, Kao H-Y (2018) R2-D2: color-inspired convolutional neural network (CNN)-based android malware detections. In: 2018 IEEE international conference on big data (Big Data). IEEE
8. Hao S et al (2020) Multisensor bearing fault diagnosis based on one-dimensional convolutional long short-term memory networks. Measurement:107802
9. Anthony D (2019) Androguard documentation. Release 3.4.0
10. Arp D, Spreitzenbarth M, Hubner M, Gascon H, Rieck K, Siemens CERT (2014) Drebin: effective and explainable detection of android malware in your pocket. In: Yan J, Yong Q, Qifan R (eds) NDSS, LSTM-based hierarchical denoising network for Android malware detection. Security and communication networks 2018, vol 14. pp 23–26
11. Chiossi, R. 2014. "A deep dive into DEX file format. https://elinux.org/images/d/d9/A_deep_dive_into_dex_file_format--chiossi.pdf

A Definition of Covering Based Decision Table and Its Sample Applications

Thanh-Huyen Pham, Thi-Cam-Van Nguyen, Thi-Hong Vuong, Thuan Ho, Quang-Thuy Ha, and Tri-Thanh Nguyen

Abstract Covering based rough set, an extension of the traditional rough set theory, which uses the cover set of the universe set instead of the partition of the universe, has proven to be both theoretical and attractive in terms of applications. Corresponding to the decision table in traditional rough set theory, the concept of covering decision system has been defined. In this paper, we propose a decision table type based on covers, including the condition lattice of covers, and the decision lattice of covers. Two tasks on covering based decision table are also introduced. We also demonstrate the applications of the covering based decision table in collaborative filtering that corresponds to the classification in the traditional decision table, and in constraint based association rule mining to indicate this covering decision table concept has a potential application.

Keywords Covering based rough set · Covering based decision table (system) · Condition lattice · Decision lattice

T.-H. Pham · T.-C.-V. Nguyen · T.-H. Vuong · Q.-T. Ha (✉) · T.-T. Nguyen
VNU-University of Engineering and Technology, Ha Noi, Viet Nam
e-mail: thuyhq@vnu.edu.vn

T.-H. Pham
e-mail: phamthanhhuyen@daihochalong.edu.vn

T.-C.-V. Nguyen
e-mail: vanntc@vnu.edu.vn

T.-H. Vuong
e-mail: hongvt57@vnu.edu.vn

T.-T. Nguyen
e-mail: ntthanh@vnu.edu.vn

T.-H. Pham
Halong University, Quang Ninh, Viet Nam

T. Ho
Institute of Information Technology (VASC), Ha Noi, Viet Nam
e-mail: hothuan@ioit.ac.vn

175

1 Introduction

Proposed by ZI Pawlak in the early 1980s [1–3], rough set theory deals with the uncertainty or ambiguity of an object's real nature (e.g., the possibility that a person is infected COVID-19). Rough set theory and fuzzy set theory refer to the uncertainty or ambiguity about a property in human perception of an object (for example, the high/ low property of a person). They are two important branches of knowledge science today. Theoretically, rough set is fundamentally important in artificial intelligence and cognitive sciences. On the application side, many methods based on rough sets have a wide application in several real life projects [4]. Collaborating with granular computing and formal concept analysis, rough set theory is a component of the conceptual framework of concept analysis for three-way decision thinking [5]. Moreover, a wide range of generalized rough set models have been developed providing very useful and flexible means to data analysis. Covering based rough set is one of most emerging models among binary relation based rough set, i.e., neighborhood rough set, rough fuzzy set, and fuzzy rough set, probabilistic rough set, decision-theoretic rough set, three-way decisions with rough set, soft rough set, rough soft set, soft fuzzy rough sets, etc. [6].

The covering based rough set theory focuses on using the covers of the universe (by tolerance relations) for describing the approximation space instead of partitions of the universe (by equivalent relations in the traditional rough set theory). Starting from the concept of covering based approximation space proposed by Zakowski [7], covering based rough set theory has been rapidly developed [8–10] to show potential applications, of which collaborative filtering is a typical one [11–13].

Corresponding to the decision table concept in traditional rough set theory, the covering decision system has been defined [14] in the same time with the reduction task, which is one of the most typical tasks in the covering decision system [14–18].

This paper aims to propose a new definition of the decision table in the covering rough set and provide its typical application domains.

This paper has following main contributions: Firstly, we propose a new definition of a covering-based decision table, in which the condition attribute set is used instead of the condition lattices (a lattice of covers of the universe), and the decision attribute set is used instead of the decision lattice. There are some studies focusing on the relations between rough set models and the concept of lattices, such as [5, 9, 19–21]. However, there hasn't been any definitions of covering based decision table with the lattice of covers of the universe. Secondly, two problems of concern for covering based decision table are represented. Finally, a typical application of the proposed a covering rough set, i.e., collaborative filtering, is introduced to show its potential application.

The remainder of this paper is organized as follows. Next section presents an evolution history from a traditional rough set decision table to a covering based decision table, some examples, and two typical problems of a covering based decision table. Section 3 presents a comparison between a traditional decision table and a covering based decision table, a classification problem using covering based collaborative-filtering, together with an experimental example. Some related studies are explored in Sect. 4. The last section presents the conclusion of the paper.

2 An Extension of Covering Based Decision Table

2.1 Traditional Rough Set Decision Table

Pawlak [3] systematized the concepts in rough set theory, i.e., the fundamental concepts like *information system, indiscernibility relation, approximation space, rough set, the approximations of set*. The *accuracy of approximation* has been introduced since the beginning of 1980s [1, 2, 22].

Let U (called the universe) be a non-empty, finite set of objects in a certain domain. Throughout this paper, U is fixed.

Definition 1 (*Information system* [1]) An information system is a 4-tuple $IS = \langle U, A, V, \rho \rangle$, where U is the universe; A is a finite set of attributes, i.e., $A = \{a_i | i = 1..k\}$; $V = \bigcup_{a \in A} V_a$, where V_a (called the range of attribute a) is the set of all values for attribute a and $card(a) > 1$; ρ is a function from $X \times A$ into V.

In [3], information system is also called *knowledge representation system* (KR-system/KRS), or the *attribute-value system*.

Definition 2 (*Indiscernibility relation*) [3] For any subset B of A, i.e., $B \subseteq A$, there exists an equivalence relation $IND(B) = \{(x, y) \in U \times U | (x, a) = (y, a), \forall a \in B\}$. $IND(B)$ is called an indiscernibility relation of B. The set of equivalence classes by $IND(B)$ is denoted as U/B.

Definition 3 (*Approximation space*) [3] The pair (U, R) is called an approximation space, where U is the universe, and R is an equivalence relation on U.

Definition 4 (*Set approximation*) [3] Let (U, R) be an approximation space. For every subset $X \subseteq U$, the lower approximation set of X, denoted as $\underline{R}X$, is defined as $\underline{R}X = \cup\{Y \in U/R : Y \subseteq X\}$; and the upper approximation set of X, denoted as $\bar{R}X$, is defined as $\bar{R}X = \cup\{Y \in U/R : Y \cap X \neq \emptyset\}$.

For $\forall X \subseteq U$, if $\underline{R}X \neq \bar{R}X$ then X is called a rough set.

Let C, D be two subsets of A, the C-*positive region* of D, denoted by $POS_C(D)$, is determined as $POS_C(D) = \bigcup_{X \in U/D} \underline{C}X$. It is clear that $POS_C(D)$ is the union of all equivalent classes $IND(C)$, and it is contained within an equivalence class with the relation $IND(D)$.

An attribute $a \in B$, where $B \subseteq A$, is dispensable in B if $IND(B) = IND(B - \{a\})$, otherwise a is indispensable in B. The core of A is defined as the collection of all indispensable attributes in A. A subset $B \subset A$ is said to be independent on A if every attribute in B is indispensable in B. If a subset $B \subset A$ is independent, and $IND(B) = IND(A)$, then B is called a reduct in A. The set of all the reducts in A is denoted as $RED(A)$. The reduction task is to find some reducts in A, i.e., $RED(A)$.

Definition 5 (*Decision table* [3]) Decision Table is a special information system $TB = \langle U, C \cup D, V, \rho \rangle$, where the set of attributes A is divided to two disjointed subsets C and D, called *condition attributes* and *decision attributes*, respectively.

In some cases, a decision table is also called a *decision system*.

As mentioned above, there exist two indiscernibility relations $IND(C)$ and $IND(D)$.

Definition 6 (*Dependency in decision table*) [3] Let $TB = \langle U, C \cup D, V, \rho \rangle$ be a decision table. We define that D depends on C at a degree $k(0 \le k \le 1)$, denoted by $C \to_k D$, and

$$k = \frac{|\text{POS}_C(D)|}{|U|} \tag{1}$$

The reduction task in decision tables is to find a reduct in C.

2.2 Covering Based Rough Sets and Covering Decision Systems

The definition of Zakowski [7] on an approximation space, i.e., the pair (U, T), where U denotes an arbitrary, nonempty set; T denotes a cover of U, is the preliminary definition of *covering based rough set*.

Definition 7 (*Covering approximation space* [23]) Let U be a universe, C be a covering of U. We call the ordered pair (U, C) a *covering approximation space*.

Definition 8 (*Minimal description* [23]) Let (U, C) be a covering approximation space, xU:

$$Md(x) = \{K \in C | x \in K \wedge (\forall S \in C \wedge x \in S \wedge S \subseteq K \Rightarrow K = S)\} \tag{2}$$

is called the minimal description of x.

Definition 9 (*family of sets bottom approximation* [23]) Let (U, C) be a covering approximation space. For any subset $X \subseteq U$, the family of sets

$$C_*(X) = \{K \in C | K \subseteq X\} \tag{3}$$

is called the family of sets bottom approximating the set X.
The set

$$X_* = \cup C_*(X) \tag{4}$$

is called the lower approximation of the set X. The set

$$X_*^* = X \backslash X_* \tag{5}$$

is called the boundary of the set X.
The family of sets

$$Bn(X) = \cup \{Md(x) : x \in X_*^*\} \tag{6}$$

is called the family of sets approximating the boundary of the set X. The family of sets

$$C^*(X) = C_*(X) \cup Bn(X) \tag{7}$$

is called family of sets top approximating the set X. The set

$$X^* = \cup C^*(X) \tag{8}$$

is called the upper approximation of the set X.
The set X is said exact when

$$C_*(X) = C^*(X) \tag{9}$$

Otherwise, it is said inexact, i.e.,

$$C_*(X) \neq C^*(X) \tag{10}$$

The partition of the set X into sets X_*, and X_*^* is called an approximating partition of the set X.

Definition 10 ([23]) Let (U, C) be a covering approximation space. The equivalence class of the relation $\sim_C$ is defined for any sets $Y, Z \subseteq U$ by the following equivalence:

$$Y \sim_C Z \Leftrightarrow C_*(Y) = C_*(Z) \wedge C^*(Y) = C^*(Z) \tag{11}$$

Definition 11 (*the induced cover of a cover* [14]) Let $C = \{K_1, K_2, \ldots, K_n\}$ be a cover of U. For every $x \in U$, let $C_x = \cap\{K_j \in C | x \in K_j\}$. The set $Cov(C) = \{C_x | x \in U\}$ is, then, also a cover of U, and called *the induced cover of C*.

Definition 12 (*induced cover of a family of covers* [14]) Let $\Delta = \{C_1, C_2, \ldots, C_m\}$ be a family of covers of U. For every $x \in U$, let $\Delta_x = \cap\{C_{ix} \in Cov(C_i) | x \in C_{ix}\}$. The set $Cov(\Delta) = \{\Delta_x | x \in U\}$ is, then, also a cover of U, and called the *induced cover of Δ*.

For every $X \subseteq U$, the lower and upper approximations of X with respect to $Cov(\Delta)$, the computation formulas for the positive, negative and boundary regions of X relative to Δ are defined.

Definition 13 (*covering decision system* [14]) A covering decision system is an ordered pair $S = (U, \Delta \cup D)$, where U is the universe, Δ is a family of covers of U, and D is a decision attribute.

Covering decision system is also called covering decision information system [17].

Assume that there exists an information function of $D : U \rightarrow V_D$ where V_D is the value set of D. As above mentioned, $IND(D)$ is the equivalence relation of D and U/D is the set of equivalence classes by $IND(D)$.

Definition 14 (Δ-**positive region of** D [14]) Let $S = (U, \Delta \cup D)$ be a covering decision system, and $Cov(\Delta)$ is the induced cover of Δ. The Δ-positive region of D is calculated as $POS_\Delta(D) = \bigcup_{X \in U/D X \in U/D} \underline{\Delta}(X)$, where $\underline{\Delta}(X) = \cup\{\Delta_x | \Delta_x \subseteq X\}$.

Covering decision system is also called covering decision information system [17].

Definition 15 ([14]) Let $\Delta = \{C_i | i = 1, \ldots, m\}$ be a family of covers of U; D is a decision attribute; U/D is a decision partition on U. If for $\forall x \in U, \exists D_j \in U/D$ such that $\Delta_x \subseteq D_j$, where $\Delta_x = \cap\{C_{ix} | C_{ix} \in Cov(C_i), x \in C_{ix}\}$, then, the decision system (U, Δ, D) is called a *consistent covering decision system*, and denoted as $Cov(\Delta) \leq U/D$. Otherwise, (U, Δ, D) is called an *inconsistent covering decision system*.

2.3 Covering Based Decision Table

2.3.1 A Definition of Covering Based Decision Table

Our idea to put the lattice aspect in a definition of *covering based decision table* based on two reasons. Firstly, lattices as an aspect have been brought into rough sets in many works [5, 9, 19–21, 24]. Degang et al. [14] investigated the covering rough set within the framework of a *complete, completely distributive* (CCD) lattice, and proposed lower and upper approximations on a CCD lattice, which can improve the definitions of upper approximation. Tan et al. [20] stated that the approximation space of a covering and the concept lattice of a formal context are correspondent. And then they introduced high-level approximation operators of formal concept analysis (concept lattice) into covering-based rough sets. Wang et al. [21] investigated lattice representation of definable sets in generalized rough sets based on the relations and relationship between distributive lattices. And the lower definable lattices in generalized approximation spaces based on reflexive and transitive relations. Yao [5] proposed a definition of four types of formal concepts in rough-set concept analysis, which leads to four lattice concepts. Secondly, the idea of using the item lattice for the constraint-based association rule mining is very meaningful. Especially, both the antecedent and the consequent are represented in the form of lattices [25]. Our definition is described as follows.

Let U be the universe. Assume that there exists a partial order relation denoted "$\leq$" in the set of all cover set of U.

Definition 16 (*the lattice of covers*) L is defined as a lattice of covers of U if and only if L is a set of covers of U $(C_1, C_2, \ldots, C_n)$, and for every C_1, C_2 belong to L, there exist Y_1, Y_2 also belong to L such that $Y_1 \leq C_1, Y_1 \leq C_2$ and $C_1 \leq Y_2, C_2 \leq Y_2$.

Definition 17 (*the top-cover and the bottom-cover of a lattice of covers*) Since the universe U is finite, there exist C_{top}, C_{bottom} that C_{top}, C_{bottom} belonging to L, and $C \leq C_{top}(C_{bottom} \leq C)$ for every C in L.

Definition 18 A *covering based decision table* (CDT) is a triple $CDT = \langle U, CL, DL \rangle$, where CL and DL are two lattices of covers of the universe U; CL and DL are called *condition lattice* and the *decision lattice*, respectively.

Based on Definition 11, we define the induced cover of the CL and DL.

Definition 19 (*the induced cover of CL*) Let $CDT = \langle U, CL, DL \rangle$ be a covering based decision table; $TopCL = \{S_1, S_2, \ldots, S_n\}$ be a top-cover of CL. For every $x \in U$, let $TopCL_x = \cap\{S_j \in TopCL, x \in S_j\}$, then, the set $Cov(CL) = \{TopCL_x | x \in U\}$ is also a cover of U and it is called *the induced cover* of CL.

$Cov(DL)$ and the induced cover of DL are also defined in the same way.

Definition 20 (CL-*positive region of* DL). Let $CDT = \langle U, CL, DL \rangle$ be a covering based decision table. The CL-positive region of DL is calculated as

$$\text{POSCL}(DL) = \bigcup_{x,y \in U} TopCL_x \left(\overline{TopDL_y} \right) \tag{12}$$

Definition 21 Let $CDT = \langle U, CL, DL \rangle$ be a covering based decision table. We say that DL depends on CL at a degree $k(0 \leq k1)$, denoted by $CL \rightarrow_k DL$, if

$$k = \frac{|\text{POS}_{CL}(\text{DL})|}{|U|} \tag{13}$$

The DL's dependency on CL is an indicator of the performance that supports decisions of covering based decision table.

2.3.2 Application Examples

We consider the task of discovering constraint based association rules in the form of $X \rightarrow Y$ in transaction databases. In a rule, the antecedent X and the consequent Y are pre-determined itemsets [Zaki15]. Let TD be the set of all transactions, I be the set of all items. For every subset $B \subset I$, there exists a lattice of items $LB = (2^B, \subseteq)$, where 2^B is the set of all subsets of B (including the empty set); "$\subseteq$" is the relation "subset of". The antecedent X's lattice (denoted as LX) and the antecedent Y's lattice (denoted as LY) can play the role of the condition lattices (the decision lattices). Thus, we have a covering based decision table $CDT = \langle TD, LX, LY \rangle$.

We also have covering based decision tables in recommender systems [11–13]. Let U be the set of all users, I be the set of all items. As above mentioned, for every itemset BI, there exists the lattice LB. For an active user U, denote T as the set of items considered to filter for recommendation. We have a covering based decision table $CDT = \langle U, IL, TL \rangle$, where IL is the lattice of I, and TL is the lattice of T.

Two Tasks of Covering Based Decision Table

Let $CDT = \langle U, CL, DL \rangle$ be a covering based decision table, where CL, DL are the condition and decision lattices, respectively. We consider two tasks related to the decision problem. Firstly, we introduce the sub-lattice definition in a covering based decision table.

Definition 22 (*sub-lattice in a covering based decision table*) Let L be a lattice of covers. A cover BL on U is called by a sub-lattice of L if and only if: the set of covers in BL is a subset of the set of covers in L; and if a cover C in BL, and for all cover X in L such that $X \leq C$, then X is also in BL.

The Definition 22 of the sub-lattice in the covering based decision table is appropriate for association rule mining, where the lattice of an itemset X is a subset of a lattice Y such that $X \subset Y$.

Definition 23 (*reduction in the covering based decision table*) Let $CDT = \langle U, CL, DL \rangle$ be a covering based decision table. The task of condition reduction in CDT is to find sub-lattice CLC of CL such that $POS_{CLC}(DL) = POS_{CL}(DL)$.

The meaning of the condition reduction task is the same as the meaning of the reduction task in covering based rough decision systems [14–16, 18].

Moreover, Wei and Qi [26] investigated the relationship between the reduction in a traditional information system $\langle U, A, F \rangle$ and conceptual reduction in the formal context $\langle U, A, I, \rangle$ where U is the universe set; A is the property set; F is the information function; and I is a relationship between U and A. The authors indicated the condition for an attribute subset to be a reduct of a rough set is also a reduct of a concept lattice (this is stricter than the reduct of rough set). Such results have the potential to be applied to reduce concept lattice in a covering based decision table.

Definition 24 (*decision fitting in the covering based decision table*) Let $CDT = \langle U, CL, DL \rangle$ be a covering based decision table, $\sigma > 0$ be a threshold. The task of decision fitting in CDT is to find a sub-lattice DLD of DL such that: the dependency of $CDTD = \langle U, CL, DLD \rangle$ is not lower than the threshold σ, i.e., $CL \rightarrow_k DLD$ with $k \geq \sigma$; the induced cover $Cov(DLD)$ is the biggest according to the relation "$\leq$".

Deciding the recommendation list for an active user is an important problem in a recommender system. A good solution to the decision fitting problem in the covering based decision tables is potential to be applied to the above recommendation problem.

3 Covering Based Decision Table in Collaborative Filtering and Recommender System Application

One of the famous applications of the traditional rough set theory is the reduction of knowledge, and the approximate classification of objects based on traditional decision tables [2, 3]. Collaborative filtering methods based on covering based decision systems are getting more and more increased, such as [11–13]. In this section we discuss about applying covering based decision tables for collaborative filtering in recommender systems.

3.1 Covering Based Decision Table for Collaboration Filtering

Figure 1 describes the differences between classification and collaborative filtering. In classification (for which the traditional rough set decision table is designed for), there are distributions of independent variables and dependent variable(s), training rows and test rows in the classification. However, there's no such distribution in collaborative filtering.

In order to apply covering based decision table for user collaboration filtering, we set: the user set as the universe; set of all (or subset of) items [Aggrawal15] as the

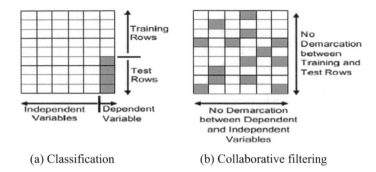

(a) Classification   (b) Collaborative filtering

Fig. 1 Comparison between the traditional classification problem and collaborative filtering [28]

conditional lattice; the lattice of potential recommendation items for the active user as the decision lattice.

3.2 Application in Recommender System

Covering based rough set theory provides two concepts: the minimal description, and the neighborhood of x belonging to the universe [23]. This is a good base for an effective solution to determine the set of neighbors for an active user in a cold start in a recommender system. The minimal description $Md(x)$ of an object x, i.e., the active user, in the universe is described in Definition 8. The neighborhood of an object is defined as follows:

Definition 23 (*the neighborhood of an object* [9, 23, 27]) Let (U, C) be a covering approximation space, x is a member of U, i.e., $x \in U$ then

$$N_C(x) = \cap\{K \in C | x \in K\} \tag{14}$$

is called the neighborhood of x with respect to (U, C). For the sake of being short, we write $N_C(x)$ as $N(x)$, and we have $N(x) = \cap Md(x)$ [23].

For user-based collaborative filtering (UBCF) in recommender systems, Zhang et al. [11, 12] proposed solutions to build an active user's neighborhood based on a covering reduction procedure in a covering-based rough set (step 5 in a seven-step algorithm [11], the STCRA algorithm [12]). Zhang et al. [13] continued the approach of covering reduction solutions in covering-based rough sets to address the new user cold-start problem. An algorithm to determine the decision class of the new user nu was proposed. Then, a covering reduction algorithm (CRA) in covering-based rough set was applied to determine an new user's neighborhood.

For user based collaborative recommender systems, the covering based decision table is represented as:

$$CDT = \langle U, CLItems, DLItems \rangle$$

where $CLItems$ is the lattice of the input itemset; $DLItems$ is the lattice of items to be filtered for recommendation to a new user in partial cold start problems. The conditions for an attribute subset to be a reduct of a rough set as well as a reduct according to the concept lattice [26] need to be investigated to apply to the conditional and decision lattice.

4 Related Work

As above mentioned, Degang et al. [14] defined the covering decision system as a tri-tuple $\langle U, \Delta, D \rangle$, where U is the universe; Δ is a family of convers; and D is a decision attribute. The consistent (inconsistent) covering decision systems have been defined, and the conditions for attribute reduction of two kinds of covering decision systems have been investigated. Based on the definitions of the intersection of covers (the induced cover of Δ), a new method to reduce redundant covers in a covering decision system and an algorithm for computing all of the reductions has been proposed. Our definition uses the lattice of condition covers and the lattice of decision covers instead of the set of covers and the decision attribute in [14], respectively. Because of using the top sub-lattice, we do not determine the induced cover of Δ.

Li and Yin [16], Wang et al. [18], Tan et al. [17], Chen et al. [15] focused on solving the reduction problem for the covering decision system f $\langle U, \Delta, D \rangle$, which is the same as the covering decision system in [14]) in two cases, i.e., the system is consistent and inconsistent.

As mentioned in subsection 2.3, Tan et al. [20] introduced high-level approximation operators of formal concept analysis (concept lattice) into covering-based rough sets, and proposed several reduction algorithms for a cover that could also be used to compute the reducts of a formal context. A. Tan et al. did not mention about covering based decision table, however, the results of representing the structures of coverings by concept lattices should be considered in covering based decision table.

5 Conclusions and Future Work

The paper proposed a definition of covering based decision table, in which both of the two components are the lattices of the covers of universe set. Two fundamental problems, i.e., conditional lattice reduction and decision fitting, were also introduced. In addition, two application domains, i.e., association rule mining with the itemset constraints in the antecedent and consequent; and user-oriented collaborative filtering, were introduced to show the potential application of the proposal.

The properties of the covering based decision table; the condition reduction and decision fitting problems; as well as the representation of covering based decision table in correspondent with covering based rough set and formal concept analysis need to be further studied.

References

1. Pawlak Z (1981) Rough sets. ICS PAS Reports, vol 431
2. Pawlak Z (1982) Rough sets. Int J Comput Inf Sci 11(5):341–356
3. Pawlak Z (1991) Rough sets: theoretical aspects of reasoning about data. Springer, Netherlands
4. Skowron A, Dutta S (2018) Rough sets: past, present, and future. Natural Computing, pp 1–22
5. Yao Y (2020) Three-way granular computing, rough sets, and formal concept analysis. Int J Approx Reason 116:106–125
6. Zhan J, Zhang X, Yao Y (2020) Covering based multigranulation fuzzy rough sets and corresponding applications. Artif Intell Rev 53(2):1093–1126
7. Zakowski W (1983) Approximations in the Space (U, Π). Demonstratio Mathematica 16(3):761–769
8. Yao Y, Yao B (2012) Covering based rough set approximations. Inf Sci 200:91–107
9. Zhao Z (2016) On some types of covering rough sets from topological points of view. Int J Approximate Reason 68:1–14
10. Zhu W (2007) Basic concepts in covering-based rough sets. vol 5. ICNC, pp 283–286
11. Zhang Z, Kudo Y, Murai T (2015) Applying covering-based rough set theory to user-based collaborative filtering to enhance the quality of recommendations. In: IUKM, pp 279–289
12. Zhang Z, Kudo Y, Murai T (2017) Neighbor selection for user-based collaborative filtering using covering-based rough sets. Ann Oper Res 256(2):359–374
13. Zhang Z, Kudo Y, Murai T, Ren Y (2020) Improved covering-based collaborative filtering for new users' personalized recommendations. Knowl Inf Syst 62(8):3133–3154
14. Degang C, Wang C, Qinghua H (2007) A new approach to attribute reduction of consistent and inconsistent covering decision systems with covering rough sets. Inf Sci 177(17):3500–3518
15. Chen J, Lin Y, Lin G, Li J, Zhang Y-L (2017) Attribute reduction of covering decision systems by hypergraph model. Knowl Based Syst 118:93–104
16. Li F, Yin Y (2009) Approaches to knowledge reduction of covering decision systems based on information theory. Inf Sci 179(11):1694–1704
17. Tan A, Li J, Lin Y, Lin G (2015) Matrix-based set approximations and reductions in covering decision information systems. Int J Approx Reason 59:68–80
18. Wang C, He Q, Chen D, Qinghua H (2014) A novel method for attribute reduction of covering decision systems. Inf Sci 254:181–196
19. Lirun S, Zhu W (2017) Closed-set lattice and modular matroid induced by covering-based rough sets. Int J Mach Learn Cybernet 8(1):191–201
20. Tan A, Li J, Lin G (2015) Connections between covering-based rough sets and concept lattices. Int J Approx Reason 56:43–58
21. Wang Z, Feng Q, Wang H (2019) The lattice and matroid representations of definable sets in generalized rough sets based on relations. Inf Sci 485:505–520
22. Pawlak Zd (1981) Information systems-theoretical foundations. Inf Syst 6(3):205–218
23. Bonikowski Z, Bryniarski E, Wybraniec-Skardowska U (1998) Extensions and intentions in the rough set theory. Info Sci 107(1–4):149–167
24. Degang C, Zhang W, Yeung DS, Tsang ECC (2006) Rough approximations on a complete completely distributive lattice with applications to generalized rough sets. Inf Sci 176(13):1829–1848
25. Aggarwal CC (2015) Chapter 5. association pattern mining: advanced concepts. In: Aggarwal CC (ed) Data mining: the textbook. Springer, pp 135–152

26. Wei L, Qi JJ (2010) Relation between concept lattice reduction and rough set reduction. Knowled-Based Syst 23(8):934–938
27. Zhu W (2007) Topological approaches to covering rough sets. Info Sci 177(6):1499–1508
28. Aggarwal CC (2016) In: Recommender systems: the textbook. Springer

Image-Based Prediction of Respiratory Diseases Including COVID-19 Using Convolutional Neural Networks

Parsa Yousefi and Yu-Fang Jin

Abstract Respiratory diseases such as COVID-19, Pneumonia, SARS, and Streptococcus have caused severe worldwide public health concerns. Specifically, COVID-19, as an emerging worldwide pandemic, imposed the most critical challenge to all scientists and researchers for prognosis, diagnosis, and treatment of COVID-19 infection. This study aims to predict the aforementioned 4 respiratory diseases and normal people with chest X-ray and CT scan images using convolutional neural networks. A total of 1,156 images has been collected from 3 published databases. The combined dataset was enriched by empowering augmentation techniques and visual filters such as rotation and lung segmentation. The noises for augmentation include Gaussian and Speckle noises with zero mean and variance of 0.05, 0.10, and 0.20, and Salt and Pepper noise with 50% and 75% ratio. The customized convolutional neural network reached a prediction accuracy of 94% in classifying the test images into the normal and 4 disease categories, and 92%, 93%, and 92% as average precision, recall, and F1-score over all categories, respectively.

Keywords COVID19 · Respiratory diseases · X-ray and CT scan · Convolutional neural networks · Augmentation · Segmentation

1 Introduction

During the past 30 years, the world has suffered from multiple respiratory disease outbreaks including Cryptosporidiosis in the 1990s, SARS in 2003, H1N1 since 2009, and the outbreak of the world pandemic COVID-19 caused by SARS-CoV-2 virus at the end of 2019 [1–4]. With over 22.6 million infected people and about 791,676 deaths worldwide by December 4th, 2020, the pandemic has extremely

P. Yousefi · Y.-F. Jin (✉)
Department of Electrical and Computer Engineering, The University of Texas At San Antonio, San Antonio, TX 78249, USA
e-mail: yufang.jin@utsa.edu

P. Yousefi
e-mail: parsa.yousefi@utsa.edu

© The Author(s), under exclusive license to Springer Nature Singapore Pte Ltd. 2021   189
H. Kim et al. (eds.), *Information Science and Applications*, Lecture Notes
in Electrical Engineering 739, https://doi.org/10.1007/978-981-33-6385-4_18

impacted the economic growth and people's daily lifestyles [5]. Nine months post the outbreak of this pandemic, detection of COVID-19 is still not easy and getting more complicated when the infected people show no symptoms or similar symptoms caused by other respiratory diseases. This research focused on the prediction of respiratory diseases using image-based deep learning algorithms with existing X-ray and CT scan images from people infected by COVID-19, Pneumonia, SARS, and Streptococcus, and normal people. Different research groups have implemented binary classification of X-ray images from COVID-19 infected people vs. non-infected people with deep learning methods [6–8]. Tartaglione et al. have proposed a comparison between classifying small labeled datasets for COVID-19 chest X-ray images with deep algorithms such as ResNet18 gathering 79% of F1-score in image classification with 61% Area Under the Curve [9]. Minaee et al. have proposed COVID-19 X-ray image classification with ResNet18 and SqueezeNet algorithms with transfer learning [10]. They have reached over 90% accuracy in total for the classification task on the dataset. Abbas et al. have proposed a convolutional neural network algorithm for decomposing, transfer, and composing for classifying chest X-ray images of COVID-19 infected and non-infected people. They have reached 95.12% of accuracy in total with 97.91% sensitivity and 91.87% specificity [11]. All these researches performed a good binary classification of COVID-19 patients and normal people. However, no research has been performed to predict and explore the features of COVID-19 comparing against other respiratory diseases.

To address this need, we have employed two reported datasets that include over 700 images of patients' X-ray and CT scan for multiple respiratory diseases at different stages [12–14]. To enrich images for categories other than COVID19, 2,802 X-ray images for Pneumonia infected patients from NIH have been integrated [15]. This research aims to establish a deep learning model to predict 5 groups featuring COVID-19, Pneumonia, SARS, and Streptococcus infections, as well as normal people with X-ray and CT scan images. Due to the complicated task of multi-category classification and feature-extractions in images, traditional machine learning algorithms may not converge to high accuracy results, thus, a customized convolutional neural network, VGG net (Visual Geometry Group), was developed to classify the 4 respiratory diseases and normal people. Due to the limited images resources, three different levels of noises (Gaussian noise, Salt and Pepper noise, and Speckle noise) with different values of mean and variance ($\mu = 0, \sigma^2 = 0.05, 0.10, and\ 0.20$) and noise ratio are added to rotated images ($\pm 10°$) to generate a sufficient number of images samples for the neural network to converge. Also, this project has manually segmented (masked) the lungs from images for maximizing feature extraction from the lungs. The established model has achieved 94% of total prediction accuracy in classifying segmented and augmented images of chest X-ray and CT scan images by adding rotation and different noise levels to enrich the datasets. To interpret the results, a detailed investigation of normal and COVID-19 features was performed to illustrate the patterns learned by the model at different layers.

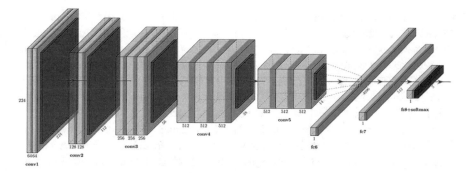

Fig. 1 VGG model architecture including 5 Conv layers and 3 fully connected layers

2 Methodology

2.1 Model Architecture

The architecture of the customized CNN based on VGG16 was shown in Fig. 1. It includes 13 layers of convolution + ReLU activation function as well as max-pooling layers. The head of the network is connected to 3 fully connected layers with a dropout of 0.5, and input of $224 \times 224 \times 3$ (RGB) for images and output layer of 5 classes. The kernel size for Conv layers is of dimension (3×3) and max-pooling windows are of size (4×4). The network has 14,747,845 trainable parameters.

2.2 Dataset

Table 1 shows the details of three datasets used in this research, leading to a combined

Table 1 Attributes of 3 datasets used in this study

Dataset Attribute	Chest X-ray	NIH X-ray and CT scan	COVID19 CT dataset
Resource	[12]	[15]	[14]
Generation date	04/27/2020	02/21/2018	03/03/2020
Number of images	COVID19: 206 SARS: 16 Streptococcus: 17	Normal: 1341 Pneumonia: 2802	COVID19: 203 No finding: 321
Number of patients	383	999	216
View	X-ray: AP, PA, and AP Supine CT: Axial	X-ray: PA CT: Axial	Axial

version of those three datasets only for categories of "COVID-19", "Pneumonia", "SARS", "Streptococcus", and "Normal (no-finding)". It's worth mentioning that the Chest X-ray dataset does not have images from a normal group. The NIH X-ray and CT Scan dataset were established in 2018, therefore, we assume the normal group has no COVID-19 cases due to the outbreak of COVID-19 was at the end of 2019. Also, no finding group in COVID-19 CT Dataset was classified as normal people [12, 14, 15]. A total of 763 images were collected from recent COVID-19 studies as shown in columns 1 and 3 in Table 1. Also, 1,341 images from normal people and 2,802 images from Pneumonia patients were collected from the NIH X-ray and CT Scan dataset. The combined dataset is pretty large but biased with significant number of images from Pneumonia patients. To have unbiased training for CNN, we only chose partial data from NIH X-ray and CT Scan dataset so the number of images for pneumonia and normal people would not overwhelm the other categories.

2.3 Preprocessing Images

Figure 2 shows an example of X-ray images for all categories studied in this research. These images from different views PA (posteroanterior), AP (anteroposterior), and AP Supine for X-ray, and Axial for CT Scans) are labeled with different attributes (categories and views) and resources. Since the images have variable resolutions, they all are resized to (224 × 224) to be fed to the CNN.

2.4 Augmentation

Due to the limited number of images for COVID-19, SARS, and Streptococcus, three different levels of noise have been implemented on each image to enrich the dataset. Three different types of noises, Gaussian, Salt and Pepper, and Speckle, with zero-mean and different variances or ratios, were augmented to the original images. Gaussian noise augmentation just adds the noise to each pixel of the original images. A zero-mean Gaussian noise was shown in Eq. (1) and the value of pixels in the output image as shown in Eq. (2) as below,

$$p_G(z) = \frac{1}{\sigma\sqrt{2\pi}} e^{-\frac{x^2}{2\sigma^2}}, \tag{1}$$

$$output_{image} = input_{image} + noise. \tag{2}$$

Salt and Pepper noise augmentation randomly selects pixels from original images and replaces them with 0 (0 in RGB) and 1 (255 in RGB) with a defined ratio. Speckle noise augmentation is based on Gaussian distribution but with the generation method shown in Eq. (3). For both Gaussian and Speckle, the mean has been chosen 0 as the

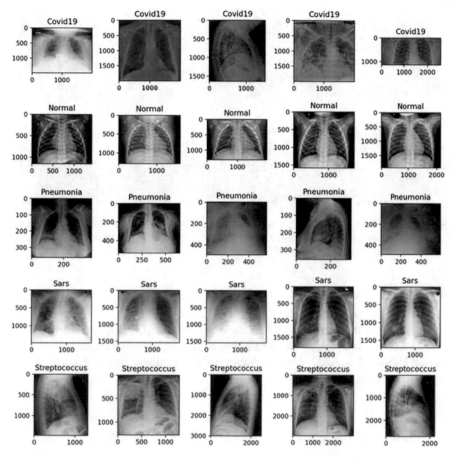

Fig. 2 Examples of X-ray for an initial dataset for 5 categories; row 1: COVID19, row2: Normal, row 3: Pneumonia, row 4: SARS, and row 5: Streptococcus

images would lose features (saturation) with increasing absolute value of the mean.

$$output_{image} = input_{image} + noise \cdot input_{image}. \tag{3}$$

In this paper, the value of $\mu = 0$ and $\sigma^2 = 0.05, 0.10,$ and 0.20 were used for Gaussian and Speckle noise and the ratio of 0.25 (25% of randomly selected pixels get a value of 0 and 75% of randomly selected pixels get a value of 255), 0.5 (50% of randomly selected pixels get a value of 0 and 50% of randomly selected pixels to get a value of 255), and 0.75 (75% of randomly selected pixels get a value of 0 and 25% of randomly selected pixels get a value of 255) were used. Figure 3 showed the generated images with different augmentations in color and grayscale mapping. To gain more enrichment, all noises were added to rotated images ($\pm 10°$) as shown in Fig. 4.

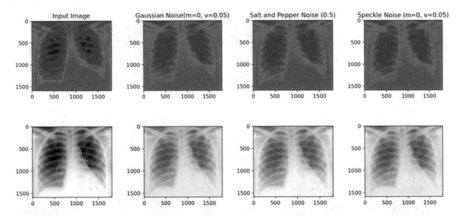

Fig. 3 Results of augmentation with noise on a sample image (color and grayscale view), column 1: input image, column 2:Gaussian noise augmentation, column 3: Salt and Pepper noise augmentation, and column 4: Speckle noise augmentation

2.5 Lung Segmentation (Masking)

To maximize feature extraction from lungs in X-rays and CT Scans, the lungs in the images were masked (segmented) to binary masks (value of 0 where it is not lung, and 255 where it is lung). Figures 5 and 6 showed binary masks and augmented masks respectively.

2.6 Training Parameters

Training the network has been performed on the original, augmented, and augmented + segmented datasets with a distribution of 80% for training and 20% for tests. It is essential to mention that the original images and augmented images are distributed uniformly in the training-test datasets. About 20 epochs for each training procedure were performed with the categorical loss function shown in Eq. (4). Adam optimizer with learning rate $1e - 3(10^{-3})$ and decay $5 \times 10^{-5} \left(\frac{learning rate}{number of epochs} \right)$ were adopted for CNN,

$$L(y, \hat{y}) = - \sum_i y_i . \log \hat{y}_i. \tag{4}$$

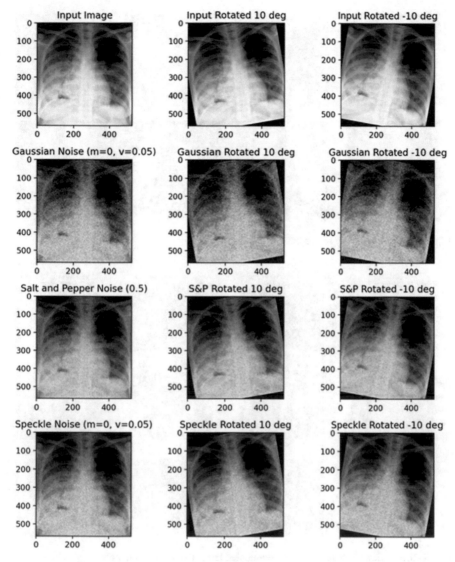

Fig. 4 Augmented images with different rotation angles

3 Results

A total of 7 different scenarios were studied in this project:

(a) Original Images (PA View only for X-rays and Axial for CT)
(b) Augmented Images with $\mu = 0 \, and \, \sigma^2 = 0.05$ for Gaussian and Speckle noise, and $r_{s\&p} = 0.5$ for Salt and Pepper noise only on PA Views (X-ray)

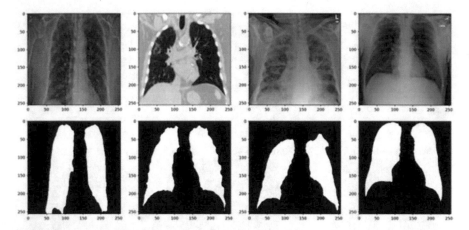

Fig. 5 Samples of binary segmentation of lung images

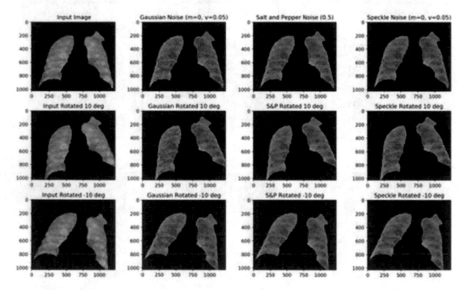

Fig. 6 Sample images for augmenting segmented lungs from original images

(c)　Augmented Images with $\mu = 0$ and $\sigma^2 = 0.05$ for Gaussian and Speckle noise, and $r_{s\&p} = 0.5$ for Salt and Pepper noise on all views (PA, AP, and AP Supine for X-ray and Axial for CT)

(d)　Augmented Images with $\mu = 0$ and $\sigma^2 = 0.10$ for Gaussian and Speckle noise, and $r_{s\&p} = 0.5$ for Salt and Pepper noise on all views

(e)　Augmented Images with $\mu = 0$ and $\sigma^2 = 0.20$ for Gaussian and Speckle noise, and $r_{s\&p} = 0.5$ for Salt and Pepper noise on all views

Table 2 Number of images in each scenario used for model training and test procedures

Categories	COVID19		Normal		Pneumonia		SARS		Streptococcus	
Scenarios	Train	Test	Train	Test	Train	Test	Train	Test	Train	Test
(a)	208	42	195	39	196	39	16	3	17	3
(b)	208	41	195	39	196	39	192	38	204	41
(c)	406	81	1382	276	1241	248	192	38	204	41
(d)	406	81	1382	276	1241	248	192	38	204	41
(e)	406	81	1382	276	1241	248	192	38	204	41
(f)	1392	278	1092	218	1236	247	192	38	204	40
(g)	1392	278	1092	218	1236	247	192	38	204	40

(f)   Augmented + Segmented Dataset with $\mu = 0$ and $\sigma^2 = 0.10$ for Gaussian and Speckle noise, and $r_{s\&p} = 0.5$ for Salt and Pepper noise on all views

(g)   Augmented + Segmented Dataset with $\mu = 0$ and $\sigma^2 = 0.20$ for Gaussian and Speckle noise, and $r_{s\&p} = 0.75$ for Salt and Pepper noise on all views

With this research design, scenario (a) was used as a reference to see the effects of different views of X-ray, augmentation, and segmentation. Scenarios (b) and (c) would allow us to further examine the effects of prediction from different views of X-ray images after augmentation. Scenarios (c), (d), (e), (f), and (g) provided a comprehensive comparison of different augmentation and segmentation effects.

The number of images used for each model training and test were shown in Table 2, leading to a total of 1,156 raw images for all 7 scenarios. For scenario (a) (failure of model convergence due to not a sufficient number of images) and other successful scenarios b to g, the results were shown in Fig. 7.

Interestingly, the best-obtained result is for scenario (f) where the network learns the features of augmented + segmented images with noise variance of 0.10, indicating the enriched data with augmentation and segmentation may be useful for prediction with small variances. The results also showed that increasing the variance in noises would decrease the accuracy since the noises may damage the features. It's worth mentioning that the augmentation and segmentation might be used on adversarial purposes and a careful study on the effect of a threshold of variances on prediction accuracy needs to be performed in the future. The performance of the CNN model and the confusion matrix of the best-case scenario was provided in Tables 3 and 4. Specifically, though both COVID-19 and normal images were collected from two different studies as shown in Table 1, the confusion matrix in Table 4 showed that our prediction is disease-specific and was independent of data sources. This was also confirmed with the precision and recall results in Table 3.

Figure 7 showed feature maps of First (initial), Thrid (middle), and Fifth (final) convolutional layers of two examples, one from COVID19 class, and the other from Normal class. The first convolutional layer mostly highlighted the general features from the input image in the single-channel resolution of (224×224). The middle convolutional layer focused on detecting edges from features in input images.

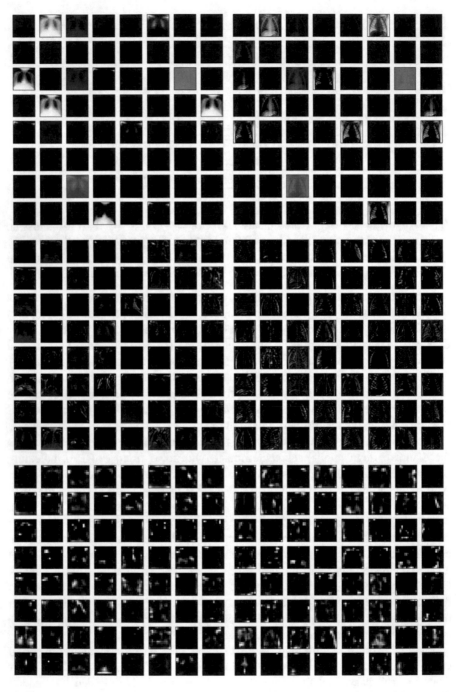

Fig. 7 Feature maps of 1st (row 1), 3rd (row 2), and 5th (row 3) convolutional layer of the network for a true positive of COVID19 class (left column) and Normal class (right column)

Table 3 Performance of best-case scenario (*f*)

	Precision	Recall	F1-Score
COVID19	0.95	0.94	0.94
Normal	0.93	0.92	0.93
Pneumonia	0.94	0.94	0.94
SARS	0.92	0.92	0.92
Streptococcus	0.86	0.93	0.89

Table 4 Confusion matrix of best-case scenario (*f*)

		Predicted				
		COVID19	Normal	Pneumonia	SARS	Streptococcus
Labeled	COVID19	261	7	7	1	2
	Normal	7	201	7	1	2
	Pneumonia	6	6	233	1	1
	SARS	1	1	0	35	1
	Streptococcus	1	1	1	0	37

The features were extracted from X-ray and CT images with a lower resolution as (56×56). The Fifth convolutional layer, on the other hand, focused on important regions in the input image which results in the classification task. Those images had a resolution of (14×14) and showed the most important regions for prediction in lungs.

4 Conclusion

In this study, an image-based CNN model was established to predict multiple respiratory diseases with augmented and segmented X-ray and CT scan images. Deep neural networks were chosen to fulfill the task, and a custom VGG16 CNN was used as the core model. Also, due to the lack of enough images for all categories, augmentation filters such as different levels of noise and rotation were used as well as lung segmentation (masking). The best result of this project was obtained by implementing augmentation on segmented images with zero mean and 10% variance for Gaussian and Speckle noise, also 50% ratio for Salt and Pepper noise. This project was able to reach a total accuracy of 94% for the best case in predicting patients with 4 respiratory diseases (COVID-19, SARS, Pneumonia, Streptococcus) and normal people. Based on the best result obtained by the CNN model, further interpretation was performed to examine the features of the images from COVID-19 infected and normal people at different layers of the CNN model.

References

1. Gharpure R, Perez A, Miller AD, Wikswo ME, Silver R, Hlavsa MC (2019) Cryptosporidiosis Outbreaks–United States, 2009–2017. In: MMWR Morb Mortal Wkly Rep 2019, vol 68. pp 568–572. [Online] Available https://www.cdc.gov/mmwr/volumes/68/wr/mm6825a3.htm
2. CDC (2009) H1N1 Pandemic (H1N1pdm09 virus). Centers of Disease Control and Prevention. https://www.cdc.gov/flu/pandemic-resources/2009-h1n1-pandemic.html.Accessed 20 Aug 2020
3. Winter K, Carol G, James W, Kathleen H (2020) Pertussis Epidemic—California, 2014. In: Morbidity and mortality weekly report (MMWR), vol 2020 [Online]. Available https://www.cdc.gov/mmwr/preview/mmwrhtml/mm6348a2.htm
4. CDC (2020) Get the facts about coronavirus. Centers for Disease Control and Prevention. https://www.cdc.gov/coronavirus/2019-ncov/index.html. Accessed 20 August 2020
5. Worldometer (2020) COVID-19 Coronavirus Pandemic. https://www.worldometers.info/coronavirus/. Accessed 20 August 2020
6. Maguolo G, Nanni L (2020) A critic evaluation of methods for covid-19 automatic detection from x-ray images. arXiv preprint arXiv:2004.12823
7. Selvan R, Dam EB, Rischel S, Sheng K, Nielsen M, Pai A (2020) Lung Segmentation from chest X-rays using variational data imputation. arXiv preprint arXiv:2005.1005
8. Cohen JP et al (2020) Predicting covid-19 pneumonia severity on chest x-ray with deep learning. arXiv preprint arXiv:2005.11856
9. Tartaglione E, Barbano CA, Berzovini C, Calandri M, Grangetto M (2020) Unveiling COVID-19 from chest X-ray with deep learning: a hurdles race with small data. arXiv preprint arXiv:2004.05405
10. Minaee S, Kafieh R, Sonka M, Yazdani S, Soufi GJ (2020) Deep-covid: predicting covid-19 from chest x-ray images using deep transfer learning. arXiv preprint arXiv:2004.09363
11. Abbas A, Abdelsamea MM, Gaber MM (2020) Classification of COVID-19 in chest X-ray images using DeTraC deep convolutional neural network. arXiv preprint arXiv:2003.13815
12. Dataset G (2020) COVID19 Chest X-ray. https://github.com/ieee8023/covid-chestxray-dataset. Accessed 20 August 2020
13. Cohen JP, Morrison P, Dao L, Roth K, Duong TQ, Ghassemi M (2020) Covid-19 image data collection: prospective predictions are the future. arXiv preprint arXiv:2006.11988
14. Zhao J, Zhang Y, He X, Xie P (2020) COVID-CT-Dataset: a CT scan dataset about COVID-19. arXiv preprint arXiv:2003.13865
15. Dataset K (2020) NIH Chest X-ray." https://www.kaggle.com/nih-chest-xrays/data. Accessed 20 August 2020

Determination of Muscle Power Using RMS of Electromyography for Stroke Survivors

Choon Chen Lim, Kok Swee Sim, Shing Chiang Tan, and Zi Sen Yeoh

Abstract Stroke is a threatening disease. Rehabilitation exercises can rehabilitate impaired limbs. Instead of using a manual assessment method of Medical Research Council (MRC) scale in muscle strength measurement, electromyography (EMG) is utilised to measure muscle signal. A two-channel surface EMG sensor is employed to measure biceps and arm muscles power. In this study, ten stroke patients and ten healthy persons were recruited to join the project. The patients were exposed to the EMG measurement throughout all the rehabilitation sessions. Overall root mean square (RMS) was used to determine the strength of the muscles. The result showed that the overall RMS value of impaired upper limb was averagely improved by 4.22% at the end of rehabilitation process. In the muscle power measurement, EMG signal measurement presented evident improvement (+2.43%) as compared to the MRC scale measurement (+0%). Therefore, EMG signal measurement can precisely describe the muscle power of the patients.

Keywords EMG · Muscle strength · Stroke rehabilitation

1 Introduction

Stroke is categorised among the five leading causes of death. It is also one of the main reasons for hospitalisation in Malaysia [1]. According to the World Stroke Campaign

C. C. Lim (✉) · K. S. Sim · Z. S. Yeoh
FET, Multimedia University, Melaka, Malaysia
e-mail: junqinglim@gmail.com

K. S. Sim
e-mail: kssim@mmu.edu.my

Z. S. Yeoh
e-mail: yeoh.jason97@gmail.com

S. C. Tan
FIST, Multimedia University, Melaka, Malaysia
e-mail: sctan@mmu.edu.my

under the World Stroke Organization, there are 17 million people worldwide affected by stroke, with 5.5 million deaths and 26 million survivors within two years.

Stroke leads to health problems such as fingers motor dysfunction, upper limb impairment, balancing impairment and so on. As a consequence, the quality of life is deteriorated [2, 3]. However, stroke can be treated through visual-based rehabilitation exercises. It aims to help stroke patients in regaining their lost skills. The key of recovering is repeating the task for times. The repetitive process will help to rebuild muscle [4].

Muscle power has to be measured throughout the rehabilitation sessions to determine the improvement in muscle strength. Many physiotherapists employ manual muscle testing method to assess the muscular paresis, which is known as the Medical Research Council (MRC) scale [5]. MRC testing procedures are quick and straightforward to conduct. It can determine the muscle strength without any equipment since the muscle power of stroke patient is assessed manually by physiotherapist. MRC scale applies for numbers between 0 and 5 to determine the muscle strength [5]. MRC scale with a grade of 0 means no muscle contraction, whereas grade 1 signifies only flicker of muscle contraction. Besides, grade 2 implies the muscle can actively move (with gravitational effect eliminated) whereas grade 3 means the muscle can actively move (against the gravitational effect). Grade 4 also indicates the muscle can actively move (against the gravitational effect and resistive force) whereas grade 5 reflects normal power.

Nevertheless, the implementation of numbers from 0 to 5 in the MRC scale is inappropriate since it practices equidistant points along a continuous scale [6]. The scale number only results in the whole number of muscle strength. For instance, there is no decimal grade of 4.5 if the muscle strength falls within the range of grades 4 and 5. Besides, the grade presented by the method is vague [6]. Grade 5 is termed as having normal muscle power. However, it is hard to determine whether the normal power is the power possessed by normal people. It presents the problem that the stroke patients with grade 5 scale have less muscle strength when compared to the normal people with grade 5 scale. Therefore, a computerised method is introduced to assess and quantify the muscle strength possessed by stroke patients.

Electromyography (EMG) is a diagnostic method that is employed for evaluating the health of muscles and the nerve cells that control motor neurons [7]. When the muscle is contracting or relaxing, the motor neurons will transmit electrical signals. An EMG will translate these signals into numbers or graphs. It can ease the doctors in constructing a diagnosis. Surface EMG electrode is opted as a sensor device to detect the EMG signal. It is a non-invasive technique that applied to the skin surface at the muscle area. It is relatively easy to implement without any need of medical certification. Thus, it is recommended especially for those physically disabled patients.

The EMG measurement is employed to obtain the muscle power in terms of root mean square. The EMG signal is first analysed in the MATLAB. Several signal processing steps are used to remove the unnecessary features of the signal. In the end, the muscle power of the stroke patients can be assessed in terms of root mean square value of the signal after the signal processing steps. EMG measurement technique describes the muscle strength with a precise and decimal numerical value; hence it provides better muscle evaluation than the MRC method.

2 Methodology

2.1 Signal Processing Steps

The normal EMG raw signal contains both muscle signal and noise. The dominant sources of noise are the fluctuating of signal during the relaxation of muscles and the changes in the conductivity of tissue properties [8]. The noise will interfere with the result and lead to an inaccurate result. Thus, filtering out the noise is the main priority. To remove unnecessary signal features, the raw signal has to go through signal shifting, filtration, rectification, smoothing and finally calculation of root mean square (RMS).

First, the raw signal is shifted to the center of the axis. The signal is then passed through a notch filter which is normally set to be 50 Hz. It aims to filter out the ambient interferences on the EMG raw signal. Next, the EMG signal must pass through a band-pass typically ranged around 20–450 Hz because around 80% of the muscular energy is concentrated around this range [9]. Although there are very small changes between the filtered and raw signals which are not visible by human eyes, the values in the signal do affect the result.

Signal rectification is the next signal processing step, where it converts all the negative values in a signal to positive by adding it to the positive side. Full-wave rectification is performed on the filtered signal by using Eq. (1). By implementing this step, it provides visual aid in which the user can clearly see that which burst has greater power. It also eases the calculation of mean, peak value and the area under the curve.

$$X = |EMG| \tag{1}$$

where X means the absolute value of the filtered EMG signal. Then, signal smoothing can filter out the extremes, and it will create a linear envelope to the signal. Signal smoothing is an extra step to smoothen the signal since the linear envelope will cut out the non-reproducible part of the signal. There is a way to capture the EMG envelope by computing the root mean square (RMS) value of the signal within a window which "slides across" the signal. Sliding root mean square is one of the smoothing techniques for the EMG signal. It is a technique of performing the root mean square within a window that "slides across" the entire signal. Window size is normally limited within the range of 50–100 ms. The EMG data recorder is able to receive 1000 data in 1 s. Hence, a window size of 100 ms is chosen. Next, the sliding RMS smoothing technique is performed for every 100 data and the process is repeated until the last data. In the sliding root-mean square, the value of the calculation is plotted in the middle to avoid time shift of the graph. The next step is performing the calculation of overall root mean square (RMS) value using Eq. (2). The calculation of overall RMS targets to provide the most insight on the amplitude of the EMG signal as it is an indicator of muscle power. It is also a measure of the force produced by the muscle. The larger the muscle power, the larger the electrical signal produced by the muscle, the larger the overall RMS.

$$X_{RMS} = \sqrt{\frac{1}{N} \sum_{n=1}^{N} |X_n|^2} \tag{2}$$

Where X_{RMS} refers to the sliding (RMS) value, X_n denotes the value of X at position n and N is the sample size.

2.2 Participants

Ten stroke patients were recruited in the study. Besides, another ten normal participants without any stroke condition were also invited to form a control group in the research study. The information of participants was shown in Table 1.

The patients were required to join 6 virtual rehabilitation sessions in order to recover the impaired upper limb. The virtual rehabilitation exercises performed by them were Pick and Place. Throughout the 6 rehabilitation sessions, they were exposed to the EMG measurement at the end of each rehabilitation exercise. Besides,

Table 1 Basic information of stroke patients

Characteristics	Stroke patients (n = 10)	Control (n = 10)
Age, year (mean ± SD)	55.36 ± 11.37	59.26 ± 9.64
Gender (Male/Female)	8/7	8/7
Height, cm	166.49 ± 7.42	166.23 ± 5.24
Mass, kg	62.84 ± 7.87	60.33 ± 8.69
Impaired side (Left/Right)	10/10	-
Stroke duration, months (range)	11.24 ± 10.66 (2–30)	-
MMSE (mean, range)[a]	27.72 ± 3.49 (25–28)	30 ± 0 (30)
FMA (range)[b]	22.34 ± 2.89 (19–27)	34 ± 0 (34)

[a]MMSE = Mini-Mental State Examination; 0–30; high: better
[b]FMA = Fugl-Meyer assessment; 0–34; high: better.

the normal participants (control group) who did not participate in the rehabilitation session also needed to join the EMG measurement process. Nevertheless, the control group was only exposed to one-time EMG muscle signal measurement to collect the normal muscle power possessed by normal people. The muscle power of healthy people was set as a control variable so that comparison of muscle power between stroke and the healthy group could be made. The surface EMG sensor used was a two-channel OYMotion analog EMG sensor. The EMG sensor was attached at the bicep and arm area (flexor carpi ulnaris muscle) of the impaired limb. During the EMG signal measurement process, the patients were instructed to fully grab a stress ball by using the paralysis side of hand as hard as they could. The maximum grabbing force could ensure the maximum voluntary contraction of the patients. After the motion of maximum grabbing, the hand was released, and the overall process was repeatedly carried out for 30 s. The similar process was repeated for the control group. This aimed to collect the muscle strength of the upper limb [10]. Next, 'Processing' application was applied to record the output data to the spreadsheet for analysing purpose. For that purpose, the output muscle signal was analysed in the MATLAB software by using the signal processing steps mentioned earlier to obtain the overall RMS value.

In order to make a comparison between the computed overall RMS value and the MRC scale, one of the patients was exposed to MRC scale assessment at the end of each rehabilitation exercise.

2.3 Virtual Rehabilitation Exercise

Virtual rehabilitation exercise is introduced to the stroke patients to replace the conventional rehabilitation exercise. The 3-dimensional virtual environment brings more fun to the patients so that the patients will be immersed in the virtual environment [11]. Pick and Place is a rehabilitation training that is introduced to restore upper limb impairment [12]. Leap Motion sensor is applied to detect the fingers' motion of the patients. By looking at the computer monitor screen, the patients are able to carry out the virtual upper limb training. There are a number of blocks with different sizes in the game. There are 3 levels of difficulty with Level 1 consists of 3 blocks, Level 2 comprises 4 blocks, and Level 3 consists of 5 blocks.

The patients were required to stack up the virtual blocks according to size of the largest at the bottom to the smallest at the top. The stack was considered successful when all the blocks were stacked in the correct sequence. The game would proceed the patients to the next higher level once the current level was completed. The time utilised to successfully stack the blocks was recorded. The patients were required to carry out 4 sets of training for every rehabilitation session.

3 Results and Discussion

In EMG data analysation, the raw EMG data had undergone signal processing steps of shifting, rectification and filtering. These techniques were crucial as they strongly enhanced the data. Figure 1 showed the result (fully-rectified EMG signal) of Patient C.

Each of the well-organised burst in the signal clearly represented each contraction of Patient C. Obviously, the number of contractions performed by Patient C during the 30 s was 16. It meant that the patient repeatedly grabbed and released the stress ball for 16 times within the given time. Next, the smoothing technique (sliding RMS) was employed to filter out extreme values such as huge noise interference.

Figure 2 displayed the comparison of applying the sliding RMS filter and without applying the filter. The blue signal represented the normal signal (before applying sRMS filter) and the orange signal represented the signal after applying the sRMS. The amplitude was significantly reduced by applying this technique. Nevertheless, the shape and distance of the burst were still the same as the signal before applying the technique. Therefore, this technique would only remove the extreme noise and would not change the pattern of muscle contraction.

The overall RMS value of the signal was calculated to determine the strength of the patients' muscle. Figure 3 depicted the graph of the averaged overall RMS value of 10 patients throughout 6 rehabilitation sessions.

Since the EMG two-channel sensor was used, it resulted in two muscle power values for each patient. The overall RMS values of 10 patients were averaged to determine the improvement of muscle strength over the 6 rehabilitation sessions. From the graphs, the muscle behaviour of the bicep and arm were quite similar as both increased and decreased relatively. However, the bicep showed a larger magnitude

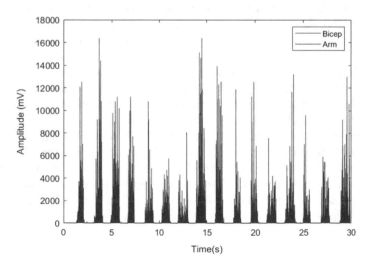

Fig. 1 Fully-rectified EMG signal of patient C

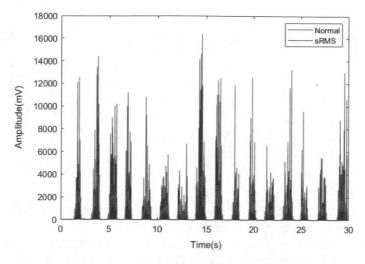

Fig. 2 Graph of comparison between EMG original rectified signal and sliding RMS signal of patient C

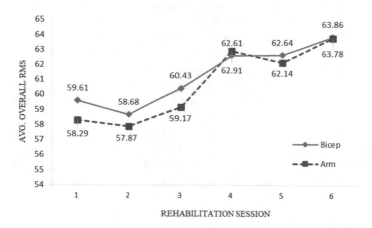

Fig. 3 Graph of averaged overall RMS value of 10 patients throughout 6 rehabilitation sessions

of RMS value than the arm area (flexor carpi ulnaris muscle). This reflected that the bicep muscles exerted more muscle force than flexor carpi ulnaris muscle during the squeezing of stress ball.

The averaged RMS value for upper limb was computed by taking the average value of averaged bicep RMS and averaged arm RMS. The muscle strength of the control group was computed to make a comparison between both groups so that any muscle improvement of the patient group could be determined. The average RMS value possessed by the control group was set as an optimum muscle power owned

Table 2 Recovery rate of the patients

Rehabilitation sessions	Averaged RMS for upper limb (mV)		Recovery (%)
	Patient group	Control group	
1	58.95 ± 12.69	115.43 ± 6.94	51.07
2	58.28 ± 11.87	115.43 ± 6.94	50.49
3	59.80 ± 12.76	115.43 ± 6.94	51.81
4	62.76 ± 13.72	115.43 ± 6.94	54.37
5	62.39 ± 13.17	115.43 ± 6.94	54.05
6	63.82 ± 12.24	115.43 ± 6.94	55.29

by a normal person. Therefore, it was set as a target value where the patients were considered fully recovered when this value was achieved.

Table 2 illustrated the recovery rate of the patients. The variance of upper limb RMS value of the patient group was large since there were a large variety of patients with different stroke condition. Some light stroke patients had great muscle power, whereas some severe stroke patients owned little muscle power. On the other hand, the variance of muscle strength of control group was lower since the participants who were invited to join the experiment came from the same range of age. The age of the control group ranged from 52 to 66. There was no powerful teenager who joined the study. Therefore, the control group's muscle strength variance was low. From the table, the increased value of recovery rate showed that the patients were on the right track of stroke recovery by conducting rehabilitation exercises. Nevertheless, the muscle strength was only slightly improved by 4.22% throughout the rehabilitation process. This was because that rehabilitation was a long process that takes up to months and some even to years. Therefore, 6 rehabilitation sessions were not long enough to see any significant improvement in the muscle strength of the patients.

At the same time, one of the patient (Patient E) was exposed to MRC scale measurement throughout six rehabilitation sessions to determine the efficiency of MRC scale in determining the upper limb power of the patients. The result of comparison between the MRC scale and EMG measurement of Patient E was presented in Table 3.

As displayed in Table 3, measurement of MRC scale and EMG muscle signal were focused at the impaired upper limb of Patient E. MRC scale of 3 implied the ability of upper limb movement against the gravitational effect. However, Patient E did not exhibit any improvement in MRC scale throughout the rehabilitation process. This reflected that he could only raise his impaired upper limb without the resistive force applied by the nurse. Thus, Patient E showed no evident muscle power improvement for the MRC scale measurement since the MRC scale of 3 remained unchanged from the beginning until the end. Nevertheless, there was a noticeable improvement of muscle strength exhibited by Patient E through the EMG measurement. The numerical value of the averaged RMS value of upper limb could precisely describe the muscle power owned by Patient E. The result of the last rehabilitation session was slightly lower than the result presented in the fifth session. This might be due to

Table 3 Comparison between the MRC scale and EMG measurement of patient E

	MRC scale (Patient E)			EMG measurement (Patient E)		
	Actual result	Optimum value	Recovery (%)	Actual result	Optimum value (mV)	Recovery (%)
1	3	5	60.00	58.09	115.43	50.32
2	3	5	60.00	57.73	115.43	50.01
3	3	5	60.00	58.77	115.43	50.91
4	3	5	60.00	60.46	115.43	52.38
5	3	5	60.00	61.23	115.43	53.05
6	3	5	60.00	60.89	115.43	52.75

the reason of tiredness of muscle after the vigorous rehabilitation exercise. Thus, it declined the capability of the muscle in force generation [13]. However, it showed an overall improvement of 2.43% at the end of the rehabilitation session. Compared to the MRC scale, EMG muscle measurement was an excellent approach to measure the muscle power of stroke patients since it could accurately and precisely describe the muscle power of the patients.

4 Conclusion

To conclude it, a two-channel surface EMG sensor is capable of assessing the muscle power of paralysed upper limb. The EMG signal must be undergone through some signal processing steps to eliminate the unnecessary signal features so that the computed overall RMS value will be accurate. The result showed that the patients' overall RMS value of the impaired upper limb was averagely improved at the end of the rehabilitation process. In the muscle power measurement, EMG signal measurement presented evident improvement (+2.43%) compared with the MRC scale measurement (+0%). Therefore, EMG signal measurement can precisely and accurately describe the muscle power of the patients. Thus, it can be applied as a muscle power measurement tool in the stroke rehabilitation field to determine the recovery rate of stroke patients.

Acknowledgements This work was supported in part by the TM R&D under Grant (MMUE/180026) for the research program with title of 'Rehabilitation Using Biofeedback System'.

References

1. Loo KW, Gan SH (2012) Burden of stroke in Malaysia. Int J Stroke 7(2):46–48

2. Steiner T, Juvela S, Unterberg A, Rinkel G (2011) In: Hand motor impairment after stroke, 3rd edn. The Canada University Press, Toronto, Canada
3. Hopkins J (2015) Stroke rehabilitationwhat to expect after a stroke. In: International conference on medical treatment method, Busan, Korea
4. Stroke Association (2020) Physiotherapy after a stroke. https://www.stroke.org.uk/life-after-stroke/physiotherapy-after-stroke. Last Accessed 16 May 2020
5. Escolar D, Henricson EK, Mayhew J, Florence J, Leshner R (2001) Clinical evaluator reliability for quantitative and manual muscle testing measures of strength in children. Muscle Nerve 24(6):787–793
6. O'Neill S, Jaszczak T (2017) Using 4+ to grade near-normal muscle strength does not improve agreement. Chiropractic Manual Therapies 25:1–9
7. Young L (2001) In: The study of electromyography (EMG), University Science, Mill Valley, California, United States, pp 323–333
8. Nazmi N, Abdul Rahman M, Yamamoto S, Ahmad S, Zamzuri H, Mazlan S (2016) A review of classification techniques of EMG signals during isotonic and isometric contractions. Sensors 16(8):301–328, Basel
9. Hermens HJ, Freriks B, Disselhorst-klug C, Rau G (2000) Development of recommendations for SEMG sensors and sensor placement procedures. Electromyograph Kinesiol 10(5):361–374
10. Vargas JF, Tarvainen TVJ, Kita K, Yu W (2017) Effects of using virtual reality and virtual avatar on hand motion reconstruction accuracy and brain activity. IEEE Access 5:23736–23750
11. Lew KL, Sim KS, Tan SC, Fazly SA (2020) 3D Kinematics of upper limb functional assessment using HTC vive in unreal engine 4. In: 12th international conference on computational collective intelligence, vol 1, no. 1
12. Sim KS, Lim ZY, Wong BWH, Kho DTK (2017) Development of rehabilitation system using virtual reality. In: 2017 International conference on robotics, automation and sciences (ICORAS), Melaka, Malaysia, pp 1–6
13. Healthline Association (2020) What Caused Muscle Fatigue?'. https://www.healthline.com/health. Last Accessed 18 May 2020

Automated Microservice Code-Smell Detection

Andrew Walker, Dipta Das, and Tomas Cerny

Abstract Microservice Architecture (MSA) is rapidly taking over modern software engineering and becoming the predominant architecture of new cloud-based applications (apps). There are many advantages to using MSA, but there are many downsides to using a more complex architecture than a typical monolithic enterprise app. Beyond the normal bad coding practices and code-smells of a typical app, MSA specific code-smells are difficult to discover within a distributed app. There are many static code analysis tools for monolithic apps, but no tool exists to offer code-smell detection for MSA-based apps. This paper proposes a new approach to detect code smells in distributed apps based on MSA. We develop an open-source tool, MSANose, which can accurately detect up to eleven different types of MSA specific code smells. We demonstrate our tool through a case study on a benchmark MSA app and verify its accuracy. Our results show that it is possible to detect code-smells within MSA apps using bytecode and or source code analysis throughout the development or before deployment to production.

Keywords Microservice · Cloud-computing · Code smells · Code-analysis

1 Introduction

Microservice Architecture (MSA) has become the preeminent architecture in modern enterprise applications (apps) [16]. MSA brings many advantages, which have led to its rise in popularity [4]. The distributed nature of an MSA-based app allows for greater autonomy of developer units. While this provides greater flexibility with respect to faster delivery, improved scalability, and benefits in existing problem domains, it also presents the opportunity for code smells to be more readily created within the app.

A. Walker · D. Das · T. Cerny (✉)
Baylor University, One Bear Place, Waco, TX 76706, USA
e-mail: tomas_cerny@baylor.edu

© The Author(s), under exclusive license to Springer Nature Singapore Pte Ltd. 2021   211
H. Kim et al. (eds.), *Information Science and Applications*, Lecture Notes
in Electrical Engineering 739, https://doi.org/10.1007/978-981-33-6385-4_20

Code smells [9] are anomalies within codebases that do not necessarily impact the performance or correct functionality of an app. They are patterns of bad programming practice that can affect a wide range of areas in a program, including reusability, testability, and maintainability. If code smells go unchecked in an MSA-based app, the benefits of using a distributed development process can be mitigated. It is, therefore, important that the code-smells in an app are properly detected and managed [9].

MSA presents a unique situation when it comes to code-smells due to its distributed nature. MSA-specific code smells often focus on inter-module issues rather than an intra-module issue. Traditional code-smell detecting tools cannot detect code smells between discrete modules, so these issues go unchecked during the development process. This paper shows that when we augment static code analysis to recognize enterprise development constructs, we can effectively detect code smells in MSA distributed apps. A case study demonstrates our approach targeting 11 recently identified code smells for this architecture. Furthermore, we share our open-source prototype code analyzer with the community that can recognize Enterprise Java platform constructs and standards, along with the MSA code smell detector recognizing the 11 code smells targeted in this paper.

The rest of the paper is as follows: Sect. 2 assesses related work. Section 3 introduces the MSA code smells. Section 4 describes the code analysis of enterprise systems. Section 5 introduces our solution for automatic MSA code smell detection, and Sect. 6 shares a case study evaluation. We conclude in Sect. 7.

2 Related Work

Code smells can be defined as "characteristics of the software that may indicate a code or design problem that can make software hard to evolve and maintain" [8]. Code smells do not necessarily impact the performance or correct functionality; they are patterns of bad programming practice that can affect a wide range of areas in a program, including reusability, testability, and maintainability. They can be seen as code structures that indicate a violation of fundamental design principles and negatively impact design quality [21].

Gupta et al. [10] underlines the need to identify and control code smells during the design and development stages to achieve higher code maintainability and quality. If developers are not invested in fixing them, code smells do matter to the overall maintainability of the software. If left unchecked, they can impact the overall system architecture [11]. Code smells can be deceptive and hide the true extent of their 'smelliness' and even carry into further refactorings of the code [6, 11]. Frequently code smells are also related to anti-patterns [18] in an app.

Code-smell correction is clearly a necessary process for developers [20], but it is often pushed aside. It was found that the most prevalent factor towards developers addressing code smells is the importance and relevance of the issue to the task worked on. Peters et al. [17] found that while developers are oftentimes aware of the code smells in their app, they do not care about actively fixing them. Code smells are often fixed accidentally through unrelated code refactoring [9].

Tahir et al. [22] studied how developers discussed code smells in stack exchange sites and found out that these sites work as an informal crowd-based code smell detector. Peers discuss the identification of smells and how to get rid of them in a specific given context. Thus, the question is not only how to detect them but also how to eliminate them in a given context. They found that the most popular smells discussed between developers are also shown to be most frequently covered by available code analysis tools. It is also noted that while Java support is the broadest, other platforms, including C#, JavaScript, C++, Python, Ruby, and PHP, are lacking in support. Concerns were also raised that there is a missing classification for how harmful smells are on a given app.

Some researchers would argue that developers do not have the time to fix all smells. For instance, Gupta et al. [10] identified 18 common code smells and identified the driving power of these code smells to improve the overall code maintainability. The effect is that developers could refactor one of the smells with higher driving power, rather than address all smells in an app, and still significantly improve code maintainability.

An attempt at automatic code smell detection [25], defined an automated code smell detection tool for Java. Since then, the field of code smell detection has continued to grow. Code smell tools have been developed for *high level design* [2], *architectural smells* [15], as well as for *language-specific code smells* [14], measuring not just code smells but also the quality [10] of the app. The field of automatic code smell detection continues to evolve with an ever-changing list of code smells and languages to cover.

It is common to identify code smells in monolithic systems using code-analysis. Anil et al. [13] recently analyzed 24 code smells detection tools (e.g, SpotBugs, PMD, etc.). While the tools correctly mapped the code smells in an app, they are limited to a single codebase, and so they become antiquated as modern software development tends towards MSA.

While extensive research has been done to define and detect code smells in a monolithic app, little has been done for distributed systems [3]. It would be possible for a developer to run code smell detection on each individual module, but this does not address any code smells specific to MSA.

In a distributed environment, in particular MSA, there have been multiple code smells identified. In one study [23], these smells include improper module interaction, modules with too many responsibilities, or a misunderstanding of the MSA. Code smells can be specific to a certain app perspective, including the *communication perspective*, or in the *development and design process* of the app. These smells can be detected manually, which usually requires assessing the app and a basic understanding of the system, but this demands considerable effort from the developers. With code analysis instruments, smells can be discovered almost instantly and automatically with no previous knowledge of the system required. However, we are aware that no tool at present can detect the code anomalies that can exist between discrete modules of an MSA app.

3   Microservice Code Smell Catalogue

In this paper, we reuse the definition of eleven MSA specific code smells from a recent exploratory study by Taibi et al. [23], which used existing literature and interviews with industry leaders to distill and rank these eleven code smells for MSA. The code smells are briefly summarized as follows:

ESB Usage (EU) An Enterprise Service Bus (ESB) [4] is a way of message passing between modules of a distributed app in which one module acts as a service bus for all of the other modules to pass messages on. There are pros and cons to this approach. However, it can become an issue of creating a single point of failure and increasing coupling, so it should be avoided in MSA.

Too Many Standards (TMS) Given the distributed nature of the MSA app, multiple discrete teams of developers often work on a given module, separate from the other teams. This can create a situation where multiple frameworks are used when a standard should be established for consistency across the modules.

Wrong Cuts (WC) WC is when modules are split into their technical layers (presentation, business, and data layers). MSA modules must be split by features, and each fully contains their domain's presentation, business, and data layers.

Not Having an API Gateway (NAG) The API gateway design pattern is for managing the connections between MSA modules. In large, complex systems, this should be used to reduce the potential issues of direct communication.

Hard-Coded Endpoints (HCE) Hardcoded IP addresses and ports to communicate between services. By hardcoding the endpoints, the app becomes more brittle to change and reduces the app's scalability.

API Versioning (AV) All Application Programming Interfaces (API) should be versioned to keep track of changes properly.

Microservice Greedy (MG) This occurs when modules are created for every new feature, and oftentimes, these new modules are too small and do not serve many purposes. This increases complexity and the overhead of the system. Smaller features should be wrapped into a larger MSA module, if possible.

Shared Persistency (SP) When two modules of the MSA app access the same database. This breaks the definition of an MSA where each module should have autonomy and control over its data and database.

Inappropriate Service Intimacy (ISI) One module requesting private data from a separate module. This likewise breaks the MSA definition, where each module should have control over its private data.

Shared Libraries (SL) If modules are coupled with a common library, that library should be refactored into a separate module to reduce the app's fragility by migrating the shared functionality behind a common, unchanging interface. This will make the app resistant ripples from changes within the library.

Cyclic Dependency (CD) Cyclic connection between calls to different modules. This can cause repetitive calls and also increase the complexity of understanding call traces for developers. This is a bad architectural practice for MSA.

To highlight the gap in MSA code smells, we assessed existing state-of-the-art architecture-specific code smell detection tools, compiled in a previous study [3], AI Reviewer, ARCADE, Arcan, Designite, Hotspot Detector, Massey Architecture Explorer, Sonargraph, STAN, and Structure 101 and verified that all are capable of detecting CD and in the case of Arcan, also HCE and SP MSA code smells. We chose these tools as they were compiled to study the existing state of the art of architecture smell detection tools and were shown to meet a minimum threshold of documentation and information about the tool.

4 Code Analysis in Enterprise Systems

The two static code analysis processes, source code analysis and bytecode analysis, ultimately create a representation of the app. This is done through several processes, including recognizing components, classes, methods, fields or annotations, tokenization, and parsing, which produce graph representations of the code. These include Abstract Syntax Trees (AST), Control-Flow Graphs (CFG), or Program Dependency Graphs (PDG) [19].

Bytecode analysis uses the compiled code of an app. This is useful in uncovering endpoints, components, authorization policy enforcements, classes, and methods. It can also be used to augment or build CFG or AST. However, the disadvantage is that not all languages have a bytecode.

We can also turn to source code analysis [5], which parses through the source code of the app, without having to compile it into an immediate representation. Many approaches exist to do this; however, most tools tokenize the code and construct trees, including AST, CFG, or PDG.

However, limits exist with these representations in encapsulating the complexity of enterprise systems. To mitigate the shortcomings of existing static code analysis techniques on enterprise systems, we must augment existing techniques with an understanding of enterprise standards [7]. A more realistic representation of the enterprise app can be constructed with aid from either source code or bytecode analysis. This primarily includes a tree representation and the detection of the system's endpoints, and the construction of a communication map. These augmented representations and metadata have been successful in other problem domains, including security, networking, and semantic clone detection.

5 Proposed Solution to Detect Code Smells

Our approach is integrable to the software development life-cycle. It uses static-code analysis for fast and easily-integrated reports on the code-smells in an system. To cover the wide variety of possible issues within an MSA app, as well as the different concerns (app, business, and data) issues that the identified smells cover, we must statically analyze a couple of different areas of an app. Our approach specifically involved the Java Enterprise Edition platform because of its rich standards for enterprise development, which include Spring Boot (https://spring.io/projects/spring-boot) and Java EE (https://docs.oracle.com/javaee). However, alternative standard adoptions exist also for other platforms. Extending for another language would be trivial since we utilize an intermediate representation for analysis, as explained below.

The core of our solution is the creation of a centralized view of the app. To begin with, we individually analyze each MSA module in the app. Once each module is fully analyzed, it can be aggregated into a larger service mesh. Then the full detection can be done on the aggregated mesh.

Our analysis process aims to generate a graph of interaction between the different MSA modules. This involves exploring each module for a connection to another module, usually through a REST API call. The inter-module communications are realized using a two-phase analysis: scanning and matching. In the first phase, we scan each module to list all the REST endpoints and their specification metadata. This metadata contains the HTTP type, path, parameter, and return type of the REST endpoint. Additionally, the server IP addresses (or their placeholders) are resolved by analyzing app configuration files that accompany system modules. These IP addresses, together with the paths, define the fully-qualified URLs for each REST endpoint. We further analyze each module to enumerate all REST calls along with request URLs and similar metadata. We list these REST endpoints and REST calls based solely on static code analysis, where we leveraged the annotation-based REST API configuration commonly used in enterprise frameworks. We match each endpoint with each REST call across different MSA modules based on the URL and metadata in the second phase. During matching, URLs are generalized to address different naming of path variables across different MSA modules. Each resultant matching pair indicates an inter-module REST communication.

Afterward, the underlying dependency management tool's configuration file, e.g., pom.xml file for maven, is analyzed for each of the different MSA modules. This allows us to find the dependencies and libraries used by each of the apps. Lastly, the app configuration, where developers define information such as the port for the module, the databases it connects to, and other relevant environment variables for the app, is analyzed. Once the processing of each module is done, we begin the process of code-smell detection. In the following text, we provide details relevant to each particular smell and its detection.

ESB Usage is detected by tallying up all of the incoming and outgoing connections within each module. We see an ESB as a module with a high, almost outlier, number of connections, and a relatively equal number of incoming and outgoing connections. Additionally, an ESB should connect to nearly all the modules.

Too Many Standards is tricky to detect since it is entirely subjective on how many standards are "too many." Additionally, there are very good reasons developers would choose different standards for different system modules, including speed, available features, and security. We tally the standards used for each of the layers of the app (presentation, business, and data). The user can configure how many standards are too many for each of the respective sections.

Wrong Cuts depends on the business logic and, therefore, nearly impossible to automatically detect without extrapolating a deep understanding of the business domain. However, we would expect to see an unbalanced distribution of artifacts within the MSA modules along with the different layers of the app (presentation, business, and data). To detect an unbalance presentation MSA module, we look for an abnormally high number of front-end artifacts (such as HTML/XML documents for JSP). For the potentially WC business MSA modules, we look for an unbalanced number of service objects, and lastly, for WC data modules, we look for an unbalanced number of entity objects. To find unbalanced MSA modules, we look for outliers in the number of the specified artifacts within each module and report the possibility of MSA module WC. We defined an outlier count of greater than two times the standard deviation away from the average count of the artifacts in each module, which is seen in Eq. 1.

$$2 * \sqrt{\frac{\sum_{i=0}^{n}(x_i - \bar{X})^2}{n - 1}} \tag{1}$$

Not Having an API Gateway is determined from code analysis alone, especially as cloud apps increasingly rely on routing frameworks such as AWS API Gateway (https://aws.amazon.com/api-gateway/). This uses an online configuration console and is not discoverable from code analysis, to handle routing API calls. In a study by Taibi [23], it was found that developers could adequately manage up to 50 distinct modules without needing to rely on an API gateway. For this reason, if the scanned app has more than 50 distinct modules, we include a warning message in the report to use an API gateway. This is not classified as an error, but rather a suggestion for best practice.

Shared Persistency is detected by parsing the app's configuration files and finding the persistence settings location for each of the submodules. For example, in a Spring Boot apps, the YAML file is parsed for the datasource URL. Then the persistence of each module is compared to the others to find shared datasources.

Inappropriate Service Intimacy can appear in a couple of different ways. First, we detect this as a variant of the shared persistency problem. Instead of sharing a datasource between two or more modules, a module is directly accessing another's datasource in addition to its own; however, once a duplicate datasource is found, if the module also has its own private datasource, then it is an instance of inappropriate service intimacy. Next, we look for two modules with the same entities. If one of those modules is only modifying/requesting the other's data, we defined it as inappropriate service intimacy.

Shared Libraries is found by scanning the dependency management files for each app module to locate all shared libraries. Clearly, some shared outside libraries will be shared among the MSA modules; however, the focus should be on any in-house libraries. Developers can then decide to extract into a separate module if necessary to bolster the app against the libraries' changes.

Cyclic Dependency is found using a modified depth-first search [24].

Hard-Coded Endpoints is found during the bytecode analysis phase of the app. Using the bytecode instructions, we can peek at the variable stack and see what parameters are passed into the function calls used to connect to other MSA modules. E.g., in Spring Boot, we took calls from RestTemplate. We link the passed address back to any parameters passed to the function or any class fields to find the path parameters used. We test for hardcoded port numbers and IP addresses as both should be avoided.

API Versioning is found in the app by first finding all the fully qualified paths for the app. To locate the unversioned paths, each API path is matched against a regular expression pattern `.*/v[0-9]+(.?[0-9]*).*`, matching the app convention. All unversioned APIs are reported back to the user.

Microservice Greedy is found by calculating a couple of different metrics for each module. This includes the counts of front-end files, service and entity objects. Then we find outliers, if any exist, as potential MG. We define outliers similarly as when finding MSA module WC using the Eq. 1. However, we focus only on those that are outliers due to being undersized, as opposed to too large.

6 Case Study

We developed a prototype open-source MSANose tool (https://github.com/cloudhubs/msa-nose) using our approach. It accepts Java-based MSA apps and performs static analysis of MSA modules. From the individual modules, it extracts the interaction patterns, combines the partial results and derives a holistic view on the distributed system. Next, it performs the smell detection and reports a list of MSA code smells with references to the offending modules and code.

Recent efforts [12] to catalog MSA testbed apps have found a lack of apps that adhere to the guidelines for testbeds outlined by Aderaldo et al. [1]. To test our app, we chose to run it on an existing MSA benchmark, the Train Ticket Benchmark [26], it is a reasonable size for an MSA app and provides a good test of all of the conditions in our app. Furthermore, it was designed as a real-world interaction model between MSA modules in an industrial environment and is one of the largest MSA benchmarks available. This benchmark consists of 41 modules and contains over 60,000 lines of code. It uses Docker or Kubernetes for deployment and relies on NGINX or Ingress for routing.

We manually analyzed the testbed for each of the eleven MSA code smells, by manual tracing of REST calls, the cataloging of entities, and endpoints within the app. We show the results of our manual assessment in column two of Table 1. Next, we ran our app on the testbed system. The app took just ten seconds to run on a system with an Intel i7-4770k and 8 GB of RAM. This includes the average time (taken over ten runs) it took to analyze the source code fully and compiled the bytecode of the testbed app. In the third column of Table 1 is a quick overview of the results from running our app on the testbed. Our tool correctly analyzed the testbed and successfully identified the MSA code smells.

Table 1 Code smells in the trainticket benchmark [26]

Smell	Manual	MSANose
EU	No	No
TMS	No	No
WC	0	2
NAG	No	No
HCE	28	28
AV	76	76
MG	0	0
SP	0	0
ISI	1	1
SL	4	4
CD	No	No

Code smells do not always break the system, but they are indicators of poor programming practice. As the testbed app has done over the past couple of years, these smells can easily work their way into the system as a system grows organically. Our tool can help developers locate code smells in enterprise MSA apps and provide a catalog of the smells and their common fix solutions.

7 Conclusions

In this paper, we have discussed the nature of code smells in software apps. Code smells, which may not break the app in the immediate time-frame, can cause long-lasting problems for maintainability and efficiency later on. Many tools have been developed which automatically detect code smells in apps, including ones designed for architecture and overall design of a system. However, none of these tools adequately address a distributed app's needs, specifically an MSA-based app. To address these issues, we draw upon previous research into defining MSA specific code smells to build an app capable of detecting eleven unique MSA-based code smells. We then run our app on an established MSA benchmark app and compare our results to manually gathered ones. We show that it is possible, through static code analysis, to analyze an MSA-based app and derive MSA-specific code smells accurately.

For future work, we plan to assess more app testbeds. Moreover, we plan to continue our work on integrating the python platform to our approach since there are no platform-specific details, and most of the enterprise standards apply to across platforms.

Acknowledgements This material is based upon work supported by the National Science Foundation under Grant No. 1854049 and a grant from Red Hat Research https://research.redhat.com.

References

1. Aderaldo CM, Mendonça NC, Pahl C, Jamshidi P (2017) Benchmark requirements for microservices architecture research. In: Proceedings of the 1st international workshop on establishing the community-wide infrastructure for architecture-based software engineering. ECASE '17. IEEE Press, New York, pp 8–13
2. Alikacem EH, Sahraoui HA (2009) A metric extraction framework based on a high-level description language. In: 2009 Ninth IEEE international working conference on source code analysis and manipulation, pp 159–167
3. Azadi U, Arcelli Fontana F, Taibi D (2019) Architectural smells detected by tools: a catalogue proposal. In: 2019 IEEE/ACM international conference on technical debt (TechDebt), pp 88–97
4. Cerny T, Donahoo MJ, Trnka M (2018) Contextual understanding of microservice architecture: Current and future directions. SIGAPP Appl Comput Rev 17(4):29–45. https://doi.org/10.1145/3183628.3183631
5. Chatley G, Kaur S, Sohal B (2016) Software clone detection: a review. Int J Control Theory Appl 9:555–563
6. Counsell S, Hamza H, Hierons RM (2010) The 'deception' of code smells: an empirical investigation. In: Proceedings of the ITI 2010, 32nd international conference on information technology interfaces, pp 683–688
7. DeMichiel L, Shannon W (2016) JSR 366: Java Platform, Enterprise Edition 8 Spec (2016). https://jcp.org/en/jsr/detail?id=342. Accessed on 27 Mar 2020
8. Fontana FA, Zanoni M (2011) On investigating code smells correlations. In: 2011 IEEE fourth international conference on software testing, verification and validation workshops, pp 474–475
9. Fowler M (2018) Refactoring: improving the design of existing code. Addison-Wesley Longman Publishing Co. Inc, USA
10. Gupta V, Kapur P, Kumar D (2016) Modelling and measuring code smells in enterprise applications using TISM and two-way assessment. Int J Syst Assurance Eng Manage 7 https://doi.org/10.1007/s13198-016-0460-0
11. Macia I, Garcia J, Daniel P, Garcia A, Medvidovic N, Staa A (2012) Are automatically-detected code anomalies relevant to architectural modularity? An exploratory analysis of evolving systems. In: AOSD'12 - Proceedings of the 11th annual international conference on aspect oriented software development, https://doi.org/10.1145/2162049.2162069
12. Márquez G, Astudillo H (2019) Identifying availability tactics to support security architectural design of microservice-based systems. In: Proceedings of the 13th European conference on software architecture, vol 2, pp 123–129. ECSA '19, ACM, New York, NY. https://doi.org/10.1145/3344948.3344996
13. Mathew AP, Capela FA, An analysis on code smell detection tools. In: 17th SC@ RUG 2019-2020, p 57
14. Moha N, Gueheneuc Y, Duchien L, Le Meur A (2010) DECOR: a method for the specification and detection of code and design smells. IEEE Trans Software Eng 36(1):20–36
15. Moha N, Guéhéneuc YG, Meur AF, Duchien L, Tiberghien A (2010) From a domain analysis to the specification and detection of code and design smells. Formal Aspects Comput 22. https://doi.org/10.1007/s00165-009-0115-x
16. NGINX Inc. (2015) The Future of Application Development and Delivery Is Now Containers and Microservices Are Hitting the Mainstream. https://www.nginx.com/resources/library/app-dev-survey/. Accessed on 27 Mar 2020
17. Peters R, Zaidman A (2012) Evaluating the lifespan of code smells using software repository mining. In: 2012 16th European conference on software maintenance and reengineering, pp 411–416
18. Reeshti Sehgal R, Nagpal R, Mehrotra D (2019) Measuring code smells and anti-patterns. In: 2019 4th international conference on information systems and computer networks (ISCON), pp 311–314

19. Roy CK, Cordy JR, Koschke R (2009) Comparison and evaluation of code clone detection techniques and tools: a qualitative approach. Sci Comput Program 74(7):470–495. https://doi.org/10.1016/j.scico.2009.02.007
20. Sae-Lim N, Hayashi S, Saeki M (2017) How do developers select and prioritize code smells? A preliminary study. In: 2017 IEEE international conference on software maintenance and evolution (ICSME), pp 484–488
21. Suryanarayana G, Samarthyam G, Sharma T (2014) Refactoring for software design smells: managing technical Debt, 1st edn. Morgan Kaufmann Publishers Inc., San Francisco, CA
22. Tahir A, Dietrich J, Counsell S, Licorish S, Yamashita A (2020) A large scale study on how developers discuss code smells and anti-pattern in stack exchange sites. Inform Software Technol 125:106333. https://doi.org/10.1016/j.infsof.2020.106333
23. Taibi D, Lenarduzzi V (2018) On the definition of microservice bad smells. IEEE Software 35(3):56–62. https://doi.org/10.1109/MS.2018.2141031
24. Tarjan R (1971) Depth-first search and linear graph algorithms. In: 12th annual symposium on switching and automata theory (swat 1971), pp 114–121
25. Van Emden E, Moonen L (2002) Java quality assurance by detecting code smells. In: Proceedings of the ninth working conference on reverse engineering (WCRE'02), p 97. WCRE '02. IEEE Computer Society, USA
26. Zhou X, Peng X, Xie T, Sun J, Xu C, Ji C, Zhao W (2018) Benchmarking microservice systems for software engineering research. In: Proceedings of the 40th international conference on software engineering: companion proceedings, ICSE. ACM, pp 323–324. https://doi.org/10.1145/3183440.3194991

On Automatic Software Architecture Reconstruction of Microservice Applications

Andrew Walker, Ian Laird, and Tomas Cerny

Abstract The adoption of Microservice Architecture (MSA) is rapidly becoming standard for modern software development. However, the added benefits of using a distributed architecture, including reliability and scalability, come with a cost in increasing the system's complexity. One way developers attempt to mitigate the effects of an overly complicated system is through Systematic Architecture Reconstruction (SAR), which creates a high-level overview of the system concerns. This is typically done manually, which takes a great amount of effort from the developers. This paper proposes a method for automatically completing SAR of an MSA application through code analysis and demonstrating it on a case study on an existing microservice benchmark application.

Keywords Microservice · Software architecture reconstruction · Code analysis · Static analysis · Information extraction · Compliance

1 Introduction

Microservice Architecture (MSA) offers many benefits for modern software development. Primarily these include benefits in scalability and reliability. With a distributed development model, different developers' teams can work separately on different modules of the application. While this offers rapid development benefits, it also allows errors to slip into the application through discrepancies between modules. Another side effect of this distributed development is an increase in the complexity of the application.

One way to mitigate these unintended errors is through Software Architecture Reconstruction (SAR). The construction of a simplified overview of the application can help developers understand the application's full scope, even beyond their

A. Walker · I. Laird · T. Cerny (✉)
Baylor University, One Bear Place, Waco, TX 76706, USA
e-mail: tomas_cerny@baylor.edu

© The Author(s), under exclusive license to Springer Nature Singapore Pte Ltd. 2021　　223
H. Kim et al. (eds.), *Information Science and Applications*, Lecture Notes in Electrical Engineering 739, https://doi.org/10.1007/978-981-33-6385-4_21

modules. One study [18] underlines that SAR is key to architecture verification, conformance checking, and trade-off analysis.

Many methods have been proposed for SAR, even for distributed systems, but none can fully automate the domain, technology, service, and operation views. In this paper, we propose a novel automation method applied to an existing distributed system SAR methodology. We then verify our automation against an existing microservice benchmark system.

The paper is organized as follows. The SAR and static-code analysis background is given in Sect. 2 and SAR state of the art in Sect. 3. Our automatic SAR method for microservice apps is described in Sect. 4. Section 5 tests our method against an existing testbed application. Section 6 concludes the paper.

2   Background

This section presents background on the SAR process as a whole, the methodology that serves as the source of our automation goals, and lastly, static-code analysis, which is the method we use to automate the SAR process.

2.1   Systematic Architecture Reconstruction

SAR has historically been defined with four distinct phases. These are as follows:

1. *Extraction*: This phase collects all of the artifacts needed during the next three phases of SAR. The artifacts collected are relevant to the "views" that are being constructed. A "view" is defined as a set of related artifacts that cover a concern of the architecture of the system.
2. *Construction*: This phase creates a canonical representation of the views and usually stores them in some form, like a database.
3. *Manipulation*: This phase combines the views to allow for the answering of more complicated questions in the next phase. How the views are combined is relevant to the specific application being reconstructed.
4. *Analysis*: This phase answers questions about a system given the overall views of the architecture constructed in the previous phases. There are almost infinite questions that can be asked, covering multitudes of domains, including networking, security, and code quality.

For this paper's purposes, we aim to demonstrate a framework for SAR analysis, so we do not focus on a specific question and thus exclude phase 4.

2.2   Views

The core of successful SAR is the construction of effective views of the architecture of a system. Care must be taken to choose views that are relevant to the questions being asked about a system. Since we are focusing on creating a general framework for SAR, instead of asking a particular question, we chose a set of views that provide good coverage of the system's concerns. We took our set of views from a previous study on SAR methodology [18]. We will briefly outline the four of them below.

- *Domain View*: This view covers the domain concerns of an MSA application. It describes the entity objects of the system as well as the datasource connections of those objects.
- *Technology View*: This view focuses on the technology aspect of an application. It describes the technologies used for microservice implementation and operation.
- *Service View*: The view focuses on service operators. It describes the service models that specify microservices, interfaces, and endpoints.
- *Operation View*: This view focuses on the ops concern of a system. It describes service deployment and infrastructure, such as containerization, service discovery, and monitoring.

Each of these views can be understood as distinct concerns within the system but also as related to the others. For example, the domain view intersects with the service view since the service view can be thought of as the intersections of data between microservices and the technology view since the technology view describes how data is stored in the application.

Another key point about the construction of these views is that ultimately each view is an aggregation of a smaller view encompassing a disparate microservice. Each microservice has a bounded-context of its concerns, but these can be aggregated into a fully centralized perspective of the system's architecture.

2.3   Static-Code Analysis

Static-code analysis is what makes an internal inspection of an application possible. It's used throughout software development but is primarily used to detect bugs in a piece of software. There are two main processes for static code analysis - source code analysis and bytecode analysis. These are used for several processes, including *recognizing* components, classes, methods, fields or annotations, *tokenization*, and *parsing*, which produce graph representations of the code. These include Abstract Syntax Trees (AST), Control-Flow Graphs (CFG) [12, 20, 25], or Program Dependency Graphs (PDG) [21, 22].

Bytecode analysis [1] uses the compiled code of an application. This is useful in uncovering endpoints, components, authorization policy enforcements, classes, and methods. It can also be used to augment or build CFG, or AST [10, 11, 13]. However, the disadvantage is that not all languages have a bytecode.

To fill this gap, we can also turn to source code analysis [3]. It parses through the source code of the application, without having to compile it into an immediate representation. Many approaches exist to do this, however, most tools toke-nize the code and construct trees, i.e., AST [21, 22], CFG [12, 20, 25], or PDG [6, 24].

Static-code analysis can also be extended beyond just the source-code and compiled artifacts to include other application artifacts. Static analysis gives the ability to access Docker images [16, 23] through the Kubernetes[1] platform. Since Kubernetes is typically the enterprise standard for service-meshes, it provides thorough coverage of the operation view for many enterprise applications.

Despite all of the usefulness of static-code analysis, limits exist with these representations in encapsulating enterprise systems' complexity. To mitigate the shortcomings of existing static code analysis techniques on enterprise systems, we must augment existing techniques with an understanding of enterprise standards [4, 14]. A more realistic representation of the enterprise application can be constructed with aid from either source code analysis or bytecode analysis. This primarily includes a tree representation and the detection of the system's endpoints, and the construction of a communication map. These augmented representations and metadata have been successful in other problem domains, including security, networking, and semantic clone detection.

3   Related Work

Most of the work done in SAR on microservices has focused on the methodology instead of the automation of the methodology. However, some works have partially automated a full SAR representation of a microservice-based system.

Ratemacher et al. [18], defined a methodology for the construction of all four of our targeted views. Their process involved the construction of a canonical representation of the data model. From there, they employed methods to fuse module views. Finally, it performed architecture analysis to answer questions about architecture implementations from the reconstructed architecture information. Unfortunately, they only provided a manual assessment of their benchmark application and included no attempt at automation. However, they highlighted the need for furthering the work in SAR to automate the process.

Alshuqayran et al. [2] conducted a manual reconstruction process that included modeling the application to derive the overall architecture while utilizing multiple module merge strategies such as data model integration and meta-model mapping rules. They do not, however, apply their strategies towards extraction or merging of domain concepts.

---

[1] https://kubernetes.io/.

Ibrahim et al. [9] considered container-based deployment configuration files to derive MSA module topology. In particular, they used Docker Compose to extract the topology of the module orchestration. In addition to topology, they generated "attack graphs", that depict actions which attackers may use to reach their malicious goal. Attack graphs help developers identify attack paths that comprise exploitable vulnerabilities in deployed services. They underlined that testers commonly construct such graphs manually, but container configuration files provide a well-structured input for automation. Their open-source tool is based on Clair [19], a vulnerability scanner for Docker containers and images. It generates image vulnerabilities linked to CVE [17] as well as connections to an attack vector for each vulnerability. An attack vector describes the conditions and effects that are connected to vulnerability.

Mayer et al. [15] propose an automatic method for extracting the domain, service, and operation information of an application. It utilizes a combination of static analysis and runtime analysis to construct a language-agnostic representation of each service and its interaction with other services. The downside to this approach is that to fully construct the representations, and not just a domain model, the application needs to be deployed. The proposed upside to this is that we no longer need a litany of parsers for the different language implementations of a heterogeneous MSA application; however since it still relies on parsing for the domain view, the parsers are needed regardless, and the dynamic approach provides extra overhead. It also creates the possibility of an incomplete view since communication paths that are not traversed during the extraction phase are absent from the final view.

MicroART [7] tool automates SAR. It extracts information about the service concern of a module (service names, ports, etc.) from source code repositories. It also performs log analysis during runtime to discover containers, network interfaces, and service interaction. The user must provide the running container's location since MicroART does not extract that information automatically. It then uses that information to generate views for the service and operations concerns. It does not consider the domain or technology views.

Zdun et al. [26] propose a methodology for the service view; however, it approaches the problem differently from the previously mentioned tools. It proposes a method for measuring MSA conformance. Because of this approach, it uses a formal model when conducting SAR. Conformance is then assessed via metrics and constraints defined by the relationships between these types. Though it considers the service view of an application, domain, technology, and operation views are not considered for its conformance measure.

We are aware of no tool at present that is capable of fully automating all four of our chosen architecture views on a microservice-based system. A breakdown of the capabilities of the tools mentioned previously is available in Table 1.

Table 1 Comparison of Modern SAR Tools

	Domain	Technology	Service	Operation	*Automated*
Alshuqayran et al.		X	X	X	No
Ibrahim et al.				X	Yes
Mayer et al.	X		X	X	Yes
MicroART			X	X	Yes
Rademacher et al.	X	X	X	X	No
Zdun et al.			X		No
Our approach	X	X	X	X	Yes

4 Proposed Method

In this section, we introduce our proposed method for automatically conducting SAR. In particular, we use static-code analysis, both source-code and bytecode, as well as other dependency and application analysis to construct bounded-contexts of the different views for each microservice. Then we aggregate them into a full-scope centralized perspective for each view, which consists of all the microservices aggregated into a mesh.

We start with the *extraction* phase of SAR. In particular, we extract Control-Flow Graphs (CFG) and Program Dependency Graphs (PDG). This allows us to construct a representation of the method calls and internal flow of data in a microservice. We also detect endpoints and their metadata, including security constraints and policies along with parameters and internal method branches, conditions, and loops. Lastly, we detect all entities within a specific module. All of this metadata is important to be able to construct a full centralized perspective of a view and offer a complete SAR overview.

The next phase in the SAR process is the *construction* phase. Once we extract the module-specific information, we represent it in a graph format. We further link the additional extracted metadata to its corresponding module. By aggregating all of the individual metadata artifacts in a cohesive and consistent way, we can easily move into the next phase.

To begin the *manipulation* phase, we use existing strategies of module fusion. Based on DDD [18], each module considers a bounded context [5], which includes a limited perspective of the data model, and often partially overlaps through certain data entities with other modules. This overlap is a key strategy in this phase of SAR. We begin with entity matching by looking for entities from distinct modules with a subset match of properties, data types, and possibly names. For this matching, we also considered natural language processing strategies (Wu-Palmer algorithm [8]). We derive the canonical data model and, through the matched entities, promote data and control dependencies.

The second perspective of static analysis we considered to merge modules is interaction. We identify all endpoints, parameter types, and metadata, and then the remote procedure calls within the methods. Next, we aim to match them and generate a complete service view of the entire application. This allows us to augment the result involving the canonical model.

We do not aggregate the technology view as each microservice is distinct in its choice of implementation; this is a hallmark of the microservice architecture.

The output of our application is a set of module metadata and the connections between them. Each individual microservice contains its bounded context domain model, while the centralized perspective contains the fused domain model. Similarly, each microservice contains its own service registry information, and the centralized perspective contains a graph that shows the connection edges between the disparate services. Each microservice contains its technology information, and an aggregate list of the technologies, broken up by layer, is in the centralized perspective. Lastly, each microservice owns its own deployment and operation information, and the centralized perspective contains a graph of connected deployments.

5 Case Study: Train Ticket

To demonstrate our framework's effectiveness, we tested it on an existing microservice benchmark, TrainTicket.[2] We chose this benchmark since it was specifically designed to emulate a real-world microservice application, consisting of 41 microservices and over 60,000 lines of code. It is written in Spring Boot and uses MongoDB for the datasource. It uses either Docker[3] or Kubernetes for deployment and either NGINX[4] or Ingress[5] for routing.

5.1 Domain View

The domain view is constructed through the analysis of the entities within each microservice. Our testbed broke from the convention with entity objects. Typically, entities can be discovered through the use of Enterprise standard annotation such as @Document for MongoDB objects or @Entity for MySQL objects. However, not all entities were annotated within the application. Because of this, we extended our definition of a system entity to include an annotated object, any object that is a field to such an object, or a POJO that matches a known entity in the system. Based on this, we could determine that object was an entity in its microservice, even if it was explicitly marked. An example merging is in Fig. 1.

---

[2]https://github.com/FudanSELab/train-ticket.

[3]https://www.docker.com/.

[4]https://www.nginx.com/.

[5]https://kubernetes.io/docs/concepts/services-networking/ingress/.

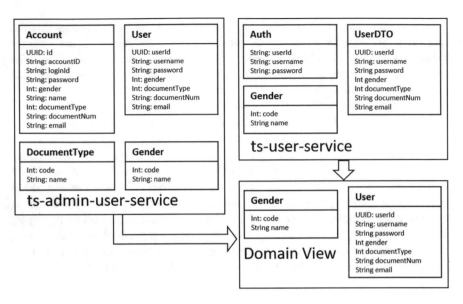

Fig. 1 Merged domain view from TrainTicket

5.2 Technology View

The technology view was broken into three sections, each representing one application layer (presentation, business, and data). For each layer, the system's underlying technologies were found by analyzing the source code files and the dependency management files (e.g., pom.xml for Maven), and system configuration files. We verified our findings against the existing TrainTicket documentation.

5.3 Service View

The service view is constructed using the generated internal communication diagram and the metadata of each module. In Fig. 2, we show a small selection of the canonical model for the service view of TrainTicket. Even a small selection from the overall service view can show some relevant observations. For example, from the documentation[6] of the TrainTicket system, it appears that the cancel service only has a dependency on the order service however through the lens of the service view, we can see that the cancel service is far more coupled to the other service. It relies on four other services, in addition to the order service.

---

[6]https://github.com/FudanSELab/train-ticket/blob/master/image/2.png.

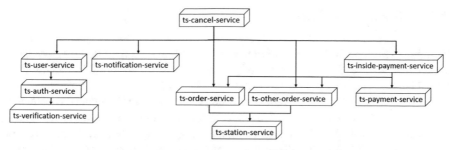

Fig. 2 Service view from TrainTicket

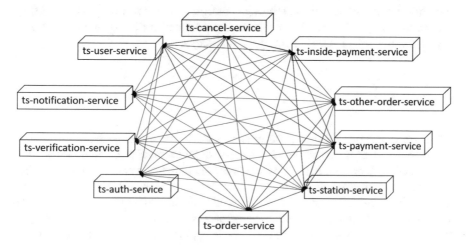

Fig. 3 Operation view from TrainTicket

5.4 Operation View

The operation view of the SAR process defines a topology of the containerization of the application. Figure 3 shows a small section of the TrainTicket topology. The TrainTicket containerization defines a singular network to connect each of the containers, which creates a graph where each node is connected to every other. This example shows our method benefit for full automation of the four application views. If only the operation view was available, it would create an incorrect/incomplete view of the application. By combining the operation view with the service view, the edges can be mapped to create the topology shown in Fig. 2.

5.5   Threats to Validity

Internal Validity: To verify our approach, we utilized the methodology we automated to extract the SAR views manually for our benchmark system. We had multiple people extract the information and construct their views.

External Validity: The effectiveness of SAR, regardless of if manual or automated, comes down to the availability of artifacts to analyze. For our application, we lack the ability to analyze artifacts such as images, diagrams, or textual descriptions of an application. We do not believe this impacts our application's overall effectiveness; however, since the concerns we automated, domain modeling, services, and containerization can be extracted through enterprise standards that do not vary from application to application. This means that our framework is capable of analyzing any application which uses the enterprise standards.

6   Conclusion

MSA is the mainstream direction for modern software development, and while extensive work has been done to define SAR methodologies on distributed systems, there is a lack of work on automation. Without automation, developers rely on expending a large amount of effort to generate the SAR artifacts manually. This paper demonstrated a framework for automatic SAR across four important views of the application, domain, technology, service, and operations. With our framework, developers can focus on answering questions about their application, phase 4 of SAR, instead of focusing all their time in phases 1–3.

Acknowledgements This material is based upon work supported by the National Science Foundation under Grant No. 1854049 and a grant from Red Hat Research https://research.redhat.com.

References

1. Albert E, Gómez-Zamalloa M, Hubert L, Puebla G (2007) Verification of java bytecode using analysis and transformation of logic programs. In: Hanus M (ed) Practical aspects of declarative languages. Springer, Berlin, Heidelberg, pp 124–139
2. Alshuqayran N, Ali N, Evans R (2018) Towards micro service architecture recovery: an empirical study. In: 2018 IEEE international conference on software architecture (ICSA), pp 47–4709
3. Chatley G, Kaur S, Sohal B (2016) Software clone detection: a review. Int J Control Theory Appl 9:555–563
4. DeMichiel L, Shannon W (2016) JSR 366: Java Platform, Enterprise Edition 8 Spec (2016). https://jcp.org/en/jsr/detail?id=342
5. Finnigan K (2018) Enterprise java microservices. Manning Publications. https://books.google. com/books?id=KaSNswEACAAJ

6. Gabel M, Jiang L, Su Z (2008) Scalable detection of semantic clones. In: Proceedings of the 30th international conference on software engineering, pp 321–330. ICSE '08, ACM, New York, NY. https://doi.org/10.1145/1368088.1368132

7. Granchelli G, Cardarelli M, Di Francesco P, Malavolta I, Iovino L, Di Salle A (2017) Towards recovering the software architecture of microservice-based systems. In: 2017 IEEE international conference on software architecture workshops (ICSAW), pp 46–53

8. Han L, Kashyap LA, Finin, T., Mayfield, J., Weese, J (2013) UMBC_EBIQUITY-CORE: semantic textual similarity systems. In: Second joint conference on lexical and computational semantics (*SEM), vol 1, Proceedings of the main conference and the shared task: semantic textual similarity. Association for Computational Linguistics, Atlanta, Georgia, USA, pp 44–52

9. Ibrahim A, Bozhinoski S, Pretschner A (2019) Attack graph generation for microservice architecture. In: Proceedings of the 34th ACM/SIGAPP symposium on applied computing, pp 1235–1242. SAC '19, Association for Computing Machinery, New York, NY, USA. https://doi.org/10.1145/3297280.3297401

10. Keivanloo I, Roy CK, Rilling J (2012) Java bytecode clone detection via relaxation on code fingerprint and semantic web reasoning. In: Proceedings of the 6th international workshop on software clones, pp 36–42. IWSC '12, IEEE Press, Piscataway, NJ, USA. http://dl.acm.org/citation.cfm?id=2664398.2664404

11. Keivanloo I, Roy CK, Rilling J (2014) Sebyte: scalable clone and similarity search for bytecode. Sci Comput Program 95:426–444. https://doi.org/10.1016/j.scico.2013.10.006, http://www.sciencedirect.com/science/article/pii/S0167642313002773

12. Kumar KS, Malathi D (2017) A novel method to find time complexity of an algorithm by using control flow graph. In: 2017 international conference on technical advancements in computers and communications (ICTACC), pp 66–68 (April 2017). https://doi.org/10.1109/ICTACC.2017.26

13. Lau D (2018) An abstract syntax tree generator from java bytecode. https://github.com/davidlau325/BytecodeASTGenerator

14. Makai M (2019) Object-relational mappers (ORMS). https://www.fullstackpython.com/object-relational-mappers-orms.html

15. Mayer B, Weinreich R (2018) An approach to extract the architecture of microservice-based software systems. In: 2018 IEEE symposium on service-oriented system engineering (SOSE), pp 21–30

16. Merkel D (2014) Docker: lightweight linux containers for consistent development and deployment. Linux J 2014(239) (Mar 2014). http://dl.acm.org/citation.cfm?id=2600239.2600241

17. MITRE: Common Vulnerabilities and Exposure (2020) http://cve.mitre.org

18. Rademacher F, Sachweh S, Zündorf A (2020) A modeling method for systematic architecture reconstruction of microservice-based software systems. Enterprise. Business-process and information systems modeling. Springer International Publishing, Cham, pp 311–326

19. Red Hat I (2020) Clair : a vulnerability scanner for Docker containers and images. https://coreos.com/clair/docs/latest/

20. Ribeiro JCB, de Vega FF, Zenha-Rela M (2007) Using dynamic analysis of java bytecode for evolutionary object-oriented unit testing. In: 25th Brazilian symposium on computer networks and distributed systems (SBRC)

21. Roy CK, Cordy JR, Koschke R (2009) Comparison and evaluation of code clone detection techniques and tools: A qualitative approach. Sci Comput Program 74(7):470–495. https://doi.org/10.1016/j.scico.2009.02.007

22. Selim GMK, Foo KC, Zou Y 2010 Enhancing source-based clone detection using intermediate representation. In: 2010 17th working conference on reverse engineering, pp 227–236 (Oct 2010). https://doi.org/10.1109/WCRE.2010.33

23. Soppelsa F, Kaewkasi C (2017) Native docker clustering with swarm. Packt Publishing

24. Su FH, Bell J, Harvey K, Sethumadhavan S, Kaiser G, Jebara T (2016) Code relatives: detecting similarly behaving software. In: ACM SIGSOFT international symposium on foundations of software engineering, pp 702–714. FSE 2016, ACM, New York, NY, USA. https://doi.org/10.1145/2950290.2950321

25. Syaikhuddin MM, Anam C, Rinaldi AR, Conoras MEB (2018) Conventional software testing using white box method. In: Kinetik: game technology, information system, computer network, computing, electronics, and control, vol 3(1), 65–72. https://doi.org/10.22219/kinetik.v3i1.231
26. Zdun U, Navarro E, Leymann F (2017) Ensuring and assessing architecture conformance to microservice decomposition patterns. Service-oriented computing. Springer International Publishing, Cham, pp 411–429

Pyclone: A Python Code Clone Test Bank Generator

Schaeffer Duncan, Andrew Walker, Caleb DeHaan, Stephanie Alvord,
Tomas Cerny, and Pavel Tisnovsky

Abstract Code clones are fragments of code that are duplicated in the codebase
of an application. They create problems with maintainability, duplicate buggy code,
and increase the size of the repository. To combat these issues, there currently exists
a multitude of programs to detect duplicated code segments. However, there are
not many varieties of languages among the benchmarks for code clone detection
tools. Without covering enough languages for modern software development, the
development of code-clone detection tools remains stunted. This paper describes a
novel tool that will take a seed of Python source code and generate Type 1, 2, and
3 code clones in Python. As one of the most used and rapidly-growing languages in
modern software development, our testbed will provide the opportunity for Python
code-clone detection tools to be developed and tested.

Keywords Code-Clone · Benchmark · Quality assurance · Testbed

1 Introduction

In many programs and code, developers are constantly trying to reduce the number
of errors, bugs, and other issues brought upon by poor coding design. One such
ingredient to bug-prone software is the inclusion of duplicated code segments or
code clones. Code clones in programs introduce problems in the forms of bugs,
unnecessarily large repositories, and difficulty in software maintenance. These can
be introduced into the program through many avenues, including developers copying
and pasting code, rather than putting in the effort to refactor code. Because of the
pervasive nature of code clones in a repository, care must be taken when modifying

S. Duncan · A. Walker · C. DeHaan · S. Alvord · T. Cerny (✉)
Baylor University, Waco, TX 76701, USA
e-mail: tomas_cerny@baylor.edu

P. Tisnovsky
Red Hat, FBC II, Purkyňova 97b, 612 00 Brno, Czech Republic

a piece of code also to modify any clones. Without a reasonable way to manage code clones, bugs can persist in the application and require significantly more developer overhead to update the repository.

A handful of useful code clone detection tools currently exist to assist developers in locating code clones in their application. However, this field is stunted by the lack of code-clone benchmark applications for the different popular industry languages. Python has recently become one of the predominant languages for applications in the industry. However, a recent mapping study on existing code clone detection tools [14] found that Python tools ranked 9th in the number of tools, with only 7 out of 67 tools able to detect Python, and no tools solely dedicated to Python code clone detection.

Given a general lack of benchmark applications, it would be beneficial to create a dynamic benchmark or test bank generator that can measure these tools' effectiveness. In this paper, we present a Python-based code clone generator, named Pyclone, that can produce a documented number of code clones to better gauge when seeded with a directory of Python code files how effective code clone detection tools are.

The rest of the paper is organized as follows. Section 2 presents background on code-clones. Section 3 presents background on existing mutation frameworks for code-clone detection tool evaluation. Section 3 presents our tool, Pyclone, and our process for the generation of the test bank. Section 4 discusses threats to validity. Finally, we conclude in Sect. 5.

2 Background

In this section, we will introduce the background to duplicated code within projects. In general, code clones or duplicate code segments are seen as poor coding practices. This is due to duplicate code being inefficient and because they introduce a level of variability, which is a breeding ground for bugs and unexpected errors. Code clones can introduce unpredictability when inconsistent changes are made to code duplicates. In general, there are four different types of code clones.

2.1 Basic Definitions

This subsection will define some general terms that will be used throughout the rest of the paper. These terms will be adapted from previous works/studies on code clones [2, 8, 9].

Code Fragment—A continuous segment of the source code, specified by (l, s, e), including the source file l, the line the fragment starts on, s, and the line it sends on, e.

Code Clone—Two code fragments that have similarities either in their syntax and/or semantics.

Table 1 Comparing/contrasting different types of code clones

Original code	Type 1	Type 2	Type 3	Type 4
int x = 5;	int x = 5;	int a = 5;	double a = 5;	int a = (10/2);
int y = 0;	int y = 0;	int b = 0;	double b = 0;	double b = 0;
// comment				
while(y <= x){	while(y<=x){y++;	while(b<=a){	while(b<=a){b+=	while(b <= 5){ b
y++; }	}	b++; }	1.0; }	+= 1.0; }

2.2 Types of Clones

In addition to the existence of a generic "code clones," these clones are broken into four sub-categories of clones, as highlighted in Table 1. In this subsection, we will define the general understandings of the four primary categories of code clones.

Type 1 code clones are said to be completely identical code fragments, disregarding comments and whitespace.

Type 2 code clones are said to be code fragments that are identical except they could have different variable and function names, along with different variable types and literals.

Type 3 code clones are said to be code fragments that are similar yet have some modifications consisting of either added/removed statements, different variable types/names, or differing function names.

Type 4 code clones are said to be code fragments that do not follow the same syntactical structure yet implement the same functionality.

Generally, these clone types are grouped by Types 1–3, which differ based on their text [1, 7] and Type 4, which is different only semantically [5].

3 Related Works

There has been a handful of studies conducted in the pursuit to evaluate the validity of code clone detection tools independently. While developers of tools typically fall back to BigCloneEval [12], several alternatives to methods of validating code clone detection tools have been proposed.

Roy and Cordy [7] proposed a unique approach to evaluating code clone detection tools. They proposed a mutation and injection automatic framework that would evaluate code clone detection tools based on code clones' editing theory. Their preliminary approach focuses on the idea of their framework being injection-based, and their approach solely relies upon textual mutation of the code. They go on to express the need for a benchmark that can be used by developers and researchers to evaluate proposed code clone detection tools. In this work, they relay the importance of verifying clones needs to be able to test results yet require minimal effort thoroughly. This aspect of easily crafting a test bank based on seed code would allow a developer more options in terms of identifying more efficient ways to code.

Svajlenko and Roy [10] also had another study done on the evaluation of code clone detection tools, this time with a focus on a clone detection tool evaluation framework named BigCloneBench [11]. In this extension of their previous work, they use BigCloneBench to validate their own mutation framework discussed above. Their proposed tool validated the effectiveness of inject-based frameworks for code-clone tool verification. Their injection framework solely aims to solve the problem of Java code clone detection tools.

3.1 Need for Pyclone

Python has been steadily rising in popularity for over a decade now and has become one of the preeminent programming languages, especially in industry. One look at the Google search trends in Fig. 1 shows that Python is at its most popular ever. The TIOBE Programming Community index [13] is a good indicator of programming language popularity. The rating bases on the number of skilled engineers world-wide, courses, and third-party vendors. For August 2020, the top was C with 16.9%, Java with 14.4%, and Python with 9.6%. Analysis of 2020 Indeed.com job postings [4] show Python, followed by Java and JavaScript. For Microservice Architecture related jobs that drive the cloud-computing field [15], 48% of companies use Java, and 21% use Python. The top languages for containers are JavaScript, Java, Python, and PHP [3].

For code clone detection, there have been various tools that have been created to test the percentage of code clones within computer programs, and while there exist a multitude of code clone detection tools for C++ and Java, comparably, there are very few available for Python. A recent analysis of open-source tools and benchmarks for code clone detection [14] made clear that there exists a significant gap for the Python language. With there being no benchmark for Python code-clone detection, it falls on these tools to test using another benchmark, in another language, and while these tools claim equal performance across their supported languages, no method exists to verify their tools.

To improve the quality of such detection tools as it pertains to the Python language, a standard set of code clones in a test bank needed to be created. The creation of clearly defined code clones would then set a benchmark for the quality of detection across the various tools.

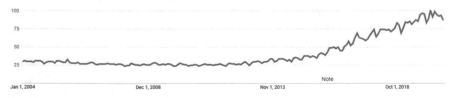

Fig. 1 Google search trends for python over time

4 Pyclone

In this section, we will detail the way Pyclone generates each type of code clone based upon its representation to the system. Having been developed in Python for testing Python code clone detection tools, Pyclone was written to create code clones based on a seed of Python files passed to it. It works based on the idea of mutation of Abstract Syntax Trees (AST) generated from the seed files.

4.1 Abstract Syntax Trees

The first step in mutating new code-clones is to scan the seed files and construct ASTs for all of the files. An AST is essentially a tree representation of abstract code structure. Each node in an AST is a different statement or condition in the code. If code is different between the two projects, then the ASTs between the two projects will also differ in the same locations, and vice versa.

With these ASTs, our tool can then modify these ASTs in order to change the underlying source code once the tool converts the AST back into Python code, giving us differing code samples.

The Python compiler uses ASTs to compile the Python code into bytecode. However, in order for Pyclone to manipulate the ASTs, it was required to find a more high-level representation of ASTs that can be manipulated within the tool itself.

Pyclone, therefore, uses a Python package called astor.[1] Astor allows for Pyclone to form ASTs based on valid Python files and modify them programmatically. These ASTs are manipulated as any other tree-based object in Python would be manipulated. Astor then handles the recreating code from the ASTs once the mutation is complete.

A key understanding of ASTs, which is used in all three type generations, is the idea of a "copyable" node in the AST. When we construct an AST, we must select a node that can constitute a fully complete and compilable code segment when taken alone. When constructing the ASTs from the seed files, we begin by utilizing the most common "copyable node" for code-clones, which is the *file-root* node. An AST containing an entire Python file is a fully-contained piece of code for code-clone detection purposes. It's worth noting that the code may not be runnable if it references classes or methods from other files or libraries, but it is fully compilable by Python.

The selection process is then further specified with a number of rules about which sub-nodes can be selected. For example, while a pure-method (one not existing as a class method) node is copyable without any additional steps, a class-method node must be copied within a duplicate of the seed class, along with any sub-methods that are called. Besides, the class must copy any necessary class fields. An example of the bare minimum needed to clone a class-method is seen in Fig. 2.

---

[1] https://pypi.org/project/astor/.

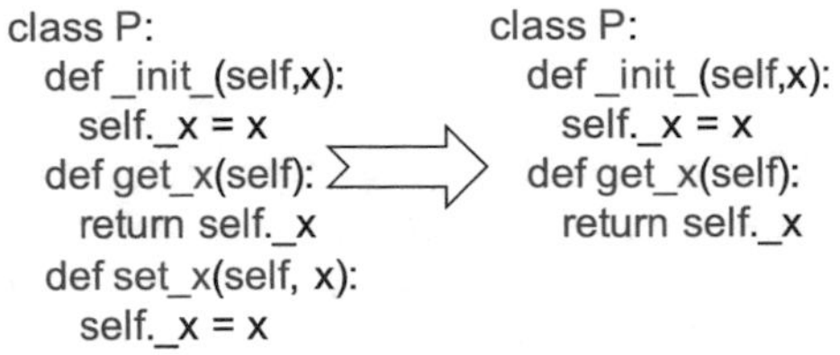

```
class P:                        class P:
    def _init_(self,x):             def _init_(self,x):
        self._x = x                     self._x = x
    def get_x(self):                def get_x(self):
        return self._x                  return self._x
    def set_x(self, x):
        self._x = x
```

Fig. 2 Cloning a class method

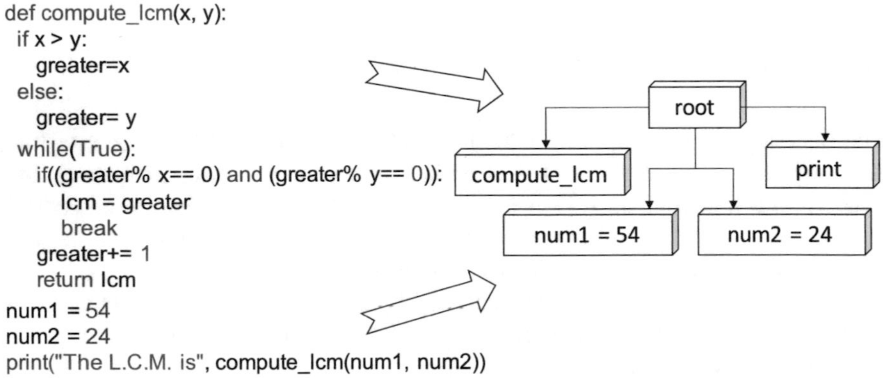

```
def compute_lcm(x, y):
    if x > y:
        greater=x
    else:
        greater= y
    while(True):
        if((greater% x== 0) and (greater% y== 0)):
            lcm = greater
            break
        greater+= 1
    return lcm
num1 = 54
num2 = 24
print("The L.C.M. is", compute_lcm(num1, num2))
```

Fig. 3 First order nodes of a python code sample

Further detailed nodes such as individual *for*, *if* or *while* blocks must be examined carefully before being copied. For example, if a *for* block calls a method within its body, then that method must either be copied along with the *for* block.

Last to be considered are individual statements. For our purposes, we only look at statements which we call "first-order" statements. These statements are any that exist as a direct child of either a file-root node or a pure-method node. An example of possible "first-order" statements is shown in Fig. 3.

Once all nodes which are copyable are identified, the process of code-clone creation can begin.

4.2   Generation of Type 1

Creating Type 1 clones is straightforward. The tool essentially selects a random number of copyable nodes to construct new clones out of and directly dumps them into a new file. Occasionally we would modify the AST before constructing the new clone to add in comment nodes. Once the new files are created, we retroactively modify their whitespace as well.

Pyclone creates a directory named 'type1clones' and puts all of the Type 1 clones with identifier file names into this directory.

4.3   Generation of Type 2

Creating Type 2 code clones was a little different from Type 1 because Pyclone has to modify the ASTs before constructing the new clone.

After selecting a new set of copyable nodes to construct the clones out of, we then proceeded to modify them by symbol translation for Type 2 clones.

The process of symbol translation involves mapping function and variable names in the existing code block. Whenever Pyclone would come across a value representing a function name or variable name, it would be inserted into a map that maps previous names to their newly generated name. When traversing future nodes, if a name that had already been mapped is discovered, it is not inserted as a new entry into the map.

After the symbol mapping phase, we begin the symbol translation phase, which involves traversing the entire tree of the node that is being copied and searching for any names that are translated in the symbol table. If the name is found, then it is mapped to its new name.

After the entire tree has been changed, the AST was then converted back to code and returned in the form of a Python code file. Additionally, Type 2 clones' generation can also include steps from the Type 1 protocol, which includes adding comments or modifying whitespace.

Pyclone creates a directory named 'type2clones' and puts all of the Type 2 clones with identifier file names into this directory.

4.4   Generation of Type 3

Creating Type 3 code clones was more similar to the generation of Type 2 clones than Type 1 clones. Like with clone types, the tool then traversed the tree, however this time it was looking for nodes that represented boolean operators, mathematical operations, or comparing operations, as modifications to these types of nodes would generate code nearly similar to the original but with different functionality.

A static list of replacement values was created for each operator type to change these nodes, and then one was chosen at random to replace the original value. Each AST could also have nodes removed. If a node is removed, any children nodes are also removed (i.e., an entire *for* block can be removed). Lastly, each AST is potentially merged with another to add lines to the code-clone. When a new AST is added in, it will have different variable names than the original AST. To mitigate this, we perform symbol realignment on the merged in section. This process is similar to the symbol translation process in Type 2 clone generation. Each symbol in each of the ASTs is mapped from its name to its data type. Then the AST to be merged in is traversed, and each variable name is mapped to a name from the original AST that matches the data type. If no suitable replacement is found, then that statement is removed, or a variable declaration is appended to the top of the code block.

After going through the entire tree, and potentially applying Type 1–2 mutations as well, the AST is then converted back to code and written to a new Python file in a directory for Type 3 clones.

Pyclone creates a directory named 'type3clones' and puts all of the Type 3 clones with identifier file names into this directory.

5   Threats to Validity

There are two main threats to the validity of Pyclone's test bank. The first is that due to the nature of injection-based frameworks, it's possible that by injecting new code lines or removing them, an accidental clone was created with an unintended code segment. The second main threat to validity is that while the code clones are compilable by Python, they may not be runnable, which is an issue with code-clone detection tools that require a running program to detect similarities [6].

The main threat to the validity of Pyclone is the development of the code clones, and whether the clone detection tools will find them. Type 1 and Type 2 clones are fairly straightforward, as they are almost identical copies of the original code with changed variable names at most. However, the detection of Type 3 or Type 4 clone may depend on the tool's implementation for finding clones, as the given tool's definition of Type 3 or Type 4 clone may not be the same as the way our tool has created them, and so the clones may be misidentified or not caught at all.

6   Conclusion

This paper presented an overview of the types of code clones and the reasons for code-clone detection tools. We further explained that there exists a large gap in the code-clone detection research field for the Python coding language, partially due to the lack of a Python code-clone benchmark. We presented an injection-based mutation framework for generating a Python code-clone benchmark from an existing seed of Python files to mitigate this. This paper concludes one of the first steps forward in creating viable means to check Python code clone detection tools.

Acknowledgements This material is based upon work supported by the National Science Foundation under Grant No. 1854049 and a grant from Red Hat Research https://research.redhat.com.

References

1. Aho AV, Lam MS, Sethi R, Ullman JD (2006) Compilers: principles, techniques, and tools, 2nd edn. Addison-Wesley Longman Publishing Co., Inc., Boston, MA
2. Bellon S, Koschke R, Antoniol G, Krinke J, Merlo E (2007) Comparison and evaluation of clone detection tools. IEEE Trans Software Eng 33(9):577–591. https://doi.org/10.1109/TSE.2007.70725
3. Datadog: 8 emerging trends in container orchestration (December 2012). https://www.datadoghq.com/container-orchestration/
4. Dowling L (2020) Top 7 programming languages of 2020. https://www.codingdojo.com/blog/top-7-programming-languages-of-2020
5. Gabel M, Jiang L, Su Z (2008) Scalable detection of semantic clones. In: 2008 ACM/IEEE 30th international conference on software engineering, pp 321–330. https://doi.org/10.1145/1368088.1368132
6. Hamerly G, Perelman E, Lau J, Calder B (2005) Simpoint 3.0: faster and more flexible program phase analysis. J Instr Level Parallelism 7
7. Roy CK, Cordy JR (2009) A mutation/injection-based automatic framework for evaluating code clone detection tools. In: 2009 international conference on software testing, verification, and validation workshops, pp 157–166. https://doi.org/10.1109/ICSTW.2009.18
8. Roy CK, Cordy JR (2007) A survey on software clone detection research. School of Computing TR 2007-541, Queen's University 115
9. Sheneamer A, Kalita JK (2016) A survey of software clone detection techniques. Int J Comput Appl 137(10):1–21. (Published by Foundation of Computer Science (FCS), NY, USA)
10. Svajlenko J, Roy CK (2014) Evaluating modern clone detection tools. In: 2014 IEEE international conference on software maintenance and evolution, pp 321–330. https://doi.org/10.1109/ICSME.2014.54
11. Svajlenko J, Roy CK (2015) Evaluating clone detection tools with bigclonebench. In: 2015 IEEE international conference on software maintenance and evolution (ICSME), pp 131–140
12. Svajlenko J, Roy CK (2016) Bigcloneeval: a clone detection tool evaluation framework with bigclonebench. In: 2016 IEEE international conference on software maintenance and evolution (ICSME), pp 596–600
13. Tiobe Software B.V. (2020) Tiobe index for august 2020. https://www.tiobe.com/tiobe-index/
14. Walker A, Cerny T, Song E (2020) Open-source tools and benchmarks for code-clone detection: past, present, and future trends. SIGAPP Appl Comput Rev 19(4):28–39. https://doi.org/10.1145/3381307.3381310
15. Zavgorodnya A (2019) RubyGarage: moving to microservices: top 5 languages to choose from. https://rubygarage.org/blog/top-languages-for-microservices

A Comprehensive Enterprise System Metamodel for Quality Assurance

Jan Svacina, Vincent Bushong, Dipta Das, and Tomas Cerny

Abstract One of the biggest challenges in code quality assurance is the amount of code that needs to be reviewed at an instance before the code is deployed on production. Reviewers need to check not only coding practices and formatting but also the meaning of the code and its compliance with requirements. Enterprise systems are notoriously known for the large codebase, challenging business logic, and advanced code constructs, which require significant resources for code review. However, enterprise systems use coding constructs that reveal aspects and constraints about the business logic, such as validation, database connection, and API. We extract these aspects and their relationships into a comprehensive metamodel. Next, we persist the metamodel into a graph database and conduct quality assurance checks via database queries. This method significantly reduces the amount of information that needs to be processed while maintaining key enterprise aspects. The method enables system administrators or project managers to discover defects and inconsistencies without reading the code.

Keywords Quality assurance · Enterprise systems · Verification

1 Introduction

Enterprise systems are vital parts of the industry that create a market share of nearly $36 billion in 2018 [17]. The enterprise systems maintenance costs steadily increased and currently create up to 90% of project cost [11]. Quality assurance is a methodology that aims to reduce the cost of finding and fixing defects. Software engineers spend from 40 to 60% of time searching for bugs in the code [7]. One way to substantially decrease the resources on maintenance is a process of code review. However, the developers need to process a large amount of a codebase, when only a small portion of code is significant and error-prone. We addressed this issue by introducing a

J. Svacina · V. Bushong · D. Das · T. Cerny (✉)
Baylor University, Waco, TX 76704, USA
e-mail: tomas_cerny@baylor.edu

© The Author(s), under exclusive license to Springer Nature Singapore Pte Ltd. 2021   245
H. Kim et al. (eds.), *Information Science and Applications*, Lecture Notes
in Electrical Engineering 739, https://doi.org/10.1007/978-981-33-6385-4_23

novel method that derives significant attributes of the enterprise systems . We represent the attributes as property code graphs, persist them into the database, and run database queries to discover issues. We validate our method on a production system and discover privacy-related issues.

The paper is structured as follows. In Sect. 2, we describe current programming techniques for enterprise applications. Next, we define current approaches to quality assurance in enterprise systems, and we describe the most significant challenges. In Sect. 4, we propose the unique metamodel for enterprise system. We conduct a case study in Sect. 5 and conclude by Sect. 6.

2 Background

Analysis of enterprise systems has some distinctive features that we are going to use in the proposed method. Enterprise systems can be written in several programming languages. Each language has a parser tool that derives from the code class names, methods, and other properties together with the relationships among them [5]. The derived code can be organized into the property code graphs, where each graph consists of nodes, edges, and properties. Edges connect nodes in a single direction called relationship. The two-way relationship is created by two single direction relationships in the opposing direction to maintain traversal algorithms. Both nodes and edges can carry data, which are called properties. Property can be of any type and often signify the variable name and type. We consider the following two examples of enterprise code. The first is a C# code snippet, which describes an entry point to the program that returns a GET HTTP request with a template as a View object. An annotation HandleError in the square brackets signifies a code's definition in another object that checks the errors' request. This abstraction enables to enhance the functionality with encapsulated constructs that give additional meaning to the code.

```
1    using System.Web.Mvc;
2
3    namespace MvcApplication1.Controllers
4    {
5        [HandleError]
6        public class HomeController : Controller
7        {
8            public ActionResult Index()
9            {
10               return View();
11           }
12       }
13   }
```
Listing 23.1 .NET code example [15]

Another example is a JAVA Spring code that handles a GET HTTP request again, but we return only a plain string this time. We also observe that the code uses annotations beginning with @, for instance, *RequestParam*. This annotation gives additional meaning to the arguments of the method. The arguments represent parameters in the HTTP Get request.

```
1   @Controller
2   public class GreetingController {
3
4       @GetMapping("/greeting")
5       public String greeting(@RequestParam(name="name",
6           required=false, defaultValue="World")
7           String name, Model model) {
8           model.addAttribute("name", name);
9           return "greeting";
10      }
11
12  }
```

Listing 23.2 Example of Java Spring Enterprise code taken from [1]

3 Related Work

Representing a code of a program began by program slicing [12]. Authors talk about selecting a reference point x, which is a base point. Then, we search for every reference to the program's base point, thus creating a subprogram. Another way is to create control flow graphs to discover security vulnerabilities [13].

Call graphs are used to analyze and discover code defects [14]. Authors in [19] used also call graphs for problems associated with insufficient validation in the code. There is an extensive research on flow-sensitive inter-procedural data flow analysis [2–4, 6, 8]. However, there is a significant gap in proposing general methods that apply to any enterprise project. There is some work on deriving trees for testers that help with manual revisions [18].

4 Proposed Method

We propose a method consisting of three consecutive phases. We start with raw source code initially, and we end up with found defects and inconsistencies in the enterprise application, as shown in Fig. 1.

In phase one, we use the static code analysis tool to derive enterprise properties from the code [5]. We derive properties common for any system, modules, class names, methods, and fields. We also derive metadata associated with these basic

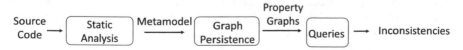

Fig. 1 Phases of the source code analysis

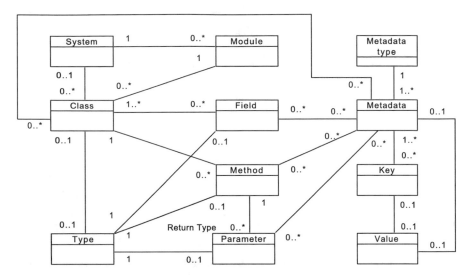

Fig. 2 A generic enterprise system metamodel

components, such as validation, API, and persistence. We organized derived enterprise properties into the novel metamodel. We depicted the metamodel as the UML domain model diagram, as shown in Fig. 2.

The metamodel consists of eleven classes and seventeen relationships. We start with the **System** domain, which describes an instance of an enterprise system that is analyzed. The name of the system is the top-level directory of the system. Each system can have one or more **Modules**, which are represented as second-level directories. Each module has one or more **Classes**. In case no modules are present, we associate classes with the system node. Each class encapsulates data and behavior. Data is represented by **Field**, which is a property of a class. The behavior is depicted by **Method**, which is a function inside the class. Each class can have any number of fields or methods. Each field has a **Type**, and each method has a return type. Both type and return type can be primitive or a class; thus, we have either a value or an association. The method has any number of **Parameters**, which are incoming data into the method.

The described structure so far can be applied to any program, even the one that is not an enterprise. However, the basic structure is enhanced by properties called **Metadata**, which is a programming construct that associates additional meaning to classes, methods, fields, and parameters. Each metadata item has a **Metadata type** association (architecture layer, validation, persistence). The metadata consists either from key-value pairs or only a value.

In the second phase, we persist the metamodel into a graph database. The graph database represents data as a property code graphs. We translate the metamodel into the graph database in the following way. We start by creating the system node and respective modules, and we connect them with *HAS_MODULE* relationship,

Table 1 Edges in the graph database system

Name	Description
HAS_MODULE	(System → Module)
HAS_CLASS	(System → Class), (Module → Class)
HAS_METHOD	(Class → Method)
HAS_FIELD	(Class → Field)
NEXT	(Method → Method)
HAS_RETURN_TYPE	(Method → Type)
HAS_TYPE	(Field → Type), (Parameter → Type)
IS_A_CLASS	(Class → Type)
HAS_(:METADATA_TYPE)	(Class → Metadata), (Field → Metadata)
	(Method → Metadata), (Parameter → Metadata)
HAS_KEY	(Metadata → Key)

as shown in Table 1. We always store the name of the system, class, module, and others as a node's property. The third column in Table 1 shows the edge's orientation, for instance, from the System node to the Module nodes. Next, we create class nodes and associate them with either System node or respective module nodes with *HAS_CLASS* method. Next, we continue with fields, methods, parameters, and types. The method can have a *NEXT* relationship to another method in case this method is called within the body of the calling method. Nodes parameter, method, field, and class are also related to the metadata. However, we use named relationship, where we name the relationship by the meaning of the relation, for instance, *HAS_ROLE* or *HAS_MAX_LENGTH*. We store the value, for instance *user* or *43*, as a property in the metadata object. In case the value of the metadata item is in the key-value format, we use the relationship to the node key, where the value is stored as a property.

The final phase is constructing queries in the graph database. Each graph database has the language for querying with its grammar, including even a library function for traversal and machine learning algorithms. The queries can check instantly, for instance, properties of methods and fields in two modules that are otherwise scattered on the enterprise system's number of places.

5 Case Study

In this section, we demonstrate detecting privacy violation issues in the enterprise system. We detected three privacy violations, hierarchy, unrelated access violations, unknown role violations, and missing security definition violations. We utilized a microservice benchmark project [9] for our analysis. The project is entirely written in JAVA using the Spring framework. We used Java parser for static code analysis, we used the Neo4j graph database as a storage of the metamodel, and we used the Neo4j [16] Java bolt connector for persistence. Neo4j uses the Cypher query language [10] that we utilized for analysis.

We focused on privacy issues that result from the incorrect usage of RBAC. In Spring Boot, some methods represent endpoints of the API. These methods can have an annotation *@RolesAllowed*, with the argument being an array of strings, where each string represents the name of the role that is allowed to use the endpoint. Roles at the endpoint are checked with the roles associated with an authenticated user. Various roles can form a hierarchy; for instance, the administrator inherits all privileges from members and moderators. We decided to detect application mutants from Walker et al. using our method.

We detected two mutant types *hierarchy access violations* and *unrelated access violations* with a single query. These violations capture situations when parts of a system are accessed from endpoints with conflicting roles. The query in the Listing 23.3 finds module *cms* with their classes, methods of respective classes, and roles to respective methods that have value user, admin, and super, admin. It makes then union operation on methods with admin and super-admin associations. The query yielded graphs that consist of related methods that use conflicting roles.

```
1    MATCH (module)-[:HAS_CLASS]->(class)
2    MATCH (method)-[:HAS_ROLE]->(role)
3    WHERE role.value="user" AND role.value="admin" \
4        AND role.value="superadmin"
5    RETURN method
6    UNION
7    MATCH (method)-[:HAS_ROLE]-(role)
8    WHERE role.value="admin" AND role.value="superadmin"
9    RETURN method
```

Listing 23.3 Cypher query for hierarchy and unrelated access violation

The *unknown role violations* happens due to misspelling the privacy role definition. We found the issues by querying all methods with other role values, then user, admin, or super admin, as shown in the Listing 23.4.

```
1    MATCH (module)-[:HAS_CLASS]->(class)
2    MATCH (method)-[:HAS_ROLE]->(role)
3    WHERE NOT role.value="user" OR role.value="admin" \
4        OR role.value="superadmin"
5    RETURN method
```

Listing 23.4 Cypher query for unknown role violations

We also identified *missing security definition violations*, where an endpoint does not have any role associated with it. We queried for classes and methods across modules and filtered methods without any relationship *HAS_ROLE*, as shown in the Listing 23.5.

Listing 23.5 Cypher query for missing security definition captionpos

```
1    MATCH (module)-[:HAS_CLASS]->(class)
2    MATCH (method)
3    WHERE NOT (method)-[:HAS_ROLE]-()
4    RETURN method
```

6 Conclusion

We proposed a unique method that creates a general metamodel from any enterprise system. We showed how to persist the metamodel as a property code graphs and persist them into the database. We also proved that we utilize the database to find defects related to privacy. We tested our solution on a production enterprise system. In future work, we want to focus on discovering other vulnerabilities and defects.

Acknowledgements This material is based upon work supported by the National Science Foundation under Grant No. 1854049 and a grant from Red Hat Research https://research.redhat.com.

References

1. 2020 VMware, I (2020) Spring web component with spring mvc. https://spring.io/guides/gs/serving-web-content/ (2020), online. Accessed 10 Aug 2020
2. Agrawal H (1994) On slicing programs with jump statements. In: Proceedings of the ACM SIGPLAN 1994 conference on programming language design and implementation, pp 302–312
3. Ball T, Horwitz S (1993) Slicing programs with arbitrary control-flow. International workshop on automated and algorithmic debugging. Springer, Berlin, pp 206–222
4. Ball T, Rajamani SK (2001) Bebop: a path-sensitive interprocedural dataflow engine. In: Proceedings of the 2001 ACM SIGPLAN-SIGSOFT workshop on program analysis for software tools and engineering, pp 97–103
5. Binkley D (2007) Source code analysis: a road map. In: Future of software engineering (FOSE '07), pp 104–119
6. Binkley D (1993) Precise executable interprocedural slices. ACM Lett Program Languages Syst (LOPLAS) 2(1–4):31–45
7. Bourque P, Fairley RE, Society IC (2014) Guide to the software engineering body of knowledge (SWEBOK(R)): version 3.0, 3rd edn. IEEE Computer Society Press, Washington, DC
8. Callahan D (1988) The program summary graph and flow-sensitive interprocedual data flow analysis. In: Proceedings of the ACM SIGPLAN 1988 conference on programming Language design and Implementation, pp 47–56
9. Cloudhubs: Tms 2020. https://github.com/cloudhubs/tms2020 (2020), online. Accessed 10 Aug 2020
10. Cypher, I.: Cypher. https://neo4j.com/developer/cypher/ (2020), online. Accessed 10 Aug 2020
11. Dehaghani S, Hajrahimi N (2013) Which factors affect software projects maintenance cost more? Acta Inform Medica: AIM: J Soc Med Inform Bosnia & Herzegovina: časopis Društva za medicinsku informatiku BiH 21:63–6. https://doi.org/10.5455/AIM.2012.21.63-66
12. Horwitz S, Reps T, Binkley D (2004) Interprocedural slicing using dependence graphs. SIGPLAN Not. 39(4). https://doi.org/10.1145/989393.989419
13. Jovanovic N, Kruegel C, Kirda E (2006) Pixy: a static analysis tool for detecting web application vulnerabilities. In: 2006 IEEE symposium on security and privacy (S P'06), pp 6, 263
14. Livshits VB, Lam MS (2005) Finding security vulnerabilities in java applications with static analysis. USENIX security symposium 14:18–18
15. Microsoft, I.: Asp .net mvc. https://docs.microsoft.com/en-us/aspnet/mvc/overview/older-versions-1/controllers-and-routing/asp-net-mvc-routing-overview-cs (2020), online. Accessed 10 Aug 2020
16. Neo4J, I.: Neo4j. https://neo4j.com/ (2020), online; accessed 10 August 2020
17. Research AM (2020) ERP software market. https://www.alliedmarketresearch.com/ERP-market (2020), online. Accessed 10 Aug 2020

18. Yamaguchi F, Lottmann M, Rieck K (2012) Generalized vulnerability extrapolation using abstract syntax trees. In: Proceedings of the 28th annual computer security applications conference. Association for Computing Machinery, New York, NY, USA. https://doi.org/10.1145/2420950.2421003

19. Yamaguchi F, Wressnegger C, Gascon H, Rieck K (2013) Chucky: exposing missing checks in source code for vulnerability discovery. In: Proceedings of the 2013 ACM SIGSAC conference on computer & communications security, pp 499–510. CCS '13, Association for Computing Machinery, New York, NY, USA. https://doi.org/10.1145/2508859.2516665

Knowledge Incorporation
in Requirements Traceability Recovery

Adnane Ghannem⊙, **Mohammed Salah Hamdi, Marouane Kessentini, and Hany Ammar**

Abstract The necessity of continuous change and adaptation in software systems makes maintenance tasks complex and difficult. One of the challenges in software maintenance is keeping requirements traceability up to date automatically. This problem can be considered as an optimization problem where the goal is to assign each requirement to one or many software artifacts (class, method, variable, etc.). However, some of the traceability links proposed using this approach do not necessarily make sense depending on the context and the semantic of the system under analysis. This paper proposes an approach that tackles this problem by adapting the Interactive Genetic Algorithm (IGA) which enables to interact with users and integrate their feedbacks into a classic GA. The proposed algorithm uses a fitness function that combines the recency of change, the frequency of change, and the semantic similarity between the description of the requirement and the software artifact, and the designers' ratings of the traceability links proposed during execution of the classic GA. Experimentation with the approach yielded interesting and promising results.

Keywords Software maintenance · Interactive genetic algorithm · Requirements engineering · Requirements traceability

A. Ghannem (✉) · M. S. Hamdi
CIS Department, Ahmed Bin Mohammed Military College, Doha, Qatar
e-mail: adnane.ghannem@abmmc.edu.qa

M. S. Hamdi
e-mail: mshamdi@abmmc.edu.qa

M. Kessentini
CIS Department, Univesity of Michigan, MI, USA
e-mail: marouane@umich.edu

H. Ammar
Lane Department of CS&Electrical Engineering, West Virginia Univesity, WV, USA
e-mail: hammar@wvu.edu

© The Author(s), under exclusive license to Springer Nature Singapore Pte Ltd. 2021
H. Kim et al. (eds.), *Information Science and Applications*, Lecture Notes
in Electrical Engineering 739, https://doi.org/10.1007/978-981-33-6385-4_24

1 Introduction

The Requirement Traceability Recovery (RTR) is defined as a process of re-/constructing traceability links between requirements and software system artifacts. RTR is known as an important factor for refining software maintenance by supporting activities such as verification, change impact analysis, program comprehension, and software reuse. Nowadays, RTR becomes more important especially, in the case of systems that change frequently and continually such as adaptive systems that need to manage the requirement changes and analyze their impact. Many reasons have made that, RTR as a practice, is not always adopted by developers such as developers dereliction, lack of time, etc. [1]. However, the automation of the RTR process is of great necessity because performing the RTR task manually is extremely tedious, error-prone, and time consuming.

To support RTR, several tools (e.g., REquirements TRacing On-target RETRO [2], ReqSimile [3], ReqAnalyst [4], and ADAMS Re-Trace [5])) have been developed using Information Retrieval (IR) techniques. Most of the proposed tools are time-consuming and semi-automatic. Therefore, we need to generate traceability links automatically through all requirements frequently.

In this paper, we present an approach aiming at automating the RTR process and helping designers update the system appropriately by extracting links between requirements and software artifacts (class, method, variables, etc.). To do this, we have considered the RTR process as a search problem using the interactive genetic algorithm (IGA). The approach takes a software system and a set of requirements needed to adapt to changes in the environment as input, and generates a set of traceability links (solution) between the requirements and artifacts (class, method, variable, etc.). A solution consists of assigning each requirement to one or many artifacts of the system taking into consideration the *recency of change (RC)* (elements that changed recently, are more likely to change now, i.e., are more likely to be related to the new requirement at hand), *the frequency of changes (FC)* (elements that change frequently, are more likely to change now, i.e., are more likely to be related to the new requirement at hand), *the semantic similarity* between the requirement documents and the software artifacts and *designer/developer feedbacks*. Although heuristics search has been applied to many software engineering problems [6], to the best of our knowledge, it is the first time that RTR is considered as an optimization problem. The main contribution of this paper is that we introduce a new automatic RTR approach based on the use of heuristic search using IGA. We evaluate the approach using two object-oriented open source projects.

The paper is structured as follows: Sect. 2 is dedicated to background information needed to understand the approach. Related work is discussed in Sect. 3. Section 4 describes the adaptation of the IGA on the RTR problem. Section 5 presents and discusses the experimental results. Conclusions and future directions of work are presented in Sect. 6.

2 Background

Generally, "Traceability" is defined as the process of establishing links between different elements manipulated. Specifically, in the case of software system, identifying traceability links between requirements and code elements is called Requirements Traceability Recovery (RTR) that consists of tracing requirements throughout a software system during and after implementation which is a complex and challenging task. Two types of RT have been identified in [7]: (1) traditional and (2) non-traditional RT. Traditional RT deals with the traceability issue during the software development process, however, the non-traditional RT deals with it at run-time to guarantee that the system is implemented according to the requirements. The ability to automatically identify the links during the software development step or after software development (i.e., at runtime) is very interesting and useful. Dealing with the second type (i.e., non-traditional RT) is more challenging than dealing with the first type (i.e., traditional RT). In this paper, we propose a search-based RT approach that can be applied to both traditional and non-traditional RT.

Interactive Genetic Algorithm (IGA) combines GA [8] with users' feedbacks (human evaluation). The main idea of GA consists of exploring the search space by making a population of candidate solutions (individuals) evolve toward a "good" solution of a specific problem. The GA process consists of evaluating each individual of the population using a fitness function that determines a quantitative measure of its ability to solve the target problem. The search space is explored by selecting individuals of the current population having the best fitness values and evolving them using genetic. This process is repeated iteratively, until a stopping criterion is met. This criterion usually corresponds to a fixed number of generations. IGA can be used to solve problems that cannot be easily solved by GA. Variety of application domains have been tackled using IGA (e.g., music composition systems [9]; Humanoid dancing [10]; software re-modularization [11]; GUI [12]). For more detail see [13]. One of the key elements in IGAs is the management of the number of interactions with the user and the way an individual is evaluated by the user.

3 Related Word

Several RTR tools were proposed in the literature [2–5]. Many IR techniques have been widely adopted as core technologies of semi-automatic tools to extract requirements traceability links such as algebraic or probabilistic models [14], data mining [15], machine learning [16], cross referencing schemes [17], scenarios [18], key-phrase dependencies [19], etc. (see [20] for an in-depth survey).

For example, ADAMS Re-Trace is a traceability recovery tool that has been proposed by De Lucia [5] based on Latent Semantic Indexing (LSI) that was developed in the fine-grained artifact management system ADAMS. ADAMS Re-Trace aims at supporting the software engineer in the identification of traceability links between artifacts of different types. Another trust-based traceability recovery approach, called Trustrace [1] that combine IR and Data mining methods in order to establish traceability links between requirements and source code. Trustrace consists of three parts, namely, Histrace, Trumo, and DynWing. Histrace consists of creating the links from requirements to source code by mining the software repositories. Trumo aims at re-ranking the recovered links obtained by Histrace using the web trust model. DynWing is a weighting technique to dynamically assign weights to experts giving their opinions about the recovered links. The approach presented by Grechanik et al. [21] combines program analysis, run-time monitoring, and machine learning to the process of recovering traceability links between types and variables in Java programs and elements of the requirements model based on Use Case diagrams. Chen and Grundy [22] presented an approach that combines three supporting techniques, namely, Regular Expression, Key Phrases, and Clustering, with a VSM in order to improve the performance of automated traceability between documents and source code. Another prototype tool, called ReqAnalyst [4] aims at testing different traceability recovery approaches and LSI methods in particular. The tool allows the user to change the settings of the parameters in the LSI reconstruction. Once executed, ReqAnalyst provides the user with the reconstructed traceability matrix with the possibility of generating different requirement views. With this tool, it is possible to obtain continuous feedback on the progress of ongoing software development or maintenance projects.

To conclude, most of the approaches that tackled RTR problem using several techniques are time-consuming. The RTR problem as an optimization problem suppose, to some extent, that a traceability link is appropriate when it optimizes the fitness function (FF). However, RTR is context-sensitive that requires some knowledge of the system to be validated. Indeed, the fact that the values of some measures composing the fitness function were improved after some iterations does not necessarily mean or ensure that all traceability links obtained in the best solution make sense. This observation is at the origin of the work described in this paper as described in the next section.

4   Interactive Genetic Algorithm Adapted to RTR Problem

This section introduces the basic concepts required to comprehend IGAs starting by the representation of the individual followed by the genetics operators, the evaluation function (fitness function), how we collected and integrated feedbacks from designer. Finally, we present a high-level pseudo-code of IGA adapted to the RTR problem.

4.1 Individual Representation

To apply IGA, we represented an individual (i.e., a candidate solution) as a list in which each element is a pair (R, A), where R is a requirement description selected from the set of requirements at hand and A is an artifact (e.g., class, method) selected from the software system under analysis.

We run IGA using the **rank selection** technique [23], **two-point crossover** technique that consists of randomly selecting two points from the parents I1 and I2, then everything between the two points is swapped between the parents, creating two children I1' and I2'. Given a selected individual, **mutation** operator randomly selects one or more elements in the list corresponding to the selected individual (solution). Then these elements will be replaced by elements (pairs (R, A)) chosen randomly from the base of examples.

4.2 Evaluation of Individual

To evaluate an individual, two factors have been defined:

Semantic Similarity To define the semantic similarity, we supposed that vocabulary used in the requirements' description is similar to the vocabulary used in the related artifacts (software elements). Thus, finding the trace between the artifacts and requirements is based on the calculus of the semantic similarity between the requirement text and the artifact text found in the names of classes, methods, fields, variables, parameters, types, etc. To do this, we used the well-known technique in the IR area which is the cosine similarity measure that aims to estimate the semantic similarity between documents (texts of requirements and artifacts) [24]. The documents are represented as vectors in a v-dimensional space where v is the number of the different terms in the vocabulary (all different terms appearing in the documents). The vector for each document is obtained by assigning a weight to each dimension (representing a specific term) of the term space. The weight w_{ij} corresponding to the term k_i and document d_j is computed using the Term Frequency – Inverse Document Frequency (TF-IDF) method known in the field of IR [24] as shown in Eq. (1). tf_{ij} is the term frequency factor. idf_i is the inverse document frequency factor. $freq_{ij}$ is the raw frequency of term k_i in document d_j. $max_l freq_{lj}$ is the maximum term frequency in document d_j. N is the total number of documents and n_i is the document frequency of k_i, i.e., the number of documents in which the term k_i appears, where $i = 1, 2, ..., v$ and $j = 1, 2, ..., N$.

$$w_{ij} = \text{tf}_{ij} \cdot idf_i = \left(\frac{freq_{ij}}{max_l freq_{lj}} \right) \cdot \log \left(\frac{N}{n_i} \right) \quad (1)$$

The similarity between a requirement R and an artifact A is computed by determining the cosine of the angle between the vector $\vec{R}$ representing the requirement

R and the vector $\vec{A}$ representing the artifact A, i.e., inner product normalized by the vector lengths, as shown in Eq. (2):

$$CosSim\left(\vec{R}, \vec{A}\right) = \frac{\vec{R} \cdot \vec{A}}{\left|\vec{R}\right| \cdot \left|\vec{A}\right|} \tag{2}$$

We called *SemSim*, the function that computes the semantic similarity for an individual I equal to the average of the semantic similarity value for the pair (R_i, A_i) over all pairs in the individual I as shown by Eq. (3), where m is the number of pairs in the individual (individual size).

$$SemSim(I) = \frac{1}{m} \sum_{i=1}^{m} CosSim\left(\vec{R}_i, \vec{A}_i\right) \tag{3}$$

Maintenance History We defined the maintenance history based on two measures:

Recency of change measure (RC). RC measure is computed based on information extracted from the history of change collected from the maintenance process of the software system. The intuition behind introducing the RC measure is that artifacts (classes, methods, etc.) that changed more recently than others are more likely to change now, i.e., are related to the new requirements at hand. The RC measure for a given artifact is obtained by looking to the time at which the artifact in the software system has undergone the last change. The RC measure for an artifact A is given by Eq. (4), where LTC stands for "Last Time of Change". The value of the fitness function RecOfChange that computes the recency of change for an individual I is computed as the average of the recency of change value for the artifact A_i over all artifacts in the individual I. This average is then normalized by the current time to ensure that the value is in the interval [0, 1]. This is shown by Eq. (5), where m is the number of artifacts, i.e., the number of pairs, in the individual (size of the individual).

$$RC(A) = LTC_A \tag{4}$$

$$Rec\ Of\ Change(I) = \frac{\frac{\sum_{i=1}^{m} RC(A_i)}{m}}{Current\ Time} \tag{5}$$

Frequency of change measure (FC) FC measure is computed based on information extracted from the history of change collected from the maintenance process of the software system. The intuition behind introducing the FC measure is that artifacts (classes, methods, etc.) that change more frequently than others are more likely to change now, i.e., are related to the new requirement at hand. The FC measure for a given artifact is obtained by looking to the number of times the artifact in the software system has undergone a change in the history of change. The FC measure for an artifact A is given by Eq. (6), where NC stands for "Number of Changes". The value of the fitness function FreqOfChange that computes the frequency of change for an individual I is computed as the average of the frequency of change value for the artifact A_i over all artifacts in the individual I. This average is then normalized by the maximum number of changes (over all artifacts in the individual) to ensure that the value is in the interval [0, 1]. This is shown by Eq. (7), where m is the number of artifacts, i.e., the number of pairs, in the individual (size of the individual).

$$FC(A) = NC_A \tag{6}$$

$$FreqOfChange(I) = \frac{\frac{\sum_{i=1}^{m} FC(A_i)}{m}}{max_l FC(A_l)} \tag{7}$$

In practice, the evaluation of an individual should be formalized as a mathematical function called "**fitness function (FF)**". In this work, we defined three fitness functions: (SemSim, RecOfChange and FreqOfChange). IGA is a mono-objective optimization technique. For this reason, we combine the three fitness functions within one fitness function. We decided for the Formula shown by Eq. (8) based on trial and error. Thus, we assigned 50, 25 and 25% as weight to SemSim, RecOfChange and FreqOfChange fitness functions respectively.

$$FF(I) = \frac{SemSim(I) + \frac{RecOfChange(I) + FreqOfChange(I)}{2}}{2} \tag{8}$$

4.3 Collecting and Integrating Feedbacks from Designers

RTR is a context-sensitive operation whose responsible is either the designer or the developer or both. In addition, depending on the semantic of the system under analysis and the requirement description, a traceability link between requirement and artifact system proposed by any optimization technique (e.g., classic GA, NSGA-II, etc.) can be considered as accurate or no sense by a designer and/or developer. Consequently, we use IGA to partly tackle this problem by interacting with designers and/or developer and getting their feedbacks on a number of the proposed traceability links. To do so, we adopted a binary scale to rate the proposed traceability link; i.e., we distinguish two types of rating that a designer can assign to a detected traceability link: (1) yes (value = 1): it is mandatory to consider the proposed traceability link; and (2) no (value = 0): the traceability link has no sense.

4.4 Interactive Genetic Algorithm Adaptation

The approach proposed in this paper exploits maintenance history, a heuristic search technique and the designer's feedback to automatically detect traceability links. Algorithm 1 shows the pseudo code of IGA adaptation to the RTR problem.

```
ALGORITHM 1:   High-level   pseudo-code   for   IGA   adapta-
tion to RTR problem
Input:    set_Of_Requirements;    set_Of_Artifacts;    mainte-
nance_History;
Percentage (P%); Max_Nbr_Of_Iterations; Nbr_Of_Interactions
1:   for i=1 ... NbrOfInteractions, do
2:      Evolve GA for NbrOfIterations
3:      Select P% of best solutions from the current population
4:   for each selected solution, do
5:        Ask the designer wether each traceability link within the
          selected solution makes
6:        Update the fitness function of the selected solution to
          integrate the feedback
7:      end
8:      Create a new population using the updates solutions
9:   end
10:  Continue GA evolution until it converges or it
     reaches Max_Nbr_Of_Iterations
```

5 Experiments

To evaluate the efficiency of our approach for the generation of traceability links using IGA algorithm, we aimed at answering two research questions: **RQ1**: To what extent can the IGA approach generate correct traceability links? **RQ2**: To what extent does our IGA approach outperforms a classic GA approach applied on the RTR problem?

5.1 Experimental Settings

After running our tool several times, we have set the IGA parameters as follow: Maximum number of iterations (stopping criterion) parameter to 10,000, Maximum size of the population parameter to 20, Crossover probability parameter to 0.9, and Mutation probability parameter to 0.1. The execution time for the generation of affected artifacts (10,000 iterations) was less than 10 min including feedback integration (10 interactions with the designer). This indicates that the approach is reasonably scalable. However, the execution time depends on the time spent by the designer to introduce his feedbacks, the number of requirements and the length of the text in requirements.

To validate our approach, we used two medium-sized open-source projects: (1) Albergate [14] which is a Hotel management system developed in Java and (2) eTour [25] which is a tour guide system. Two measures were used to answer our two research questions, which are Precision and Recall.

5.2 Results and Discussion

Figures 1 and 2 show respectively the obtained precision and recall values (y axis) for 31 executions (x axis). The average values for (Precision, Recall) are as follows: (95.5%, 95.37%) for Albergate, and (91.73%, 91%) for eTour. These results allow us to answer positively the first research question **RQ1**.

To answer the second research question **RQ2**, we compared our results to those obtained by Ghannem et al. [26] that proposed an RTR approach using classic GA based on the similarity of nouns between requirements and artifacts' names calculated by using Rita tool [27]. According to the results obtained and shown by Figs. 3 and 4, our approach outperforms Ghannem et al. [26] approach. IGA approach presented in this paper has a higher average for precision and recall than GA approach presented in Ghannem et al. [26] for the two projects under analysis. This will allow us to say that the impact of designer/developer is positive on the obtained results.

Fig. 1 Precision results for 31 executions of IGA approach

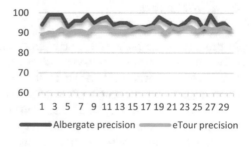

Fig. 2 Recall results for 31 executions of IGA approach

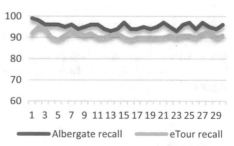

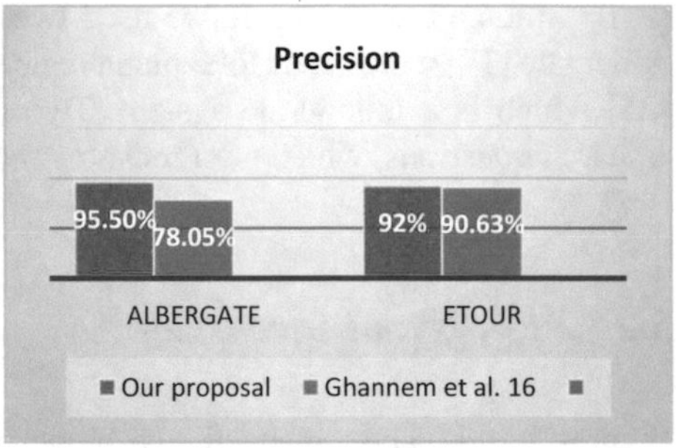

Fig. 3 Precision comparison between our proposal and Ghannem et al. [26]

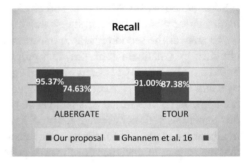

Fig. 4 Recall comparison between our proposal and Ghannem et al. [26]

6 Conclusion

In this paper, we presented a new approach aiming at identify traceability links between requirements documents and artifacts system. To do so, we adapted Interactive Genetic Algorithms (IGAs) to build an algorithm which integrates the designer's knowledge during the search process for requirements traceability links. We implemented the approach as a plugin integrated within Eclipse platform and we performed multiple executions of the approach on two medium-sized open source projects. The results of our experiment have shown that the approach is stable regarding its correctness, completeness. IGA has significantly reduced the number of meaningless traceability links in the optimal solutions for these executions. While the results of the approach are very promising in terms of precision and recall, we plan to extend it in two different ways: (1) expand our study to additional data sets (e. e., industrial systems) and (2) make further analyze the traceability links that are correctly suggested.

References

1. Ali N, Guéhéneuc YG, Antoniol G (2013) Trustrace: mining software repositories to improve the accuracy of requirement traceability links. IEEE Trans Software Eng 39(5):725–741

2. Hayes J et al (2007) REquirements TRacing on target (RETRO): improving software maintenance through traceability recovery. Innov Syst Softw Eng 3(3):193–202
3. Natt och Dag J et al (2005) A linguistic-engineering approach to large-scale requirements management. IEEE Softw 22(1):32–39
4. Lormans M, van Deursen A (2006) Can LSI help reconstructing requirements traceability in design and test? In: Proceedings of the 10th european conference on software maintenance and reengineering (CSMR'06)
5. De Lucia, A et al (2005) ADAMS Re-Trace: a traceability recovery tool. In: 9th European conference on software maintenance and reengineering (CSMR'05)
6. Harman M, Mansouri SA, Zhang Y (2012) Search-based software engineering: trends, techniques and applications. ACM Comput Surv 45(1):1–61
7. Andersson J et al (2009) Modeling dimensions of self-adaptive software systems. In: Cheng BC et al (eds) Software engineering for self-adaptive systems. Springer, Berlin, Heidelberg, pp 27–47
8. Goldberg ED (1989) In: Genetic algorithms in search optimization and machine learning. Addison-Wesley Longman Publishing Co., Inc. 372
9. Fukumoto M, Hatanaka T (2016) Parallel distributed interactive genetic algorithm for composing music melody suited to multiple users' feelings. In: 2016 IEEE/ACIS 15th international conference on computer and information science (ICIS)
10. Manfré A et al (2016) Exploiting interactive genetic algorithms for creative humanoid dancing. Biolog Inspired Cognitive Architect 17:12–21
11. Yuniarti A, Anggara S, Amaliah B (2016) Resize my image: a mobile app for interactive image resizing using multi operator and interactive genetic algorithm. In: 2016 International conference on information and communication technology and systems (ICTS)
12. Sorn D, Rimcharoen S (2013) Web page template design using interactive genetic algorithm. In: 2013 International computer science and engineering conference (ICSEC)
13. Sun X-Y et al (2010) Grid-based knowledge-guided interactive genetic algorithm and its application to curtain design. In: 2010 Second World congress on nature and biologically inspired computing (NaBIC)
14. Antoniol G et al (2002) Recovering traceability links between code and documentation. IEEE Trans Softw Eng (TSE'02) **28**(10): 970–983
15. Zhang Y et al (2008) Ontological approach for the semantic recovery of traceability links between software artefacts. IET Softw 2(3):185–203
16. Spanoudakis G et al (204) Rule-based generation of requirements traceability relations. J Syst Softw (JSS'04) 72(2):105–127
17. Evans MW (1989) In: The software factory, Wiley
18. Bouillon E, Mäder P, Philippow I (2013) A survey on usage scenarios for requirements traceability in practice. In: Doerr J, Opdahl A (eds) Requirements engineering: foundation for software quality. Springer, Berlin, Heidelberg, pp 158–173
19. Jackson J (1991) A keyphrase based traceability scheme. In: IEEE Colloquium on tools and techniques for maintaining traceability during design
20. Gotel OCZ, Finkelstein ACW (1994) An analysis of the requirements traceability problem. In: Proceedings of the first international conference on requirements engineering (RE94)
21. Grechanik M, McKinley KS, Perry DE (2007) Recovering and using use-case-diagram-to-source-code traceability links. In: Proceedings of the the 6th joint meeting of the European software engineering conference and the ACM SIGSOFT symposium on The foundations of software engineering. ACM: Dubrovnik, Croatia pp 95–104
22. Xiaofan C, Grundy J (2011) Improving automated documentation to code traceability by combining retrieval techniques. In: 26th IEEE/ACM International conference on automated software engineering (ASE'11)
23. Baker JE (1985) Adaptive selection methods for genetic algorithms. In: Proceedings of the 1st international conference on genetic algorithms. L. Erlbaum Associates Inc. pp 101–111
24. Baeza-Yates RA, Ribeiro-Neto B (1999) In: Modern information retrieval. Addison-Wesley Longman Publishing Co., Inc. pp 513

25. Challenge T (2011) In: Proceedings of the 6th international workshop on traceability in emerging forms of software engineering. Waikiki, Honolulu, HI, USA, ACM
26. Ghannem A et al (2018) Search-based requirements traceability recovery. Springer International Publishing, Cham
27. Howe CD (2009) RiTa: creativity support for computational literature. In: Proceedings of the 7th ACM conference on Creativity and cognition. ACM: Berkeley, CA, USA. pp 205–210

On Log Analysis and Stack Trace Use to Improve Program Slicing

Vincent Bushong, Jacob Curtis, Russell Sanders, Mark Du, Tomas Cerny, Karel Frajtak, Pavel Tisnovsky, and Dongwan Shin

Abstract Program slicing is a common technique to help reconstruct the path of execution a program has taken. It is beneficial for assisting developers in debugging their programs, but its usefulness depends on the slice accuracy that can be achieved, which is limited by the sources of information used in building the slice. In this paper, we demonstrate that two sources of information, namely program logs, and stack traces, previously used in isolation to build program slices, can be combined to build a program slicer capable of handling more scenarios than either method individually. We also demonstrate a sample application of our proposed slicing approach by showing how our slicer can deduce integer inputs that will recreate the detected error's execution path.

V. Bushong · J. Curtis · R. Sanders · M. Du · T. Cerny (✉)
Baylor University, Waco, TX 76706, USA
e-mail: tomas_cerny@baylor.edu

V. Bushong
e-mail: vincent_bushong1@baylor.edu

J. Curtis
e-mail: jacob_curtis1@baylor.edu

R. Sanders
e-mail: Russell_Sanders1@baylor.edu

M. Du
e-mail: Mark_Du1@baylor.edu

K. Frajtak
Computer Science, FEE, Czech Technical University, Karlovo Nam. 13, 121 35 Prague, Czech Republic
e-mail: frajtak@fel.cvut.cz

P. Tisnovsky
Red Hat, Brno, Czech Republic
e-mail: ptisnovs@redhat.com

D. Shin
Computer Science, New Mexico Tech, Socorro, NM 87801, USA
e-mail: dongwan.shin@nmt.edu

Keywords Code analysis · Log analysis · Program slicing

1 Introduction

Finding the root cause of an error is one of the most time consuming and tedious tasks a developer must face. As a result, many methods have been explored to improve the debugging process. One such method depends on program slicing. Program slicing is the process of narrowing down a program's source code to an increasingly-narrow selection. When applied to the problem of debugging, the goal of program slicing is to narrow the code down to the section of code that was actually executed during the erroneous run of the program. This reduces the amount of code that must be analyzed to find the error source, either through manual analysis or through other automated techniques.

While program slices can be constructed using the source code alone, in order to slice a program based on an actual given execution of the program, a source of additional information about the runtime data is needed. One such source of information is the stack trace a program produces when an unhandled runtime exception occurs. The stack trace provides a definite endpoint for the program slice and a cross-section of the functions that were currently executing, so a method based on stack trace analysis already has a good start for building its slice. Another source of data that can be utilized for program slicing is the application log. Logging statements are commonly used in code to provide information about the program's execution by printing errors or other interesting events in a place where developers can read and interpret the messages. Log messages can also be analyzed as part of an automated process, and, if the analyzer has access to the source code that generated them, the log messages can serve as checkpoints in the code. Using the presence of log messages compared to the log statements in the code, a program slicer can more accurately identify which paths were possible for a given execution. Developers already use this technique manually, not knowing it is program slicing; they read log messages and search for its location and meaning in the code to narrow down the problem's location.

Program logs and stack traces have both been used as a source of data to improve the accuracy of program slicing [1, 2, 6, 6, 10, 11, 14, 15, 15, 16]. However, no existing work has combined the two techniques to use simultaneously. In order to narrow down a program slice as much as possible, every available source that can provide information about the execution should be used. We propose a method that uses both the stack trace of an exception and any log messages that have been generated up until the point the exception occurred to create a program slice more accurately than using either data source in isolation.

To demonstrate our method and a potential application for the resulting program slice, we describe an implementation for the Go language. It creates the program slice for a runtime exception and uses it in conjunction with a Satisfiability Modulo Theories (SMT) solver to suggest program inputs that produce the same path through the program as the exception, thus reproducing the error. In Sect. 2, we give our motivation. Section 3 reviews prior work. Sections 4 and 5 explain our proposed method and sample implementation, respectively. We show an example usage in Sect. 5 and offer conclusions in Sect. 6.

2 Motivation

While usually unstructured and designed for human consumption, program logs represent a rich source of information that can be tapped to learn about an application and its operation. When combined with static source code analysis, log analysis techniques can be used to extract insights about a program's performance using log timing information, the presence of anomalies, and error localization within a distributed system. One particular application of log analysis is improving the accuracy of program slicing. If the locations of log statements in the code are matched with logs collected from the application, the program slice representing a particular execution of the application can be more accurately narrowed down.

Other methods have been used to increase the accuracy of program slicing. One method is to use the information contained in stack traces. If a runtime exception occurs and no exception handler is found for it, the stack trace is usually routed to the logging mechanism to prevent the entire program from crashing. If this stack trace is available post-mortem to an analyzer, its contents can be used to narrow down the slice further, based on the functions it contains. This method has been shown to improve the accuracy of a program slice, lending itself to applications such as improving fault localization.

Both log and stack trace analysis can improve program slice accuracy. However, the two techniques have not yet been combined. Since uncaught exceptions are usually printed to the log, having access to stack traces usually means having access to the program logs as well. If the goal is to build as accurate of a program slice as possible without introducing more overhead to the program itself, every possible source of information should be utilized; if access to the program log is available, then there is no need to depend solely on the stack trace to build the program slice, and vice versa. Our goal is to show that these two techniques can be used in conjunction to build a program slice, and the resulting program slice can be used as the basis for further analysis, in this case, providing the user with program inputs that will recreate the execution path that caused the error.

3 Related Work

Automated log analysis is the process of extracting some insights from a program's logs, and it has been used for detecting performance anomalies, fault localization, and identifying security anomalies [3]. A subset of log analysis methods consists of those that map logs to the corresponding location in source code. The general approach is to traverse an Abstract Syntax Tree (AST), creating regular expressions from the logging statements in the source code to match against gathered logs once the program has run. These regular expressions are associated with their location in the code, and when matched with seen logs, give the location in source code, providing partial observation of program execution. This is a common approach [1, 2, 4, 5, 7–9, 13, 14, 16], with some variations. For example, in [2], the regular expressions augment to denote that the log occurs inside a loop or conditional branch. In [9], the regular expressions are associated with the class or method they are used in, not the source line itself.

Program slicing is a technique to narrow down a code base to a particular subset, usually a subset that can affect a particular statement, performed by creating a reachability graph between the program's statements [12]. The variant of program slicing we consider is dynamic program slicing. In this method, a program slice is constructed regarding a single execution of the program, resulting in the subset of the program's code that could have affected a particular line of code in a particular execution [12]. When applied to a part of a program known to have failed, this technique helps in debugging and fault localization [12].

Many works [1, 2, 6, 10, 11, 14–16] utilize log analysis to assist with program slicing. The approach we adopt is inspired by Chen et al. [2]. Statements in the code are assigned labels to determine which lines were executed. The labels used in this study are: "may", "must-have", or "must-not", and are assigned on the basis of whether a certain log appeared during execution. The approach accounts for conditional execution and surrounding code having related execution status. If a log appears in a conditional branch, the entire block of statements in that branch is labeled "must-have"; any code in the opposing branch is consequentially labeled as "must-not" have been executed. Similarly, logs that appear in code but not in the output must not have executed and labeled as such. If no logs are present in either branch, it cannot be known which branch was executed without further analysis and is labeled as "may" have executed. A similar approach is taken by [15], with different names for labels being used. The limitation of this method is that the information needed to construct the slice is entirely dependent on the frequency/location of matchable logs within the source code.

Another method used for program slicing is to analyze stack traces to generate method-level execution paths. This is fairly straightforward, as it is known from a complete stack trace a majority of the functions currently executing at the point it is printed. This information gives a good starting point for building a slice. The approach is used by several works that specifically use the information to help localize faults [6, 15]. The authors of [10] supplement this call graph generated from the stack trace with information mined from the internet to construct a complete call-graph that could be used to show more program execution information. The drawback of this approach is that the inner workings of each function's execution are not known.

While these two methods do work in isolation, our goal is to show these two sources of information can be used to enhance each others' coverage, more accurately revealing which paths through a program could have been executed. With the stack trace analysis, we gain an overview of program execution that is then supplemented by the more detailed execution trace given by the log analysis. This way, we have a detailed execution path to analyze.

4 Method

Utilizing printed logs matching statements in the source code will let us slice down the program into the paths that must have or may have executed, further allowing us to perform value analysis on the variables along the path. To meet our goal of extracting the executed path and its conditions in a program, our approach takes to input the stack trace produced by an unhandled error, the logs produced from the program before the error occurred, and the source code that produced the error and logs.

The provided stack trace is parsed to find the functions on the stack when the error occurred. This parsed information also provides the root level function for constructing the control-flow graph (CFG), which represents the different execution paths for a given function. This root function has its CFG constructed and is then expanded. To expand the CFG, we find all of the function calls inside of some function and then replace each one with the code inside of its function declaration, meaning we are in-lining the function's contained statements. We also map the arguments passed into the function call to the parameters from the function declaration. This allows us to determine where parameters used in the in-lined statements get their values from. Having the CFG expanded allows us to backtrack from the point of the exception and slice the program based on printed logs matching log templates along the paths we find.

During the backtracking phase of our method, we use the stack trace in conjunction with the provided logs to slice the program's CFG into the paths that must have, may have, or must not have executed. Our log templates are regular expressions created from a log statement in the program. We account for static and dynamic information in the log so that any message produced from that log statement will match the regular expression. Any time a printed log matches a log template from the CFG, that block is labeled as must have executed. Once a block is labeled as must have executed, any other alternative branches (i.e., conditional control flow) that do not have a log statement matching a printed log are then labeled as must not have executed. Any time we encounter a conditional control flow in the CFG, if the preceding block is labeled as must-have executed, both outgoing branches are labeled as may have executed. May have labels can be overwritten to must-have labels if a printed log matching a log template is encountered.

For the sliced execution path, we also extract the conditions for each possible branch combination within the sliced execution path. This includes expressions within a conditional statement and assignments to variables since these are relevant to the SMT solver for determining possible input values. These conditions are recorded during the backtracking phase when each path is being traversed.

Once we have the execution path along with its conditions that need to be analyzed, we use an SMT solver to perform value analysis on the variables found in the path conditions. The only variables reported to the user calling our method are those that are user input (in our method using the heuristic that any variable never assigned or is a program argument is user input). By doing this, we can report to the user the variable values necessary to recreate the erroneous path of execution, which allows us to create a more accurate program slice overall.

5   Implementation

Here we describe an implementation of our method for the Go programming language to demonstrate the effectiveness of our approach.

Stack Trace Parsing The first step in our process is to analyze the stack trace written to standard output when an exception is unhandled. This stack trace contains the functions that were currently on the stack when the exception occurred, giving a definite list of functions that were at least partially executed and a definite end point for the path (the line where the exception was thrown). Additionally, the stack trace provides the file names that the functions are located in; since the Go language does not compile to an intermediate representation (such as Java's bytecode), the source files themselves are necessary to perform the analysis, as the exact execution path through the code is not recoverable from the executable.

In Go stack traces, lines can contain either a file path and line number or a function call. We used these two types of lines to detect all of the functions that were on the stack and their corresponding file names and line numbers. We then stored the function information into a structure for later use in constructing the CFG and labeling. Once the functions from the stack trace and point of original error are found, we move on to constructing the CFG.

CFG Construction Our approach to creating a CFG for some input program is to create a wrapper type around the existing CFG package in the Go standard library. We create wrapper types around the native Go CFG because it only parses on a per-function basis, and our analysis requires a CFG that includes all function calls connected together so we can view an entire execution trace. We created a wrapper interface with two concrete implementations. The function wrapper is the entry point of a function, representing the function signature along with its entry block. The block wrapper wraps the Go CFG package's block representation and has parent and child connections to other wrapper blocks, mimicking the structure of the Go CFG so it can be traversed in a similar way, while also allowing us to store extra information.

We construct the wrapped CFG by first building the Go CFG for the function, then recursively traversing its blocks and creating a block wrapper for each. Every time we create a block wrapper, we replicate the wrapped block's children with a wrapper substitute, allowing us to keep the sequence of the original blocks within our wrappers. The parent relation is also set to the current wrapper on the successor wrapper. As a result, we have a series of blocks where we can traverse through both its ancestors and descendants.

To connect the CFG across functions, we replace function calls with parsed function itself. Beginning with the function wrapper at the root of the stack trace, we search the chain of block wrappers for function calls. Upon finding a call expression, we split the current block in two halves, and search the code for the function declaration matching the call expression. We then construct the called function's function wrapper, passing in the list of arguments found via the call expression in order to map the arguments to the function parameters. The new function is given the first half of the original block as its parent and the second half of the original block as its child. After repeating this for all function calls, a connected CFG for the given stack trace has been created.

Log Matching The next step is to find log statements within the source code and construct regular expressions to match them. We extract log statements by examining AST nodes and looking for call expressions that match the log statements' format. In our implementation, we currently detect log calls from the Zerolog library[1] as well as the standard Go logging library and generate a regular expression for the log message that matches both the static part of the message and the dynamic portions. The dynamic portions are variable values that are interspersed in a log message. For example, the dynamic portion in a string "%d is an odd number" would be %d, and the regex generated would include a '\d' to match that portion.

Our log-matching functionality is used later during the labelling phase. Given a set of observed log messages, we attempt to match the messages to each log statement's regular expression. If we identify a match between a log message and a regular expression, we can identify that log statement and the branch it is in as having executed. This gives us better filtering of execution paths that may or may not have occurred.

Path Labeling The next step is to attach labels to each of our wrappers specifying whether the block of code it represents must have or may have executed. This is accomplished using both the stack trace and log matching. Our approach labels the tree bottom-up, starting with the block containing the line in the source that caused an exception. From here, as long as there is only one parent block, we label that parent as a must because there is no deviation in execution. A block with two parents indicates its parents are part of a conditional statement, which may be arbitrarily nested. We navigate up the tree until all conditional branches have converged, reaching the topmost condition. We then traverse down each branch, labeling every branch as may-have, and stopping when the original block with two parents, mentioned above, is reached. We then resume labeling parents starting with the block above all of the branches just analyzed. If the log matching described above finds a log statement that matches a log seen in the particular execution, its branch is changed to must-have, and any opposing branches are treated as not having been executed.

The next problem we needed to address is a limitation of the Z3 solver. The Z3 solver analyzes assertions as one unit, which means order is ignored, so con-

---

[1] https://github.com/rs/zerolog.

ditions such as in this set of statements, "`if (x < 0) x = 1; if (x > 0) return`", will contradict. The solver tries to solve a system of equations where `x < 0, x = 1, x > 0`, which is unsolvable. Our solution is to convert the variables in the conditions to a single static assignment form (SSA). SSA treats each reassignment of a variable as if it were a new variable, giving it a unique name. This is necessary because every new value needs to be considered in the conditions. We accomplish this by traversing the tree again and adding numeric values to the beginning of identifiers, incrementing the value for every instance of reassignment seen for a particular variable. Our solution would transform the statements to look like this "`if (x < 0) 1x = 1; if (1x > 0) return`". This allows the solver to assert that `x < 0, 1x = 1`, and `1x > 0`, which is solvable.

SMT Analysis After the renaming, we extract the conditions along the execution paths identified by the labeling stage. Following the path of must-have blocks, we add all conditions and assignments of variables to a list, and when necessary, create a new path when a branch is reached to construct every possible route through the code. Any conditions in branches that oppose the branch taken (e.g., an "if" condition that was false when an "else if" condition was true) are negated to provide the most correct set of constraints for the path.

The extracted conditionals (Go AST expressions) along the execution path from the backtracking stage are each passed into a function that recursively converts them into a format that Z3 can recognize. After all of these are converted, they are used as assertions with Z3 to build a model.

If the Z3 solver was able to solve the constraints, we filter the returned value assignments that satisfy the conditions down to only those variables presumed to be user input. We used a heuristic to define user input as any variable that was never given a value assignment in the path conditions. If the argument variable was user input, then the parameter is treated as user input, and vice versa. Command-line arguments were also recognized as user input despite there being an assignment to the variable in the path conditions.

To demonstrate the effectiveness of our method, we show its results of operating on code samples. First example given in Listingexample1. The listing shows the logs output during the particular execution that would lead to the panic statement on line 24, as well as the simple stack trace generated by the panic. The conditions that are collected include the ones that branch to the locations of collected logs in the source code, and the negations of ones that lead to logs that were *not* collected. The following expressions are given to the Z3 Solver after SSA reassignment has been performed: `1main.num == main.num * 2, !(1main.num > 4), main.x == 1main.num * 4, main.x < 8, 2main.num == 1main.num - 2, !(2main.num < 0)`, and `2main.num < 10`. After these statements are asserted with the Z3 Solver, the variables are filtered to show only variables that are described as user input, as discussed earlier. The final output shows a possible value for the variable 'num' is 1, thus, the expected input given for this example.

Discussion We present two primary contributions. First, combining two previous approaches to program slicing creates a more robust system. The success of any

Listing 25.1 Example code

```
1   package ifelse
2
3   func main() {
4       num, _ := strconv.Atoi(os.Args[1])
5       num *= 2
6       if num > 4 {
7           log.Log().Msgf("%d > 4", num)
8       } else {
9           log.Log().Msgf("%d <= 4", num)
10      }
11      x := num * 4
12      if x < 9 {
13          log.Log().Msgf("%d < 9", num)
14      }
15      num -= 2
16      if num < 0 {
17          log.Log().Msgf("%d is negative", num)
18      } else if num < 10 {
19          log.Warn().Msgf("%d has 1 digit", num)
20          panic(errors.New("has 1 digit"))
21      } else {
22          log.Log().Msgf("%d has multiple digits", num)
23      }
24  }
25
26  Stack Trace:
27  ifelse.main()
28   /workspaces/test/test.go:32
29
30  Log Output:
31  2 <= 4
32  8 < 9
33  0 has 1 digit
34
35  Project root: /workspaces/test/
```

technique that depends on program slicing depends greatly on the accuracy of the slice that can be achieved. Using multiple sources of information increases the likelihood that the slice can be further narrowed down, as each source of information by itself may not contain the full picture. While they will not be applicable in every situation, the program logs and stack trace can be mutually beneficial to each other. Further information sources, such as static configuration files or code from third-party libraries, could be brought into the process to increase the number of scenarios the program slicer can adequately address.

Second, we demonstrate a sample application using our program slicing method. By representing the program slice's path as a series of conditionals, we were able to obtain a list of constraints that can be programmatically analyzed, in this case, by using an SMT solver. In this case, the output was relatively simple, being input values that could recreate the path taken through the program; there is room for more complex solutions to be built upon this concept, such as recommending ranges of values, as well as considering non-integer values.

Threats to Validity As a proof-of-concept prototype work, an in-depth case study has yet to be done to address several threats to the validity of our work. Our prototype's performance on large codebases has yet to be determined, though optimizations may address performance issues to the implementation, rather than the method itself. A case study or series of studies is also necessary to determine how many scenarios the combination of program logs and stack traces affects in the broader corpus of code in the industry.

There a limitation of a CFG in describing modern program execution. While analyzing conditionals as the primary program flow indicator is a good baseline, it leaves out approaches used in emerging software trends, e.g., aspect-oriented programming, interceptors, and distributed programs, all of which defy description by a traditional CFG representation of a piece of code. The new potential execution paths these alternative methods present will have to be analyzed using new tools and techniques; if they could be analyzed and merged into an intermediate representation with the rest of the code flow, common techniques (e.g., our labeling approach using logs and stack traces) could be used on the result.

6 Conclusion

In this paper, we have presented a novel approach to program slicing by combining two sources of information that have previously remained separate, program logs, and stack traces. We have shown that these proven approaches to program slicing are able to be used simultaneously to complement each others' coverage by sharing our implementation of the combined methods. We also presented a sample application of this novel technique in using an SMT solver and the paths of conditional constraints found in the program slice to assist the programmer in recreating the program's execution path.

Acknowledgements This material is based upon work supported by the National Science Foundation under Grant No. 1854049 and a grant from Red Hat Research https://research.redhat.com.

References

1. Bao L, Li Q, Lu P, Lu J, Ruan T, Zhang K (2018) Execution anomaly detection in large-scale systems through console log analysis. J Syst Software 143:172–186. https://doi.org/10.1016/j.jss.2018.05.016
2. Chen B, Song J, Xu P, Hu X, Jiang ZMJ (2018) An automated approach to estimating code coverage measures via execution logs. In: 2018 33rd IEEE/ACM international conference on automated software engineering (ASE), pp 305–316. https://doi.org/10.1145/3238147.3238214
3. El-Masri D, Petrillo F, Guéhéneuc YG, Hamou-Lhadj A, Bouziane A (2020) A systematic literature review on automated log abstraction techniques. Inform Software Technol 122:106276. https://doi.org/10.1016/j.infsof.2020.106276
4. Fu Q, Lou J, Wang Y, Li J (2009) Execution anomaly detection in distributed systems through unstructured log analysis. In: 2009 Ninth IEEE international conference on data mining, pp 149–158. https://doi.org/10.1109/ICDM.2009.60, ISSN: 2374-8486
5. Ghanbari S, Hashemi AB, Amza C (2014) Stage-aware anomaly detection through tracking log points. In: Proceedings of the 15th international middleware conference, pp 253–264.

Middleware '14, Association for Computing Machinery, New York, NY, USA. https://doi.org/10.1145/2663165.2663319

6. Jiang S, Zhang H, Wang Q, Zhang Y (2010) A debugging approach for java runtime exceptions based on program slicing and stack traces. In: 2010 10th international conference on quality software, pp 393–398. https://doi.org/10.1109/QSIC.2010.23, ISSN: 2332-662X

7. Pecchia A, Cinque M, Carrozza G, Cotroneo D (2015) Industry practices and event logging: assessment of a critical software development process. In: Proceedings of the 37th international conference on software engineering, vol 2, pp 169–178. ICSE '15, IEEE Press. https://doi.org/10.5555/2819009.2819035

8. Schipper D, Aniche M, van Deursen A (2019) Tracing back log data to its log statement: from research to practice. In: Proceedings of the 16th international conference on mining software repositories, pp 545–549. MSR '19, IEEE Press. https://doi.org/10.1109/MSR.2019.00081

9. Shang W (2012) Bridging the divide between software developers and operators using logs. In: 2012 34th international conference on software engineering (ICSE), pp 1583–1586. https://doi.org/10.5555/2337223.2337490

10. Sui L, Dietrich J, Tahir A (2017) On the use of mined stack traces to improve the soundness of statically constructed call graphs. In: 2017 24th Asia-Pacific software engineering conference (APSEC), pp 672–676. https://ieeexplore.ieee.org/document/8306000

11. Sun C, Ran Y, Zheng C, Liu H, Towey D, Zhang X (2018) Fault localisation for ws-bpel programs based on predicate switching and program slicing. J Syst Software 135:191–204. https://doi.org/10.1016/j.jss.2017.10.030

12. Xu B, Qian J, Zhang X, Wu Z, Chen L (2005) A brief survey of program slicing. SIGSOFT Softw Eng Notes 30(2):1–36. https://doi.org/10.1145/1050849.1050865

13. Xu W, Huang L, Fox A, Patterson D, Jordan MI (2009) Detecting large-scale system problems by mining console logs. In: Proceedings of the ACM SIGOPS 22nd symposium on operating systems principles, pp 117–132. SOSP '09, Association for Computing Machinery, New York, NY, USA. https://doi.org/10.1145/1629575.1629587

14. Yuan D, Mai H, Xiong W, Tan L, Zhou Y, Pasupathy S (2010) Sherlog: error diagnosis by connecting clues from run-time logs. In: Proceedings of the fifteenth international conference on architectural support for programming languages and operating systems, pp 143–154. ASPLOS XV, Association for Computing Machinery, New York, NY, USA. https://doi.org/10.1145/1736020.1736038

15. Zhang H, Jiang S, Jin R (2011) An improved static program slicing algorithm using stack trace. In: 2011 IEEE 2nd international conference on software engineering and service science, pp 563–567. https://doi.org/10.1109/ICSESS.2011.5982378, ISSN: 2327-0594

16. Zhao X, Zhang Y, Lion D, Ullah MF, Luo Y, Yuan D, Stumm M (2014) LPROF: a non-intrusive request flow profiler for distributed systems. In: Proceedings of the 11th USENIX conference on operating systems design and implementation, pp 629–644. OSDI'14, USENIX Association, USA. https://doi.org/10.5555/2685048.2685099

Data-Driven Similarity Measures for Matrimonial Application

Yong-Fang Chia, Chian-Wen Too, and Kok-Chin Khor

Abstract Marriage is a life-long commitment that completes our life. However, the marriage rate in Malaysia continues to decline due to the difficulties in seeking a suitable partner. A web-based matrimonial application was, therefore, developed to help people finding their potential partner for marriage according to their preferences. The application used similarity measures to find matches that suited user preferences and to overcome the limitations of the rule-based approach and SQL syntax. We evaluated five similarity measures, namely, Jaccard Coefficient, Cosine Similarity, Euclidean Distance, Manhattan Distance, and Minkowski Distance, through remote usability testing with eight users. We found that using Manhattan Distance returned the most satisfying search results to them.

Keywords Similarity measures · Web-based matrimonial · Usability testing

1 Introduction

Marriage is one of the most important events in one's life. It is a bond that unites two souls together into one. Traditional offline dating is very time-consuming and based on spatial proximity. Nowadays, nearly 60% of the seven billion people in the world can access the Internet [1]. With the Internet, people can interact with anyone from different corners of the globe almost instantaneously. Hence, a new phenomenon has emerged, where people are becoming more interested in finding their life partner online. According to the study by [2], the traditional offline ways of meeting partners in U.S. have all been declining sharply since the year 2000.

Y.-F. Chia · C.-W. Too (✉) · K.-C. Khor
Lee Kong Chian Faculty of Engineering Science, Universiti Tunku Abdul Rahman, Bandar Sg. Long, 43000 Kajang, Malaysia
e-mail: toocw@utar.edu.my

Y.-F. Chia
e-mail: chia5484@1utar.my

K.-C. Khor
e-mail: kckhor@utar.edu.my

Late marriage is a recent trend in Malaysia, particularly in the non-Muslim community [3]. The women are facing difficulties in finding suitable partners, according to the Malaysia Population Research Hub [4]. It is because good opportunities for education and employment had led women to postpone marriage [5]. As women are getting better qualification for jobs and more economically independent than before, they tend to have higher expectations for choosing their marriage partner. Besides, it is not easy to find a partner that meet their expectations as the circle of suitable candidates may become smaller when they get older.

Currently, there are a lot of matrimonial applications exist in the market. Some of the existing matrimonial applications implement a rule-based approach or Standard Query Language (SQL) as the matching method. However, there are some limitations to the rule-based approach and SQL. The rule-based approach uses a series of rules that are usually expressed as "if–then" clauses to derive actions [6]. Using the rule-based approach may cause a combinatorial explosion as the classification of data often contains a considerable number of rules [7]. Generating all the rules and conditions for a complex system is sometimes difficult and time-consuming. On the other hand, almost all applications that work with databases manage relational data through SQL. Usually, SQL is excellent at finding exact matches. However, it may return zero results when no record in the database meets the SQL statements.

In this paper, we implemented five similarity measures for our newly developed web-based matrimonial application. The application performed matching through similarity measures based on the preferences set by users. The similarity measures were used to overcome the limitations of the rule-based approach and SQL. There are five similarity measures used, namely Jaccard Coefficient, Cosine Similarity, Euclidean Distance, Manhattan Distance and Minkowski Distance. We then performed an empirical evaluation to determine which similarity measure is useful in finding matches that suit user preference.

We organised the paper as follows. Section Two provides a review of the similarity measures and usability testing. In Section Three, we present the methodology that was used to determine the best similarity measure. Section Four discusses the analysis of results obtained from the usability testing and the evaluation of the similarity measures. We conclude the paper and discusses the limitations in Section Five.

2 Literature Review

2.1 Existing Applications

According to the study by [8], there are several matching mechanisms adopted by the online dating systems in the market, such as search/sort/match, personality-matching systems, and social network. Search/sort/match is the most common mechanism adopted by many online dating systems. Sites like Match.com allow users to set their match criteria; the system will find the user profiles that exactly match the criteria set by the users. eHarmony.com, on the hand, implements the personality-matching mechanism. Personality tests are provided to the users, and the system will match the users based on the compatibility of their personality features. There are also sites like Facebook Dating adopting social network mechanism. It has a function that enables the users to pair with people from events or groups they have joined via Facebook.

2.2 Similarity Measures

The similarity measure is the measure of the relation between a pair of data objects [9]. It determines how much identical two data objects are. Two objects that have a high degree of similarity are usually closed to each other in the distance. We shall discuss five popular similarity measures the following subsections.

Jaccard Coefficient. Jaccard Coefficient measures the similarity of two sets of data. To measure the similarity between two data sets through Jaccard Coefficient, the division between the size of the intersection ($\cap$) and the size of the union ($\cup$) of two data sets (A and B) is calculated. The Jaccard coefficient takes a value between [0,1] with 1 indicating the two data sets are entirely similar and 0 indicating otherwise. The mathematical representation of Jaccard Coefficient is as follows.

$$J(A, B) = \frac{|A \cap B|}{A \cup B} = \frac{|A \cap B|}{|A| \cup |B| - |A \cap B|} \tag{1}$$

Cosine Similarity. Cosine similarity measures the normalised dot product of the two attributes by finding the cosine angle between the two vectors. The outcome of the cosine similarity is between [0,1]. If the angle between two vectors is 0°, the two vectors will have a similarity value of 1. On the other hand, the two vectors at 90° have the similarity value of 0, independent of the magnitude. The larger the angle between two vectors, the smaller their similarity. The mathematical representation of Cosine Similarity is as follows.

$$\text{Similarity } (A, B) = \cos(\theta)$$
$$= \frac{A \cdot B}{\|A\| \times \|B\|} = \frac{\sum_{i=1}^{d} A_i \times B_i}{\sqrt{\sum_{i=1}^{d} A_i^2} \sqrt{\sum_{i=1}^{d} B_i^2}} \tag{2}$$

Euclidean Distance. Euclidean Distance is the widely used distance function in many applications. The Euclidean distance measures the straight-line distance between two points by following the Pythagorean rule. The result of Euclidean Distance is usually greater than or equal to zero, where zero indicates that two points are identical. The higher the value, the lesser the similarity. The mathematical representation of Euclidean Distance is as follows.

$$d(A, B) = \sqrt{\sum_{i=1}^{d} (A_i - B_i)^2} \qquad (3)$$

Manhattan Distance. Manhattan Distance is a measure to obtain the difference between two points along axes at right angles. It calculates the absolute sum of the difference between their Cartesian coordinates which are the x-coordinates and y-coordinates. The result of Manhattan Distance is similar to Euclidean Distance, where zero indicates that two points are identical. The higher the value, the lesser the similarity. The mathematical representation of Manhattan Distance is as follows.

$$d(A, B) = \sum_{i=1}^{d} |A_i - B_i| \qquad (4)$$

Minkowski Distance. Minkowski Distance is a generalisation of the Euclidean and Manhattan distances. It is a similarity measurement between two points in the normed vector space. The order of Minkowski metric, λ can be manipulated to calculate the distance in three different ways. When λ is equal to one, it is Manhattan Distance. When λ is two, it is then Euclidean Distance. When λ is ∞, it is Chebyshev Distance. The mathematical representation of Minkowski Distance is as follows.

$$d(A, B) = \sqrt[\lambda]{\sum_{i=1}^{d} |A_i - B_i|^{\lambda}} \qquad (5)$$

2.3 Usability Testing

Usability testing is the evaluation of an application by testing it with real users. It measures how easy an application is to use and how easy an application is for users to achieve their usability goal. During a usability test, participants are required to complete typical tasks. Then, their performance will be observed and recorded for evaluation purpose. Usability testing can be carried out remotely even though the researcher and the participants located in different geographical locations. Therefore, it provides an opportunity to connect with participants from various geographical regions. It also helps to save time and cost, particularly the cost of hiring a professional usability lab.

The System Usability Scale (SUS) can be used to evaluate the usability of a system [11]. It consists of ten questions with five response choices. Among the ten questions, odd-numbered items worded positively, whereas even-numbered items worded negatively. Participants will rate each question from 1 to 5, which ranges

Table 1 The SUS score grading scale

SUS score range	Grade	Percentile range
84.1–100	A+	96–100
80.8–84	A	90–95
78.9–80.7	A−	85–89
77.2–78.8	B+	80–84
74.1–77.1	B	70–79
72.6–74	B−	65–69
71.1–72.5	C+	60–64
65–71	C	41–59
62.7–64.9	C−	35–40
51.7–62.6	D	15–34
0–51.7	F	0–14

Adopted from the study by [10]

from strongly agree to strongly disagree, based on their degree of agreement [12]. The SUS score will be range from 0 to 100, while 68 is the average SUS score. If the score is below 68, it indicates that there would be some problems with the system usability. SUS scores can be translated into letter grades, as shown in Table 1.

3 Methodology

3.1 Data Preparation and Pre-processing

Figure 1 shows the methodology we used in this study to determine the best similarity measure for our newly developed matrimonial application. We collected data from 40 males and 45 females through questionnaires for creating their member profiles in our application. They provided their personal information, such as demographic info, background and lifestyle. Before storing their data into the application, we transformed the data from strings to numerical values. For example, "Male" in

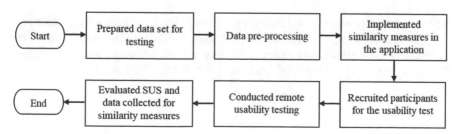

Fig. 1 The methodology to determine the best similarity measure

the gender field was represented by 1, while "Female" was represented by 2. Such transformation was necessary so that we can apply similarity measures to the data later.

3.2 Similarity Measures Implementation

There were 12 search preferences implemented in the application. They were gender, age range, height range, living state, marital status, posture, religion, mother tongue, education level, the field of study, smoking habit, and desire for a child.

We evaluated five similarity measures on the collected data, namely, Jaccard Coefficient, Cosine Similarity, Euclidean Distance, Manhattan Distance and Minkowski Distance. When the users set their preferences, their preferences will be converted into index numbers. However, the large-scaled preferences may dominate the similarity measures. Thus, the index numbers were normalised to values range between 0 and 1. After that, the similarity of the member data and the preferences set by users were measured.

3.3 Remote Usability Testing

The targets members for the matrimonial application are the individuals who are single with age between 22 and 26. Thus, we enrolled eight participants (six males and two females) who fulfilled the age requirements for the usability test. A usability testing on the matrimonial application was carried remotely with these eight participants. We prepared a user satisfaction survey form for the participants. The survey form included the question of the System Usability Scale (SUS), and also several questions regarding the accuracy of a similarity measure in finding the matches. The usability testing was conducted through one-to-one meetings using Microsoft Team.

3.4 Results Evaluation

We evaluated the data collected through the survey. The SUS scores were calculated using the method developed by Brooke [13]. The score for odd-numbered items was deducted by 1, while the score for even-numbered items was deducted by 5. Then, the sum of the scores will be multiplied by 2.5 to get the final usability score.

4   Results and Discussion

Figure 2 shows the interface of the newly developed matrimonial application. This interface will show the search results after users setting their match preferences. In the application, six matching methods can be chosen by users to find their partners, which are the exact matching (using SQL) and the mentioned five similarity measures. Exact matching will only display the search results that exactly match user preference. For the similarity measures, the top 10 search results with the highest similarity values will be displayed. The similarity values will be converted into percentages and displayed beside members' name. A higher percentage of similarity indicates that a member is closer to the user preference.

Jac Jaccard Coefficient, *Cos* cosine similarity, *Euc* Euclidean distance, *Man* Manhattan distance, *Min* Minkowski distance.

Figure 3 shows the overall user satisfaction for the matrimonial application. The SUS scores collected from each participant ranged from 70 to 92.5. The findings demonstrated that the mean of the SUS score was 80.5. According to the grading scale defined as shown in Table 1, the SUS score between 78.9 and 80.7 can be interpreted as grade A−. Grade A—reflects that the users had a good experience with the application.

During the usability test, we requested the participants to select the similarity measure method that can help them to find the potential matches close to their preferences. From the top 5 and top 10 search results returned by each similarity measure, they needed to select the profiles that suit their preferences. Table 2 shows the results obtained from the participants during the test. For the top 5 search results, half of

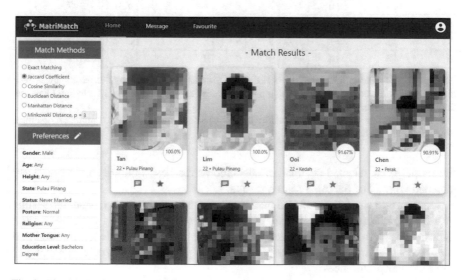

Fig. 2 The application interface that displays a user's search result

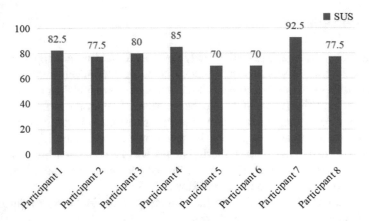

Fig. 3 The overall user satisfaction for the matrimonial application

Table 2 The user feedback on the similarity measures

Participant	Similarity measures									
	Top 5 search result					Top 10 search result				
	Jac	Cos	Euc	Man	Min	Jac	Cos	Euc	Man	Min
1	2	3	2	**5**	1	5	6	3	**7**	3
2	**3**	2	2	2	2	**5**	2	2	3	2
3	4	4	**5**	4	4	8	8	8	**9**	**9**
4	3	3	3	**5**	3	**6**	**6**	5	**6**	5
5	**4**	2	2	2	2	4	4	2	**7**	4
6	1	**3**	1	2	2	5	**7**	4	4	3
7	3	**5**	**5**	**5**	**5**	8	6	7	7	**9**
8	4	4	4	**5**	4	7	6	6	**9**	6

The numbers in bold indicate that the best similarity measures selected by the participants. Take the example of Participant 1, out of the top 5 search results returned by the Manhattan Distance, the participant agrees with all of them
Jac jaccard coefficient; *Cos* cosine similarity; *Euc* euclidean distance; *Man* manhattan distance; *Min* Minkowski distance

the participants selected Manhattan Distance as the best similarity measure. They satisfied with all of the top 5 search results.

On the other hand, there were 5 participants satisfied with the top10 search results of Manhattan Distance. The number of search results they agreed with was higher as compared with the other similarity measures. In conclusion, this test showed that Manhattan Distance is useful in helping the participants to find the potential matches close to their preferences.

The result of the exact matching method is not displayed in Table 2 as no user select this method.

5   Conclusion

In this project, we developed the matrimonial application that enables an individual to find their potential matches for marriage based on their preferences. We evaluated five similarity measures that allow a user to search for a potential partner based on his/her preference. These five similarity measures are, namely Jaccard Coefficient, Cosine Similarity, Euclidean Distance, Manhattan Distance and Minkowski Distance. We carried out a usability test remotely with eight participants to figure out which similarity measure is best suited for the application. From the study conducted, Manhattan Distance was found to be more appropriate for the application. As compared with the other similarity measures, Manhattan Distance is useful in helping an individual to find a potential partner that suits his/her preference.

However, there are rooms for improvement. Firstly, we found that the participants tended to focus more on member appearance during the usability test. To ensure the accuracy of the results, the participants had been advised to make their selection solely based on the member data. But it was unable to deny that physical attraction is a significant factor in selecting a partner. Some participants also mentioned that the first impression is crucial while choosing a partner that matches their preferences. Secondly, the data set collected for usability testing is small since it only consists of 40 male and 45 female data. Hence, we will collect larger data and include the physical appearance into the measurement for our future work.

References

1. Internetworldstats.com (2020) World internet users statistics and 2019 world population stats. (online) Available at: https://www.internetworldstats.com/stats.htm. Accessed 4 Feb 2020
2. Rosenfeld M, Thomas R, Hausen S (2019) Disintermediating your friends: how online dating in the United States displaces other ways of meeting. Proc Natl Acad Sci 116(36):17753–17758
3. Department of Statistics Malaysia (2019a) Marriage and divorce statistics, Malaysia, 2019. (online) Available at: https://newss.statistics.gov.my/newss-portalx/ep/epFreeDownloadCont entSearch.seam?cid=198734. Accessed 15 Feb 2020
4. Malaysia Population Research Hub (2019) Malaysian getting married later in life. (online) Available at: https://mprh.lppkn.gov.my/2019/04/09/malaysian-getting-married-later-in-life/. Accessed 15 Feb 2020
5. Yuen MK (2019) Feature: Malaysia's shrinking families. (online) The Star Online. Available at: https://www.thestar.com.my/news/nation/2019/11/10/malaysia039s-shrinking-families. Accessed 15 Feb 2020
6. Kwasny SC, Faisal KA (1990) Overcoming limitations of rule-based systems: an example of a hybrid deterministic parser. In: Dorffner G (eds) Konnektionismus in artificial intelligence und Kognitionsforschung. Informatik-Fachberichte, vol 252. Springer, Heidelberg, pp 48–57
7. Liu B, Ma Y, Wong CK (2001) Classification using association rules: weaknesses and enhancements. In: Grossman RL, Kamath C, Kegelmeyer P, Kumar V, Namburu R (eds) Data mining for scientific and engineering applications. Massive computing, vol 2. Springer, Boston, pp 591–605
8. Fiore AT, Donath JS (2004) Online personals: an overview. In: Conference on human factors in computing systems, CHI 2004. ACM, New York, pp 1395–1398

9. Polamuri S (2015) Five most popular similarity measures implementation in python. (online) Available at: https://dataaspirant.com/2015/04/11/five-most-popular-similarity-measures-implementation-in-python/. Accessed 1 Mar 2020

10. Sauro J, Lewis JR (2012) Quantifying the user experience, 1st edn. Morgan Kaufmann, United State

11. Usability.gov (2020b) System usability scale (SUS) | Usability.Gov. (Online) Usability.gov. Available at: https://www.usability.gov/how-to-and-tools/methods/system-usability-scale.html. Accessed 19 Mar 2020

12. Lewis JR, Sauro J (2009) The factor structure of the system usability scale. In: Kurosu M (eds) Human centered design: first international conference, HCD 2009. Lecture notes in computer science, vol 5619. Springer, Berlin, pp 94–103

13. Brooke J (1996) SUS: a quick and dirty usability scale. In: Jordan PW, Thomas B, Weerdmeester BA, McClelland IL (eds) Usability evaluation in industry. Taylor & Francis, London, pp 189–194

Proposal of Situation Estimation System Using AI Speaker

Yuki Kitasako and Tsutomu Miyoshi

Abstract We propose a system that uses the sound acquired by voice AI to guess the user's situation and provide services according to the situation. The system analyzes environmental sounds before the user's command, and records and collects the characteristic environmental sounds as symbol sounds, the type and time of occurrence. We thought we could analyze the pattern in which a particular command was executed each time a particular symbol string occurred at the same time interval. In Experiment 1, we investigated the characteristic sounds that were candidates for symbol sounds from the recording of daily environmental sounds. In Experiment 2, we selected three characteristic sounds, converted 120 recorded data into spectrogram images, and obtained a Convolutional Neural Network (CNN) model. We built a model, trained it, identified unidentified data in the trained model, and verified whether the prediction accuracy can identify it.

1 Introduction

In recent years, the number of devices equipped with voice AI is increasing in each home. In this research, we propose a system that can estimate and share the user's situation using the environmental sound acquired by voice AI. The feature of this system is to guess the user's situation by connecting the command and the environmental sound. We believe that the operation by voice recognition, which only responds passively, will be improved, and it can be expected to be applied to a wide range of fields such as providing services according to the user's situation and building an indoor watching system.

Assuming a command with a relatively high frequency in which commands are repeated daily, analyze the environmental sounds before the user's command at that time and classify the characteristic environmental sounds into several types of symbol sounds. Every time a command or a symbol sound with a certain degree of accuracy

Y. Kitasako (✉) · T. Miyoshi
Department of Media Infomatics, Ryukoku University Graduate School, 1-5 Yokotani, Seta oe-cho, Otsu Shiga 520-2194, Japan
e-mail: t19m057@mail.ryukoku.ac.jp

or more is generated, the type and time of occurrence of the command or symbol sound is recorded and collected using a device equipped with multiple voice AIs installed in a house By using the collected data, it is possible to analyze patterns such that a specific command is executed each time a specific symbol sound sequence occurs at similar time intervals. Recognizes a pattern of occurrence of a specific symbol sound found by machine learning and proposes to the user whether to give a specific command. The pattern that the proposed system uses to infer the situation of the user. As an experiment related to the identification of the symbol sounds used in matching, two types of experiments were conducted to verify (1) what can be used as the symbol sounds and (2) whether different symbol sounds can be identified.

2 Methods

In Experiment 1, in order to verify (1), characteristic sounds that were candidates for symbolic sounds were examined from the recorded daily environmental sounds. Recording was performed for 3 weeks and 6 h or more per day. A characteristic sound, which is a candidate for a symbol sound among the environmental sounds by listening, was summarized in a list. In Experiment 2, in order to verify (2), machine learning was performed from the symbol sound candidates examined in Experiment 1. Then, we verified whether the identification was correctly performed. We selected three characteristic sounds, that is, the sound of opening the key from the outside, the sound of opening and closing the door, and the sound of closing the key from the inside, and 120 of each of the three characteristic sounds. The recorded data was converted into image data by performing a short-time Fourier transform so that the vertical axis represents frequency and the horizontal axis represents time spectrogram. The sound of opening the lock, the sound of opening and closing the door, and the sound of closing the key are labeled for each image data, and 80 of the 120 images are used as training data and 20 are used as verification data to build a CNN model for learning. The structure of the CNN model is shown in Table 1. At this time, the accuracy of identification and the loss value were graphed as the learning result of CNN. We identified 20 unconfirmed data each in the trained model and verified whether the identification was possible from the prediction accuracy.

3 Results and Discussion

3.1 Experiment 1

In Experiment 1, we investigated the characteristic sounds that are candidates for symbol sounds from the environmental sounds recorded at the entrance and the

Table 1 CNN model structure

Layer	Filter size
Input	
Conv	3*3
Maxpool	2*2
Conv	3*3
Conv	3*3
Maxpool	2*2
Conv	3*3
Maxpool	2*2
Flat	
(Dopout)	(0.5)
Relu	
(Dropout)	(0.5)
Sigmoid	

living room.The characteristic sounds that are candidates for symbol sounds from the environmental sounds recorded at the entrance are listed below.

- Cars and buses running outside
- Car engine sound
- Airplanes and helicopters
- Inserting a key from the outside
- Sound to unlock
- Sound of removing key from keyhole
- Sound of door opening and closing
- The sound of opening and closing the key from the inside
- Human voice·footsteps
- The sound of luggage hitting the wall
- The noise of taking off shoes
- The noise of fixing shoes
- Intercom sound.

The characteristic sounds that are candidate symbol sounds from the environmental sounds recorded in the living room are listed below.

- The sound of cars and buses running outside
- Car engine sound
- The sound of airplanes and helicopters
- Human voice
- Human footsteps
- Rubbing sound
- Belt metal noise
- Metal sound of a key or key bunch

- The sound of opening and closing the doors of balconies and corridors
- Sound for opening and closing sliding doors
- TV audio
- Squealing sound of chair
- Sound of objects hitting the table
- Dryer sound
- Water sound of wash basin
- The sound of washing things in the kitchen
- The sound of opening food packaging
- The sound of hitting a cutting board with a knife
- The sound of opening the refrigerator
- Operation sound of air conditioner
- Sound using home appliances such as microwave ovens
- Crying and footsteps of cats
- Sound to turn on electricity
- Sound to turn off electricity
- Sound about alarm clock.

3.2   Experiment 2

The results of Experiment 2 are shown in Figs. 1, 2, where the vertical axis of Fig. 1 is the model identification accuracy, the vertical axis of Fig. 2 is the model loss value, and the horizontal axis is epoch (the unit of processing of the entire data set. The number of iterations), the line graph is the training data, and the dotted line is the verification data.

Fig. 1 The accuracy of CNN

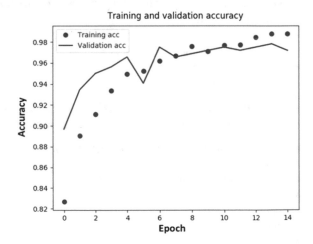

Fig. 2 Loss value of CNN

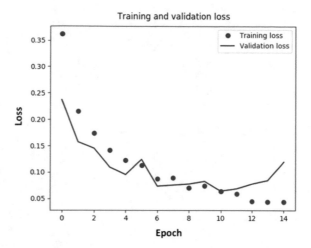

4 Conclusions

From the results of Experiment 1, when we examined the characteristic sounds that are candidates for symbol sounds from everyday environmental sounds, we found that there were many candidate candidates for symbol sounds.

At first glance, some of the characteristic sounds listed as candidates for symbol sounds seem to be noise, but the symbol sounds, which are the clues for the pattern analysis or machine learning performed by the proposed system, have characteristics that are not regular to humans. Even with sounds, artificial intelligence may be able to find regularity during long-term analysis, so it is desirable to have as many symbol sound candidates as possible. Therefore, cohesive sounds are considered to be candidates for symbolic sounds.

In Experiment 2, we verified whether the characteristic sounds listed as candidates for the symbol sounds in Experiment 1 could actually be identified as different symbol sounds. 80 learning data and 20 verification data for each sound. As a result of training with individual spectrogram images, it was found that the classification accuracy of the CNN model was 97.8% on average for unconfirmed data, and classification was possible with high accuracy.

From the graphs in Figs. 1 and 2, it can be seen that there is a little overlearning of the image data for the verification data, but because of the high identification accuracy for unconfirmed data, the learning performed with 80 images is sufficient for learning. Since it was found that individual tones could be identified, it became possible to treat the symbol tones as a sequence of symbol data. In the future, it will be necessary to verify whether it is possible to find a symbol pattern similar to the symbol pattern contained in the environmental sound before the user's command from the symbol data string.

RPL-Based Adaptive Multi-metric Routing Mechanism

Hung-Chi Chu and Xing-Dai Liao

Abstract Due to the rapid development of communication technology and embedded technology, tiny electronic devices can be embedded with diversified functions such as precision sense, calculation, and communication for use in environments that are difficult to monitor around life. However, wireless sensors are hardware with limited energy and memory performance. In order to operate in limited resources, its transmission method is extremely important. Therefore, RPL (IPv6 Routing Protocol for Low Power and Lossy Networks) is designed to provide routing protocols for devices with limited resources. However, the traditional RPL routing mechanism considers a single metric, which leads to a reduction in the energy of a specific node and gradually cannot meet the needs of users. Therefore, this paper proposes an adaptive multi-metric routing mechanism that considers Expected Transmission Count, Energy Consumption, and Hop Count to extend network lifetime. Simulation results show that the proposed method can effectively prolong the network lifetime by 15.91%, maintain a packet delivery ratio of more than 99%, and reduce the total delay by 6.6% compared with the traditional method.

Keywords RPL · LLN · Routing

1 Introduction

With the Internet of Things (IoT) [1, 2], a large number of different devices are connected to the Internet in wired or wireless ways to achieve identification, automation and management purposes. Many application technologies have been developing under the IoT concept. One of the important wireless network techniques in IoT is discussed in the working group of IEEE802.15, which is mainly used in Wireless Personal Area Network (WPAN). The wireless low-power transmission standard, IEEE802.15.4 corresponds to Low-Rate Wireless Personal Area Networks (LR-WPANs), which are commonly used in Wireless Sensor Networks (WSN) platform.

H.-C. Chu (✉) · X.-D. Liao
Department of Information and Communication Engineering, Chaoyang University of
Technology, Taichung, Taiwan, ROC
e-mail: hcchu@cyut.edu.tw

The WSN is composed of multiple sensing nodes, and the sensing data is sent to the data collection center in a single-hop or multi-hop routing method through wireless communication between the sensor nodes. WSN practical applications include environmental monitoring, military [3] and smart city [4] and other applications. Since the nodes have the ability to self-organize, the network topology and connections can be automatically formed after the sensor nodes are arranged. Therefore, the monitoring function can be achieved according to the requirement of the user.

There are still many problems to be solved under the IEEE802.15.4 protocol [5], including transmission traffic, path selection [6], network lifetime [7], scalability and insufficient IPv4 addresses. Note that the problem of insufficient IPv4 addresses makes it impossible to deploy a large number of sensors. In order to solve this problem, the Internet Engineering Task Force (IETF) proposed the IPv6 Internet standard specification to replace the traditional routing with insufficient IPv4 addresses as the IPv6 over Low-Power Wireless Personal Area Networks (6LoWPAN) standard, which is based on openness and standardization. IPv6 is connected to the Internet so that a large number of sensor nodes can connect to the Internet. Let 6LoWPAN have more operability, flexibility and multi-function in the Low-Power and Lossy Network (LLN) environment, and propose the IPv6 routing protocol suitable for LLN environment, namely IPv6 Routing Protocol for Low-Power and Lossy Network (RPL) [8] to improve routing and greatly to reduce the use of network resources. In the LLN environment, there is an inseparable relationship between packet transmission, equipment, and users, so that routing protocols can evaluate the balance between applications under limited resources and find the best configuration. In order to improve the performance of the overall network without reducing the lifetime of the network in the LLN environment, this will be the subject of this paper.

2 Related Works

2.1 Routing Metrics

According to the protocol of wireless networks, the routing metrics of wireless networks have been diversified to obtain different information. However, the design of routing metrics is not arbitrary, because it has a significant impact on the operation of routing agreements. In [9], it proposed a new security solution for the IoT to enhance the security and credibility of the routing protocol, and it conforms to the design of the wireless routing path weight structure and the criteria for the specific attributes that the routing metrics must have to ensure consistency, optimality and loop-free of routing.

The traditional routing mechanism can be divided into two types of methods: additive [10] and lexical [11]. The former is to give weight values to perform routing metric combination operations. The latter is to set specific conditions for routing metrics based on priority. The additive and lexical methods can be used in routing metrics such as HC, ETX, EC and Link quality level (LQL) to carry out one or more routing metric combinations [12]. A 4-tuple in (1) shows that it conforms to the nature of strictly increasing and transition.

$$(S, \preceq, \omega, \oplus) \tag{1}$$

Note that S is the set of all paths in the environment. ω is expressed as a function, and the path using this function is mapped to its weight. $\oplus$ is the action of continuously adding links between nodes. $\preceq$ is the weight order of all paths relationship.

2.2 RPL

RPL is an IPv6 routing protocol developed based on the 6LoWPAN standard. It is suitable for use in LLN environments with limited processing power, memory and power, and it is not suitable for a large number of periodic route maintenance methods. RPL helps nodes in peak packet loss rates, low-rate transmissions and unstable environments to connect to the Internet and establish data streams such as Point-to-Point, Point-to-Multipoint, and Multipoint-to-Point. RPL uses the Distance Vector (DV) routing protocol, so that each node does not need to store the overall network routing status, establishes a Directed Acyclic Graph (DAG) and based on routing information and objective function (OF) establishes a Destination-Oriented Directed Acyclic Graph (DODAG). The operation of RPL and the definition of symbols used are as follows.

Control Message
In order to establish and maintain the network topology, RPL defines four Control messages.

- DODAG Information Object (DIO)
  The main message is used to establish and maintain the network topology, and assist each node to join the DODAG and select the basis of the parent.
- DODAG Information Solicitation (DIS)
  Used for nodes that have not joined DODAG and seek DIO messages from neighbor nodes.
- Destination Advertisement Object (DAO)
  When the node selects the parent, it will send the DAO messages to the parent in the path until it is transmitted to the root to update the Routing Table in real time.
- Destination Advertisement Object Acknowledgement (DAO-ACK)
  When the DAO message is received, the sender has received the response.

Objective Function
OF provides optimization standards and defines how RPL nodes select the best route and form DODAG under resource constraints. OF is a combination of one or more

routing metrics and routing constraints. OF has currently defined Objective Function Zero (OF0) [13] and The Minimum Rank with Hysteresis Objective Function (MRHOF) [14] two protocols. OF0 uses the number of Hop Count (HC) as the routing consideration, and MRHOF uses the Expected Transmission Count (ETX) as the routing consideration. One or more routing metrics must be set in the RPL network to let the node know how to choose the parent.

2.3   RPL Simulation Tools

Contiki OS
Contiki OS [15] is an open source operating system for the Internet of Things, which can connect small, low-cost and low-power microprocessors to the IoT, and is used as a powerful toolbox for building complex wireless systems. Contiki supports fully standard IPv6 and IPv4, as well as recent low-power wireless standards: 6LoWPAN, RPL and CoAP (Constrained Application Protocol) agreements.

Cooja Simulator
Cooja [16] is a network simulator that supports 6LoWPAN technology. Cooja simulator runs on the Contiki operating system and provides a simulation environment that can imitate Motes and supports radio medium simulation and integration with external tools, providing additional functions for applications.

3   Adaptive Multi-metric Routing Mechanism

The proposed method is modified by the traditional RPL method. Therefore, the traditional RPL and the proposed method is presented in the following subsection.

3.1   Routing Metrics of Traditional RPL

RFC6551 [17] proposes a variety of routing metrics that can be applied to RPL. The ETX, HC and EC (Energy Consumption) are possible routing metrics in traditional RPL. However, traditional RPL routing mechanism just only can trigger one of them (i.e. single metric) to perform routing task.

Expected Transmission Count (ETX)
ETX is the number of times that a data packet needs to be transmitted successfully to the destination node, which is calculated by Eq. (2).

$$ETX = \frac{1}{D_f * D_r} \tag{2}$$

D_f is the forward delivery ratio, that is, the success rate of a data packet successfully delivered to the receiving end. D_r is the receiving rate (Reverse delivery ratio), which refers to the success rate of the sender successfully receiving the returned packet.

Hop Count (HC)

HC is the number of nodes passed between the node and the root, which can be used to avoid the path caused by the loop state, and its calculation formula is shown in Eq. (3)

$$HC = \left(\frac{Rank_{self}}{Rank_{root}} - 1 \right) \tag{3}$$

$Rank_{self}$ is the number of nodes passed between itself and the root. $Rank_{root}$ is the value of the root.

Energy Consumption (EC)

EC is the total energy consumption of the node. There are three types of energy consumption that considered in this paper: Low Power Module (LPM), Radio Listening Mode (RX) and Radio Transmission Mode (TX). The total EC can be obtained by Eq. (4).

$$EC = \left(\sum_{n,m} P_{m,n} \cdot T_{m,n} \right) \tag{4}$$

m represents all modes of operation. n represents the state (open or closed) in the m operation mode. $P_{m,n}$ is expressed as the power in the current operating mode. $T_{m,n}$ presents the running time in the current operating mode.

3.2 Weighted Evaluation Function

In this paper, the routing metrics of ETX, HC, and ECP are considered simultaneously.

The energy consumption with Punishment (ECP) is EC plus P, and it is listed in Eq. (5). If the network is stable and the routing is unchanged, then P is zero. On the other hand, if the routing is changed, it needs to consume energy to change the route, then P is the energy consumption of the route change operation. P is used to prevent the node from excessively exchanging the parent to consume too much energy, as shown in Eq. (6).

$$ECP = (EC + P) \tag{5}$$

$$P = k \times p \tag{6}$$

k is the number of times DIO is sent, p is the energy consumed by multicasting a DIO message.

In order to make each metric value can be measured concretely, the original value of each metric will be normalized. The large value of the metric indicates its high importance. The normalized ETX, ECP, and HC are listed in Eqs. (7), (8) and (9).

$$ETX_{Nor} = (-1) \times \frac{ETX}{ETX_{max}} \tag{7}$$

$$ECP_{Nor} = (-1) \times \frac{ECP}{ECP_{max}} \tag{8}$$

$$HC_{Nor} = (-1) \times \frac{HC}{HC_{max}} \tag{9}$$

ETX_{max} is the maximum values of ETX, HC_{max} is the maximum values of HC, and ECP_{max} is the total energy of the node.

The routing evaluation function (Eva) of the proposed method is listed in Eq. (10). α, β and γ are the weight values corresponding to its normalized routing metrics ETX_{Nor}, ECP_{Nor}, and HC_{nor}, respectively. Among them, α, β and γ must be greater than zero and $\alpha + \beta + \gamma$ is one.

$$Eva = \alpha \times ETX_{Nor} + \beta \times ECP_{Nor} + \gamma \times HC_{Nor} \tag{10}$$

3.3 Adaptive Multi-metric Routing Algorithm

The pseudocode of the proposed Adaptive Multi-metric Routing Algorithm is shown in Algorithm 1. Lines 1–22 are the DODAG construction and maintenance process. In detail, lines 1–2 are the root starts to send DIO to establish RPL topology. Lines 3–15 are the operation of the node after receiving DIO. Lines 4–6 are after the node receives the DIO message, if it judges that it has not joined the DODAG, it will join the DODAG and calculate the evaluation value of the parent. Lines 7–14 are the DIO message sent by the neighbor node after the node has joined the topology. The node will calculate the routing evaluation function (Eva_n) of the neighbor node and judge with the routing evaluation function of the parent (Eva_p). If E_p is greater or equal to Eva_n, then the source parent will be kept. Otherwise, replace the parent (i.e. Eva_p is less than Eva_n). Line 12 is when the node changes its parent node, the trickle timer will be remade. Lines 16–21 are that the node has not received DIO message after a period of time after startup, the node will send DIS message to its neighboring nodes to request DIO message. Lines 24–30 are the process of routing evaluation function.

The 25th line is to set the weight values α, β and γ corresponding to ETX, ECP and HC.

Algorithm 1 : Construct and Maintain DODAG
1. root start sending DIO message for neighbor node
2. construct DODAG topology
3. **if** (node receive DIO message) **then**
4.   **if** (node have not yet joined DODAG) **then**
5.   node join DODAG
6. $Eva_p \leftarrow$ Calculate_Evaluation( )
7.   **else**
8. $Eva_n \leftarrow$ Calculate_Evaluation( )
9.   **if** ($Eva_p \geq Eva_n$) **then**
10.   parent remain unchanged
11.   **else**
12.   reset trickle timer
13.   neighbor node becomes parent of node
14.   **end if**
15.   **end if**
16. **else**
17.   node is waiting to receive DIO
18.   **if** (node waiting time > T) **then**
19.   send DIS to DIO from neighbor node
20.   **end if**
21. **end if**
22. **end**
23.
24. Function Calculate_Evaluation( )
25. $\alpha + \beta + \gamma = 1,\ \alpha \cdot \beta \cdot \gamma \geq 0$
26. ETX $\leftarrow \dfrac{1}{D_f \times D_r}$, $Eva_{Nor} \leftarrow (-1) \times \dfrac{ETX}{ETX_{max}}$
27. ECP $\leftarrow ((\sum_{n,m} P_{m,n} \times T_{m,n}) + k \times p)$, $ECP_{Nor} \leftarrow (-1) \times \dfrac{ECP}{ECP_{max}}$
28. HC $\leftarrow \dfrac{Rank_{self}}{Rank_{root}}$, $HC_{Nor} \leftarrow (-1) \times \dfrac{HC}{HC_{max}}$
29. Eva $= \alpha \times ETX_{Nor} + \beta \times ECP_{Nor} + \gamma \times HC_{Nor}$
30. **return** Eva

---

4 Experimental Results

The experiment uses the Contiki OS operating system (version 3.0) and the Cooja simulator as the simulation platform. Table 1 shows the parameters used in the simulator. The experiment is divided into three stages. In each stage, the weight ratio of the best Network Lifetime (NL) will be obtained one by one and substituted into the subsequent experiments. In order to show the performance of the proposed

Table 1 Experimental parameters

Parameter	Describe
Network simulator	Contiki OS 3.0
Platform	Sky mote
Transmission range	50 m
Number of node	30 node + 1root
Initial energy	7500 mJ
Network topology	Randomly generate

Table 2 Performance evaluation with single metric

α	β	γ	k	NL	PDR (%)	TD
1	0	0	0	1169	87.78	28.4
0	1	0	0	1562	71.75	19.6
0	0	1	0	3423	99.96	9.9

method, the results of Packet Delivery Ratio (PDR), NL, and Total Delay (TD) are considered.

4.1 Experiment 1. Performance Analysis with Single Metric

This paper considers that the RPL routing mechanism uses a single metric to select the routing path. Therefore, this experiment will focus on the three single metrics of ETX, ECP and HC. As shown in Table 2, it can be observed that the network performance focusing on ETX and ECP is not good, while the network performance focusing on HC is the best of the three metrics. The reason is that during the initial establishment of DODAG, the focus on the routing of ETX and ECP will cause nodes to frequently exchange parent node and multicast DIO messages, resulting in excessive power consumption. In addition, the routing mechanism is used on a lossy network, causing the node to lose the packet sent by the neighbor node, which affects the PDR, NL and TD. HC will have the best performance because the initial HC is not available. If the network topology is stable after a period of time, the HC will gradually become fixed, and then the routing metric is also fixed.

4.2 Experiment 2. Adaptive Weight of ETX, ECP and HC

Considering that traditional RPL can no longer meet the needs of users for multiple services. Therefore, this experiment will focus on the results obtained in Experiment 1 and perform experiments with different weights to find a better network lifetime.

Table 3 Performance evaluation with Multi-metric

α	β	γ	k	NL	PDR (%)	TD
0	0	1	0	3423	99.96	9.9
0.05	0.05	0.9	0	3773	99.69	10.3
0.05	0.15	0.8	0	3774	99.56	10.3
0.15	0.15	0.7	0	3729	99.54	10.4
0.2	0.2	0.6	0	3834	99.58	10.4
0.4	0.1	0.5	0	3677	99.88	10.3
0.35	0.25	0.4	0	3530	99.10	11.5
0.6	0.1	0.3	0	3264	99.61	11.5

According to Experiment 1, it is known that the route metric focusing on HC will have the best network performance. Therefore, experiment 2 will use HC as the main metric to conduct experiments with different weight adjustments. To simplify, this experiment selects the best 8 sets of results from 63 sets of weights for performance analysis. In Table 3, it showed that when γ is set to 0.4 to 0.9, the network lifetime is significantly extended. However, when γ is 0.3, the network lifetime is shorter, which shows that if HC is not the main metric, the network lifetime will be shortened. And due to the consideration of the other two metrics (ETX and ECP), the routing path variation increases, which makes PDR and TD worse.

4.3 Experiment 3. Adaptive Weight of ETX, ECP and HC with k Value

According to the results of experiment 2, it showed that the weight ($\alpha = 0.2, \beta = 0.2$, $\gamma = 0.6$ and $k = 0$) can get the best network lifetime and improve the results of PDR and TD. In Experiment 3, it will change the number of transmissions k (see Eq. (6)) for $\alpha = 0.2$, $\beta = 0.2$, and $\gamma = 0.6$. The Performance evaluation with different k is listed in Table 4. It shows that regardless of k, PDR is higher than 99.18%. The NL is mostly unstable when the k is set to 1 to 12. The NL when k is between 12 and 20 is better than the NL when k is less than or equal to 1. When k is between 1 and 4, the TD is similar to the initial value. When k is set between 10 and 19, it will increase, but when k is 20, it will decrease.

5 Conclusions

In this paper, a routing mechanism with multi-metrics based on ETX, EC (or ECP), and HC was proposed. According to the experimental results, the proposed method

Table 4 Performance evaluation with different k

α	β	γ	k	PDR (%)	NL	TD
0.2	0.2	0.6	1	99.18	3825	10.4
0.2	0.2	0.6	2	99.33	3766	10.4
0.2	0.2	0.6	3	99.55	3871	10.4
0.2	0.2	0.6	4	99.86	4007	10.8
0.2	0.2	0.6	5	99.68	3737	10.3
0.2	0.2	0.6	6	99.52	3916	9.3
0.2	0.2	0.6	7	99.96	3905	9.3
0.2	0.2	0.6	8	99.86	3968	9.5
0.2	0.2	0.6	9	99.93	3829	9.5
0.2	0.2	0.6	10	99.88	3848	9.7
0.2	0.2	0.6	11	99.90	3828	9.4
0.2	0.2	0.6	12	99.93	3949	9.4
0.2	0.2	0.6	13	99.84	3897	9.3
0.2	0.2	0.6	14	99.72	3967	9.4
0.2	0.2	0.6	15	99.63	4027	9.4
0.2	0.2	0.6	16	99.71	4058	9.3
0.2	0.2	0.6	17	99.63	3978	9.6
0.2	0.2	0.6	18	99.54	3955	9.3
0.2	0.2	0.6	19	99.43	3937	9.5
0.2	0.2	0.6	20	99.58	3966	10.4

can effectively expand NL by 15.91%, keep PDR above 99%, and reduce TD by 6.6% when the weight ($\alpha = 0.2$, $\beta = 0.2$, $\gamma = 0.6$ and $k = 7$) is used. Overall, the proposed method can effectively improve the performance of RPL. In future works, other routing metrics can also be considered to meet the different data transmission requirements of IoT application systems.

References

1. Stojkoska BLR., Trivodaliev KV (2017) A review of Internet of Things for smart home: challenges and solutions. J Clean Prod 140:1454–1464
2. Ngu H, Gutierrez M, Metsis V, Nepal S, Sheng QZ (2017) IoT middleware: a survey on issues and enabling technologies. IEEE IoT J 4(1):1–20
3. Thulasiraman P, Wang Y (2019) A lightweight trust-based security architecture for RPL in mobile IoT networks. In: 16th IEEE annual consumer communications and networking conference. IEEE, Las Vegas, pp 1–6
4. Sebastian A, Sivagurunathan S (2018) Multi DODAGs in RPL for reliable smart city IoT. J Cyber Secur Mobility 7(1):69–86

5. Ghaleb B, Dubai A, Ekonomou A, Alsarhan E, Nasser A, Mackenzie Y, Boukerche AL (2019) A survey of limitations and enhancements of the IPv6 routing protocol for low-power and lossy networks: a focus on core operations. Journal 21(2):1607–1635

6. Bezunartea M, Wang C, Braeken A, Steenhaut K (2018) Multi-radio solution for improving reliability in RPL. In: 29th annual international symposium on personal, indoor and mobile radio communications, pp 129–134. IEEE, Bologna

7. Lassouaoui L, Rovedakis S, Sailhan F, Wei A (2017) Comparison of RPL routing metrics on grids. In: Zhou Y, Kunz T (eds) Conference 2017, ICST, vol 184. Springer, Cham, pp 64–75

8. Winter T, Thubert P, Brandt A, Hui J, Kelsey R, Levis P, et al (2012) RPL IPv6 routing protocol for low-power and lossy network. RFC 6550 (Proposed Standard), Internet Engineering Task Force

9. Djedjig N, Tandjaoui D, Medjek F, Romdhani I (2017) New trust metric for the RPL routing protocol, In: 8th international conference on information and communication systems. IEEE, Irbid, pp 328–335

10. Ahmed AR. Bhangwar A (2017) WPTE: weight-based probabilistic trust evaluation scheme for WSN. In: 5th international conference on future internet of things and cloud workshops. IEEE, Prague. pp 108–113

11. Cao Y, Wu M (2018) A novel RPL algorithm based on chaotic genetic algorithm. Sensors 18(11):3647

12. Park P, Coleri Ergen S, Fischione C, Lu C, Johansson KH (2018) Wireless network design for control systems: a survey. Journal 20(2):978–1013

13. Thubert P (2012) Objective function zero for the routing protocol for low-power and lossy networks (RPL). RFC 6552 (proposed standard), Internet Engineering Task Force

14. Gnawali O, Levis P (2012) The minimum rank with hysteresis objective function. RFC 6719 (proposed standard), Internet Engineering Task Force

15. Zikria YB, Afzal MK, Ishmanov F, Kim SW, Yu H (2018) A survey on routing protocols supported by the Contiki Internet of Things operating system. Future Gener Comput Syst 82:200–219

16. Hendrawan INR, Arsa IGNW (2017) Zolertia Z1 energy usage simulation with Cooja simulator. In: 1st international conference on informatics and computational sciences. IEEE, Semarang, pp 147–152

17. Vasseur JP, Kim M, Pister K, Dejean N, Barthel D (2012) Routing metrics used for path calculation in low-power and lossy networks. RFC 6551, Internet Engineering Task Force

NFC Label Tagging Smartphone Application for the Blind and Visually Impaired in IoT

Kah Yong Lim and Yean Li Ho

Abstract This paper details a mobile labelling application for visually impaired and blind user to help them identify an object's character in IoT. Visually impaired and blind people recognize items through their sense of touch. However, it is not very easy for them to know the object's character, for example, the colour of clothes. NFC technology was used to develop this project, the main reason for using NFC is because it requires the least effort to scan and automatically generates information without needing to open any third-party application. Problems arise from the existing applications which are implemented with barcode and QR code, these codes are only convenient for sighted people. If we consider the perspective of the visually impaired or blind people, they need to spend more time searching for the location of the code blindly. Moreover, the existing applications only generate output if the scanner successfully captures the code in full size. Besides that, the lack of labelling feature is another problem of the existing system, which is unable to identify an object's character which is also crucial for the visually impaired or blind users. Research testing of the usability of the mobile application was tested by 30 people, classified as sighted blindfolded participants and short-sighted users without glasses. The results show that NFC works in the simplest way and requires the least effort during the scanning process, and it is suitable for blind or visually impaired and also sighted user. NFC-enabled Smartphones will generate output when detecting any NFC tags in a range of 3 cm. Having good accessibility and usability of application not only bring benefits to sighted people, but it also helps the visually impaired and blind users enjoy this benefit.

Keywords Tagging system · Blind · Visually impaired · Smartphone · Near field communication · Mobile devices

K. Y. Lim · Y. L. Ho (✉)
Multimedia University, Jalan Ayer Keroh Lama, 75450 Bukit Beruang, Melaka, Malaysia
e-mail: ylho@mmu.edu.my

© The Author(s), under exclusive license to Springer Nature Singapore Pte Ltd. 2021
H. Kim et al. (eds.), *Information Science and Applications*, Lecture Notes
in Electrical Engineering 739, https://doi.org/10.1007/978-981-33-6385-4_29

1 Introduction

The smart phone is becoming an important aspect of our lives. The comfort and convenience certainly makes our lives much easier than ever before. Smart phones really bring so many advantages to mankind. For example, e-commerce connects people everywhere and etc.

Although the smart phone brings many advantages to mankind, it is still difficult for visually impaired or blind people to use the smart phone. Application developers should include functions and features that are able to support blind people. For instance, text-to-speech technology, sound of a clicked button, vibration and etc.

This mobile application is built to help the visually impaired or blind people to identify objects and its characteristics by scanning using a technology called Near Field Communication (NFC tag). This advanced technology requires less efforts to successful identify an object compared with other scanning mechanisms such as QR code. In other words, NFC brings convenience to different kinds of user regardless if they are visually impaired or sighted people.

In this project, the user is allowed to add or change contents with speech-to-text technology unlimited times until the tag is physically damaged. Content is stored in NFC tags independently. No matter if the smart phone is in the online or offline environment, it is still able to access the content of NFC tags. Once a user inserts content into NFC tags, the smart phone will automatically generate and display the contents without opening ant third-party application.

The NFC tag is a passive item. It operates without the need of a power source. NFC tags are powered through a process called electromagnetic induction. As the phone is placed near to the tag, the phone creates a magnetic field and the NFC receiving coil will produce a match with the electrons coming from a powered phone.

2 Existing Systems

2.1 WayAround Application

The WayAround app and the smart WayTags lets the user add helpful information to items around their home and office [1]. The simple tag-and-scan system lets the user add a custom description or more information such as expiration dates of food or washing instruction for clothes.

After tagging an item, scan it anytime with a smart phone to hear the description and details. WayAround use the technologies of NFC (Near Field Communication) in this application. WayTag are designed in 4 types, such as button, sticker, clip and magne [1].

WayAround makes information that has never before been available to people who are blind easily accessible. It is an elegant solution, which meets previously unmet needs of many who live with vision loss [1]. Strengths of this application include adaptability. Waytags have 4 types of tags such as sticker, magnet, button and clips so the user can tag anything and everything. Waytags are reusable. WayAround is affordable, each of the tags cost about a dollar each. Besides that, it is accessible because this application works with NFC technologies to store information and transmit it wirelessly [1]. All a user needs to do is hold the device close to the target WayTag.

The weakness of this application is that it only works with a smart phone that has the feature of NFC (Android phone). iPhone users should install the additional NFC application and it only allow the user to read NFC and cannot insert or modify content inside NFC tags [1]. Another limitation of NFC is it only works in short distance, in a range of 10 cm to read the NFC tag.

2.2 Digit-Eyes

There are 2 versions of this application, Digit-Eyes with full features and Digit-Eyes Lite with limited features. This is an application only available on iPhone, Digit-Eyes allows user to read Digit-Eyes QR (Quick-Response) text bar code labels [2].

They have their own website called Digit-eyes website, which allows a user to create a PDF file of Digit-Eyes QR bar codes that each contain up to 250 character of text. These codes are useful for labelling items where the content that cannot be changed and also create labels to identify clothing, CDs, canning and personal items [2].

Digit-Eyes includes a feature that enables a user to identify many groceries, CDs, and other consumer goods by scanning the UPC and EAN codes on the products. It will connect to the product database, after a matched result is returned in text form. If a user's phone has Voiceover Active, it will read aloud to the user [2].

The cons of this application are UPC and EAN codes features are only available on the full version application. The user needs to pay USD9.99 to use the full features version. Besides that, this application only supports the iPhone operating system. Based on the chart from [3], there are only 22.17% IOS users in globally, which means this application only focusses on this 22.17% of a narrow market.

The strength of this Digit-Eyes application is that it supports multiple languages including English, Danish, French, German, Italian, Norwegian Bokmal, Polish, Portuguese, Spanish, Swedish [2]. Besides that, it can create its own label code and record information that the user wants in text form and audio form. After the user successfully recorded the information, a rescan of the label code will display the information that the user recorded.

3 Layout Interface Requirement

3.1 Interface Design Requirements for the Moderately Visually Impaired User

Touchscreens became a standard feature of mobile devices such as smart phones and tablets. Touchscreens provide friendly and easy to use interface for modern mobile devices [4]. However, visually impaired people are faced with numerous difficulties using a touchscreen. There are several functions that are able to improve accessibility on smart device, including screen reading, voice input, large buttons, screen magnification and high-contrast screen [4].

Nowadays, the mobile device is designed with the function buttons of numerous items integrated into the software interface, this approach tries to minimize the production costs, but it also increases the difficulties faced by visually impaired user when they experience the touchscreen-based smartphone. Therefore, the researcher should do further exploration on the accessibility of the touchscreen interface and also focus on visually impaired user accessibility requirement and expectation [4].

There are several factors which influence accessible touchscreen interface design. The first factor is 'Tracking', which includes bold borders, touch clues and etc. Visually impaired users often collect information by touch, therefore traditional mobile phones with bumps on the keys can enhance their touch perception of an object's surface, while perceiving objects on touchscreen-based smartphone is challenging for a visually impaired user because the touchscreen surface is flat and slippery. Thus, adding a bold border around the tapped button will help moderately visually impaired user to easily locate and know related clues on the touchscreens. Therefore, providing useful tracking clues for touchscreen is crucial [4].

Second factor is 'Detection Quality', this factor helps moderately visually impaired users to perceive signals and other information on the screen. Items including information sequence arrangement on screen, screen density, vibrotactile, reading glass and sensitive click on screen are able to assist the visually impaired user to receive more information. For example, picture and text information should be in big size and arranged [4].

Furthermore, third factor 'Fuzzy Field' includes designing a related object shape that easily recognizes the dynamic effects of the item [4]. This factor is mainly for moderately visually impaired because they have some light perception and able to vaguely identify the shape of an object. For these kinds of user, although the information may be not very clear, but fuzzy object shapes able to give them a reference to guide them the right location to press on the screen.

Factor 4, 'Dominance' is used to control the capabilities of touchscreen interface, such as voice control and speed of sliding screen [4]. Some of the modern smartphones provide voice control functions. For example, Apple product (Siri) and Android product (Google Now). This is a useful assistance for user with physical disabilities. They can easily interact with smartphones by using voice control to perform daily tasks such as phone calling, listening to music and etc. Moreover, the sliding speed must be adjusted to ensure that the touch feedback was good.

Besides that, factor 5 'Colourity' is another design factor which should be considered for moderately visually impairment user [4]. A high-contrast screen can be result in easy perception of an object, visually impaired people prefer appropriate screen colourity when they are looking for information on the touchscreen. For example, black colour background with text in white colour is the best colour contrast for visually impaired user.

Factor 6 'Speech Quality' refers to the quality of screen reader and natural voice item [4]. Visually impaired users rely on auditory information to perform different mobile activities. When a user press on a block of text or information, the content must be automatically read out. Besides that, the button should announce its function, for example home button should pronounce 'home' after user press on it. Screen reader enables the visually impaired user to receive messages through sound, this will increase the speed of receiving information and also reduce the false selections on the touchscreen.

3.2 Interaction Experience of Visually Impaired People with Assistive Technology

Affect from auditory sense play a significant role for visually impaired user [5]. But on this experiment, it was observed that problems included speech is too slow, screen reader voice are not friendly and so on. Screen reader is a crucial element for visually impaired user interact with smartphones, therefore, to develop a good quality screen reader is necessary to meet the expectation of the visually impaired user [5].

Besides that, participants found it difficult to perceive spatial information through an auditory channel. For example, the screen reader expresses the information in nine sections of a screen (3 X 3) (e.g. top left, right down). Many visually impaired indicated that using clock metaphor such as direction 11 o'clock would help them easily to understand spatial structure [5].

Moreover, before beginning to access any application, the visually impaired user will memorize the location of the item or button on the screen by practicing alone or asking to someone help. However, the lack of the tactile landmark on the touchscreen is difficult for them to memorize the interface layout [5].

Furthermore, the most influential factor that affect the ability of moderately visually impaired people to use a smartphone is eye fatigue. Color contrast and transparent background are the factors that cause eye fatigue for people with low vision. Most participants preferred using black background with white text and also suitable text sizes [5].

From 4 experimental findings of unique interaction experience of visually impaired user mention on above paragraph, it is very important for the designer to understand what technologies is able to assist the visually impaired user [5].

First, it is important for visually impaired users to receive information mostly through the auditory sense. Several auditory design features can be included, such as softness of bell sound, number of default bell sounds, and steps required to adjust the volume of the bell [5].

Secondly, visually impaired user cannot estimate distances between items on the screen. Therefore, the experimenter indicated that visually impaired user preferred guidance with the clock metaphor when receiving spatial information on the touch-screen. In addition, development of a navigation system can be helpful for them perceive visual-spatial information [5].

Third, provide a consistent and simple-structured UI layout. From the experiment observation [5], they found out that users tried to memorize the layout of the smart-phones before starting to use it. The layout must be consistent and should not change according to the function. For example, (in Fig. 2) the button location inside pop-up components change frequently according to the message length. In other words, the button is not always at the same location. Hence, it was difficult for visually impaired user to know the exact location of the button (e.g. okay, cancel). In addi-tion, the layout of application should be simple and familiar. Visually impaired users prefer a familiar layout [5].

Fourth, increase configurable settings. There should be an additional settings func-tion which can improve the accessibility of application. Text size and colour should not be set arbitrarily because visually impaired users with different levels of severity have different requirements. Providing changeable color contrast, transparency of background, font shape, space between letter and lines will be helpful for them [5].

Fifth, improve speech performance of the screen reader. There is a lack of studies on improving the speech performance for assists blind people [5]. Visually impaired users rely heavily on the screen reader to receive all information. Therefore, the demand for speech performance should be further studied and applied [5].

3.3 Designing Mobile Application for Visually Impaired People

Visually impaired user faces different problems due to the rapid development of infor-mation technology has made to mobile applications. Application without features of screen readers, screen magnification, voice commands recognition is useless for visually impaired user [6]. Besides that, accessibility to the functionality of mobile devices is another problem. Modern smartphones and tablets have flat surface of the touchscreen. The touchable "home" button is not enough for a fluent navigation through the content of the screens. Next problem is visually impaired users are impos-sible download available mobile application in Google Play or App Store. Action such as selecting item from a list or cancellation in the pop-up dialog windows are difficult task for them to do [6].

Most important feature of operating system accessibility for blind people is screen reader. It interacts with user by using various gestures. For the example, a single-tap reads the label of the selected elements. Double-tap activates a selected element, similarly to double-click a mouse button on computer [6].

One of the most powerful features of mobile accessibility is the function of intelligent voice assistant. Mobile personal assistant helps user to answer question, make recommendations. Additionally, it analyses user's behavior, based on their search queries, it will predict what user may want to know, or what to do. Intelligent assistant is useful for visually impaired user because this the best method to interact with smartphones [6].

Currently most mobile operating system allow user to change text size. IOS 8.1 offers 12 sizes to choose from horizontal slider, but the changed sized only affects e-mails, contract and text messages. While Android OS 5.0 offers 4 different sizes and selected text size applied in the whole operating system. Besides that, both mobile operating system provides color inversion as the option of accessibility. The is a very important option for reducing reflects and visual fatigue. This could be helpful for those visual impairment users. This feature changes the text in black color and white in background [6].

Both IOS and Android operating system have their own built-in screen magnification. This accessibility feature is important and useful for those people with low vision. The IOS includes the ability to increase the viewable display with a three-finger double-tap. With Android OS, activate the screen magnification by triple tapping the screen with a single finger. Disabling magnification is done by triple tapping the screen again with a single finger [6].

4 Proposed System

4.1 Interface Design

The proposed system was implemented by using software Android Studio. Latest version of Android Studio support many different programming languages such as Java, C+ + , GO and Kotlin. This application was built by using Java as the programming language. Java is the official language for programming Android apps, so it could be the best languages for Android apps and there are many useful resources on internet that can be used for references during implementing this application.

Besides that, this application required an enabled NFC smartphone to run and test usability of application. Huawei P10 is the model of smartphone that used in testing and running this application.

Before start to build this application, study interface layout requirement for visually impaired user is a must. After studied literature reviews [4–6], learned knowledges was applied during designing interface layout.

Refer to Fig. 1, the insert content process was designed by using the recording

Fig. 1 Implemented application layout

mechanism rather than using the QWERTY keyboard, because visually impaired users face challenges to use QWERTY keyboard. Speech-to-text mechanism is used in the recording process. The recorded content will be stored in text form. Besides that, this application also provides a replay button used to check the content if it is captured correctly. After the user confirms that the content was captured correctly, the user is required to place the phone close to the NFC tag in order to successfully save the content into NFC tag. The range of adding content into the tag is within 3 cm distance. If the phone cannot detect any NFC tags, the screen reader will interact with user and tell the user to get closer to the NFC tag.

In addition, designed button can be single-clickable and double-clickable. The screen reader will speak out the button feature at a single-click, while double-click is used to select the button.

Moreover, this application provided explanation pop-up dialog use to explain the usability of this application. Screen reader will read aloud the content and after the content read finished, the dialog will automatic close. Furthermore, configurable settings were provided in this application to improve the accessibility of application. Users can change the language of the application to English, Chinese and Bahasa Malay, user also allow to change the voice reader.

5 Proposed System

This test was conducted to study accessibility, usability and satisfaction about the implemented mobile application. 30 people were tested, which included sighted participants who were wearing a blindfold and also near-sighted users without wearing their glasses. Participants will be classified into 2 groups, group A is for

Table 1 Interview questions about usability of application

	1	2	3	4	5
(1) Does this application effective and easy teaching process?					
(2) Are you satisfy about screen reader quality					
(3) Is the font size easy to see? (for moderately visually impaired user only)					
(4) Did you feel comfortable for Color-contrast? (for moderately visually impaired user only)					
(5) Did the icon relate with button feature? (for moderately visually impaired user only)					

those participants aged under 30, while group B is for participants aged 30 or above. Participants were given assistance if they had problems. They were required go through the tutorial part first before they start performing any task.

Secondly, they need to change the application language and screen reader type on setting page. Third, they were required to add new content into the NFC tags by using speech-to-text technology, all the processes were estimated to be completed within 15 min. After they finish the ordered tasks, they were required to give their feedback about this application. Several questions about testing usability were prepared as shown in Table 1 to find out whether this application meets the expected result and allows the user to feedback their experience after they run through everything on this application. Likert score scale is used from 1 (Very poor), 2 (Poor), 3 (Neutral), 4 (Good) to 5 (Very good) as illustrated Table 1.

6   Result

After conducting the testing of 2 groups of participants and collecting feedback from the users, result can be referred to in Table 2.

Group A average score of teaching process is 3.04 (Neutral), Screen reader quality in average score of 3.66 (Neutral), font size average score of 4.33 (good), color-contrast average score of 4.33(good), button's icon average score is 3.66 (Neutral) and overall.

experience average score is 3.83 (neutral).Group B average score of teaching process is 2.5 (Poor), Screen reader quality in average score of 2.66 (Poor), font size average score of 3.66 (Neutral), color-contrast average score of 4.66(good), button's icon average score is 3.66 (Neutral) and overall experience average score is 3.66 (neutral).

Through this test, we found out that this application still has a lot of room for improvement. Besides that, we retrieved a lot of feedback from users for future improvement. For instance, feedback from participant stating "Explanation is too complicated, if able to change sound effect better and the explanation is not complete". From this feedback, we learned that if the explanation takes too long, the

Table 2 Result of interview question (Q3 to Q5 for moderately visually impaired user only)

Name	Age	Gender	Q1	Q2	Q3	Q4	Q5	Q6
User 1	15	Female	2	3				3
User 2	25	Female	3	2				4
User 3	25	Male	4	4	4	4	4	4
User 4	55	Male	3	3				3
User 5	25	Male	2	3				2
User 6	25	Male	3	4				3
User 7	54	Female	2	3	4	5	3	4
User 8	20	Male	4	5				4
User 9	21	Female	3	4				4
User 10	21	Female	2	4				4
User 11	27	Male	4	5				4
User 12	30	Male	3	3	4	4	4	4
User 13	24	Male	3	3				3
User 14	18	Female	4	3				4
User 15	15	Male	4	5				5
User 16	20	Male	3	4				4
User 17	24	Male	4	4				4
User 18	43	Female	2	3				3
User 19	21	Female	3	3	4	4	3	4
User 20	58	Male	2	2	3	5	4	4
User 21	40	Male	3	2				4
User 22	25	Male	3	3	5	5	4	4
User 23	25	Female	2	3				3
User 24	25	Female	3	4				3
User 25	25	Male	3	4				4
User 26	26	Female	2	3				4
User 27	25	Female	3	4				4
User 28	24	Male	4	4				5
User 29	27	Male	3	4				5
User 30	25	Male	2	3				4
Total score			88	104	24	27	22	114
Total score of user in group A (age under 30)			73	88	13	13	11	92
Average score of group A			3.04	3.66	4.33	4.33	3.66	3.83
Total score users in group B (age from 30 or above)			15	16	11	14	11	22
Average score of group B			2.5	2.66	4.66	4.66	3.66	3.66

user will feel impatient to listen to the explanation till the end. The 'explanation' feature can be made simpler and easier to understand.

7 Conclusion

The growth of mobile technology is going rapidly in the last few years. As visually impaired or blind users are increasing in this market, it is necessary for the developer to build useful applications for them. Therefore, this system was proposed to enable the visually impaired or blind user to identify object correctly in an effective way and also easy to use way. This mobile application is built by using Android Studio.

The main goals of this project are to provide a simple and easy access mobile application which allow blind and visually impaired users to identify objects quickly and effectively using NFC technology.

Last but not least, after conducting usability testing, users' overall feedback is neutral, which means that this application still needs to be improved. This application could be improved in the future. For instance, the help function could be improved by making the explanation simpler and placing explanations at every page separately instead of putting all the explanations in one page. Furthermore, more options may be included in the settings page. For example, user can turn on or off the optional sound effects.

References

1. Blind InSites LLC. Home—WayAround
2. Digit-Eyes, L. L. C. Digit-eyes UPC database and iPhone bar code scanner app
3. StatCounter GlobalStats (2019) Mobile operating system market share Europa | StatCounter Global Stats
4. Huang H (2017) Blind users' expectations of touch interfaces: factors affecting interface accessibility of touchscreen-based smartphones for people with moderate visual impairment. Univers Access Inf Soc 17:291–304
5. Kim HK, Han SH, Park J, Park J (2016) The interaction experiences of visually impaired people with assistive technology: A case study of smartphones. Int J Ind Ergon 55:22–33
6. Dobosz K (2017) Designing mobile applications for visually impaired people. 19

COVID-19's Telemedicine Platform

Gyöngyi Szilágyi Kocsisné and Kocsis Attila

Abstract The fight against COVID-19 has become part of our everyday lives. In this paper, we will introduce a novel telehealth solution we developed to provide powerful help in this fight. The integrated system we have developed provides technology support for special features that have already proven their effectiveness in fighting the virus. Such key functions include social distance control, effective contact detection, continuous monitoring of people's health status, disinfection of affected areas, analysis and optimization of work processes to prevent the spread of the virus. Our health and safety solutions for COVID-19 are based on digitizing and monitoring spaces, time, motion and physiological parameters with smart IoT and artificial intelligence.

Keywords COVID-19 · Telemedicine · Social distance · Contact tracing · Indoor positioning

1 Introduction

Nowadays, the technology feasibility and efficiency of different Telemedicine systems [3] have been proven, so it is important to examine how they can be applied in the situation caused by COVID-19. In the cycle that characterizes the spread and treatment of COVID-19 [1, 2] virus, it is possible to intervene in several places to prevent its destructive power.

Some possible points of intervention: 1. The person becomes infected; 2. Asymptomatic (Invasive and non-invasive tests to detect positivity); 3. Infects through personal contact (directly from person to person, or through objects); 4. Appearance

Health and safety solutions for COVID-19 for Enterprises and citizens by digitizing and monitoring spaces, time, motion and physiological parameters with smart IoT and artificial intelligence.

G. S. Kocsisné (✉) · K. Attila
I-QRS Research and Development Inc, 677 N Washington BLVD, Sarasota, FL 34236, USA
e-mail: szilagyi@i-qrs.com

K. Attila
e-mail: kocsis@i-qrs.com

of disease symptoms (Detection and identification of symptoms, different treatment methods).

In the present study, we present a Platform that supports the effective fight against COVID-19 in all four main phases and provides a common interface for correlating and integrating the results of solutions based on invasive and non-invasive technologies. Our Health and safety solutions for COVID-19 for Enterprises and citizens is based on digitizing and monitoring spaces, time, motion and physiological parameters with smart IOT and artificial intelligence.

Our solution is built on three main pillars, including telemetric measurement devices based on smart IOT technology (position, medical and other data), data collection, analysis and visualization and Artificial Intelligence in every step.

The Platform uses special high-tech sensors and devices to continuously measure and analyze the physiological and movement parameters and orientation of people in real time, both indoors and outdoors [6]. The system's online functions controlled by artificial intelligence support real-time and continuous control, reporting, sending alerts, 3-dimensional visualization, and synchronized video analysis. Meanwhile, in the background, there is continuous machine learning guided by artificial intelligence (AI). AI ensures that other data coming from the interface can be integrated with continuous measurement results, the data can be processed using data mining methods, and more efficient solutions can be developed based on these. For example, if a user is found by a test method to be a virus carrier and the system has physiological and other (e.g. position-based) measurement data about that person, it can be automatically and very quickly localized, isolated, a list of contact persons affected by the infection can be identified retrospectively, and the areas where urgent disinfection is required can be identified (where the positive person was active in the last days).

On the other hand, automatic signal processing algorithms [6] can be used to search for special features by analyzing of physiological measurement signals that can serve as markers in terms of infection. By examining a sufficiently large number of cases, a solution could be found that eliminates the test and can only generate an alarm in the event of a probable positive case by continuously measuring and analyzing physiological signals (e.g. body temperature, ECG, heart rate variability, skin resistance, blood oxygen level change, etc.).

With this approach, we can provide effective assistance in screening during the asymptomatic period, preventing the spread of the virus by contact detection, by disinfecting the affected areas, and in the event of symptoms (such as fever) to quickly identify the virus and take the necessary measures.

We have developed smart IOT solutions that are able to perform contact detection indoors and outdoors (using local position measurement), which are capable of continuous tracking and historical management of this information. We also measure movement and physiological parameters (e.g. ECG, heart rate, blood oxygen level, skin temperature, ambient temperature, etc.) using an easy-to-wear integrated device.

Based on previous research and development, we have sophisticated expert systems for the analysis of physiological signals, which are suitable for continuous real-time measurement of skin temperature, identification of arrhythmic signals in everyday life, analysis of sympathetic/parasympathetic nervous system effects, measurement of stress levels, correlation analysis, historical analysis, setting alarms, etc.

In our research [7, 8], we correlated the results from invasive technologies with the results of noninvasive measurements, combining these with expert knowledge to develop directly usable tools and expert systems based on noninvasive technologies for the users. The various special measuring modules have been integrated into a common platform which allows information, data (such as non-invasive test results, other health and position data) from other systems to be received via interfaces, and their joint analysis, data mining processing, and correlation detection.

2 Functionalities and System Architecture of the COVID-19's Platform

The structure of COVID-19's Platform is shown in Fig. 1.

Various modules provide the operation of telemetry devices, their synchronization, data management and analysis, various information services for users, management of alarms and reports, connection to other systems.

The main modules of COVID-19's Platform and their functions are listed in Table 1.

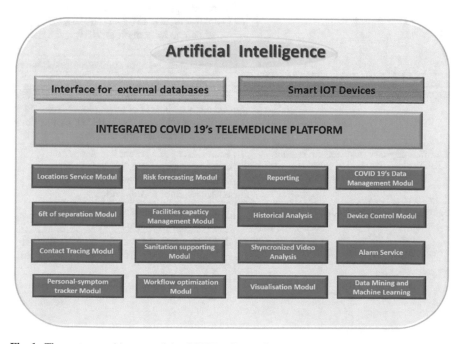

Fig. 1 The system architecture of the COVID-19's platform

Table 1 The main modules of COVID-19's platform

Modules	Functions
Locations Service Module	1. Tracking people within an office, facility, water park, a warehouse or manufacturing plant: With an I-QRS tag all person's and visitor's position can be tracked continuously
	2. Know the exact location of Employees who are being tracked
	3. Indoor wayfinding: The persons can be automatically navigated to specific places, buildings, rooms where they want to reach
	4. Tracking information on the indoor map of the building
	5. Control the ban on Prohibited Areas
6ft of separation module	Enforce social distancing: alert users (and also the call center) when they are too close to each other through sound or vibrations
Contact tracing module	Identification of the list of colleagues which were within close proximity in the past week with the confirmed case, to isolate close contacts of the confirmed case and break the chain of transmission
Personal-symptom tracker module	1. Send an immediate alert if a symptom of an infection occurs: Continuous monitoring of body temperature (fever) and ECG/HR (abnormal heart rhythm), HRV, and SpO2 level
	2. Historic analysis for employees health status monitoring
Risk forecasting module	Historical employee motion analysis to identify and analyze the risk of an employee based on the time spent in the areas affected by the virus carrier, and contact time with them
Facilities capacity management module	Facilities capacity, event and stateroom capacity management. Timing of events, scheduling the attendance of programs with respect to current virus policies
Sanitation supporting module	1. Crowd flow analysis to determine high traffic areas for sanitation
	2. Disinfection of target areas: list the areas in which infected person was active in the past days
	3. Heat maps (number of employees per time)

(continued)

Table 1 (continued)

Modules	Functions
Workflow optimization module	Analyzing and optimizing the activity of employees in space and time and performance based on location and tracking (information in percent- walking/stationary); Understand space dwell time by using activity; rearrange the workflow for contact minimization
Reporting and historical modules	The different reports and the chronological interpretation supports the sophisticated interpretation of the measurement data
Synchronized video analysis module	Camera based enter support: 1. Face recognition to check in the database whether the employee or visitor can enter or not (quarantined person, ask to stay home, unauthorized intruders) 2. Thermal Camera to check the body temperature
	3. Supports synchronized analysis of physiological and motion time series with video recording
Visualization module	Position, Tracking, activity and health status information in 3D animation on the indoor map of the building
COVID-19's Data Management and device control modules	General modules connected to the telemetric system support the administrative management of data stored in the central database Radio transmission and Distributed algorithms control the operation of the devices. The management of logs is available
Alarm service, data mining, machine learning, and AI	We use state-of-the-art methods of Artificial Intelligence in our telemetry system Intelligent processing algorithms perform the analysis of the measured signals and the data mining. Many subfields of artificial intelligence have been used in the tools, from data analysis to the development and operation of the expert systems
Interface for other service providers	We make system data available to other external security and other systems through an interface External data is integrated into the system

Fig. 2 Wearable devices:
chest belt, identification
card, smartwatch

3   The Technology Background

3.1   *Measurement Devices*

The requirements for continuously wearable devices are to have small size, and low power, preferably to integrate different functions in a single device, to provide comfortable wearability to improve patient's mobility and prolonged network lifetime.

Technologies used: High-tech hardware development, embedded systems, radio planning and protocols, distributed algorithms.

In the family of telemetry devices we have developed, only one small device needs to be worn, which measures physiological signals and indoor and outdoor position in an integrated way.

The form of wearable devices: chest belt, identification card, smartwatch (Fig. 2).

3.2   *Indoor and Outdoor Positioning*

In the current viral situation, it has become particularly important to monitor the distance of people from each other, to trace their position and to detect the persons who come into contact with the infected person, which poses a particularly high risk in the case of indoor contact (see Fig. 3.).

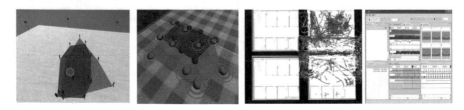

Fig. 3 Sreenshots: 3D animation and physiological state

There are many ways to apply indoor positioning [4], and many technologies to choose from. For example technologies based on signal strength (Wi-Fi, Bluetooth, RFID), and based on time-of-flight (UWB). Most systems use a combination of technologies, because each technology has its own strengths and weaknesses.

The Global Navigation Satellite System (GPS, GLONASS, Galileo) is a commonly used technology for position tracking outdoors, but they do not provide enough accurate data for some important purposes.

The key element of our COVID-19's Platform is high-tech telemedicine devices that are able to provide accurate position data both indoors and outdoors using a wireless sensor network. The wireless sensor network (WSN) [3] has important applications such as remote environmental monitoring, health monitoring and target tracking.

A wireless sensor network [5] consist of spatially distributed autonomous sensor nodes for data acquisition. Each node is able to sense the environment, perform simple computations and communicate with its other sensors or with the central unit.

The IQRS system is optimized for applications in Real Time Location Systems and Wireless Sensor Networks.

The solution we have developed integrates three different technologies: 1. GPS (outdoor tracking) 2. LPM—UWB (high accuracy positioning system for indoor and outdoor measurements [4]) 3. Inertial Measurement Unit (accelerometer, gyroscope and magnetometer for fine motion analysis).

We analyze these signals using Artificial Intelligence algorithms and integrate these signals using sensor fusion. Sensor integration makes it possible to track fine movements and determine orientation, which can also be important information in risk analysis. Currently, in the wake of the COVID-19 pandemic, our engineers and researchers have built 20–50 cm precision systems. The results of tens of millions of position measurements show reliable and robust positioning at an accuracy below 0.5 m, the observed standard deviation is about 0.25 m, depending on local circumstances (walls, metal objects, anchor geometric configuration, etc.).

3.3 Physiological Signal Monitoring

In the sensor layer wireless sensor network of wearable noninvasive sensor units are integrated [6].

Many physiological signals can be measured from individuals during their everyday activities and are used to observe the deviations of health status in the early phase or to alert automatically in special cases.

Our solution includes user friendly, flexible and comfortable mobile/wireless signal recording devices. The devices can continuously measure different physiological signals: ECG, SpO2, heart rate, body temperature and ambient temperature.

The heart that controls life of every human being is an important organ in our body. By studying the electrocardiogram (ECG) valuable information about the health status of the patient can be obtained. From the ECG can be calculated and predicted some life endangering situations.

We developed an intelligent agent for automatic, real time and continuous detection, analysis warning and predictive diagnostic. In the case of a mobile ECG device the patient can move, run, do sports using the device, so the different disturbances can be very large. In our solution we used special feature extraction and machine learning algorithms to give automatically validated diagnosis. The built-in *motion sensors* (accelerometer, gyroscope, magnetometer) provide an opportunity to examine physiological signals as a function of the intensity of physical exertion. Our Physiological Expert system has shown its effectiveness in tens of thousands of measurements.

In a special study 63 patient was examined to compare the physiological effects of an extreme physical and psychological stress. Physical stress test was carried out in an exercise physiology laboratory, where subjects completed an incremental treadmill test to full exhaustion. Psychological test was brought off in a military tactical room, where subjects met a street offence situation. ECG and blood pressure were recorded directly before, during (ECG), immediately after (BP) and 30 min after the test. Artificial Intelligence algorithm analyzed the measured signals and calculated the Heart Rate Variability parameters. Majority of HRV indices changed significantly between the baseline and post test at both stress protocols. Inverse changes and very strong significant difference ($p < 0.001$) was found when comparing the changes of pre and post stress HRV indices between the physical and psychological tests. Significant difference was found in the change of systolic and diastolic blood pressure between the test protocols, and also between the baseline and post test measurements [7]. According to the results HRV seems to be a sensitive tool to measure the impact of either physical or psychological stress.

Measuring *body temperature* is medically important. This is because many diseases are associated with characteristic changes in body temperature. The course of certain diseases can be monitored by measuring body temperature. Even in special environmental conditions or in physical activity, the temperature may rise temporarily, for example, extreme physical exertion has a special body temperature characteristic.

With the help of a high precision fast thermometer our device is able to asses the metabolic processes, as compared *to ambient temperature.*

Blood Oxygenation Measurement (SpO2) and *Heart Rate* are also measured. SpO2 or pulse oximetry is a noninvasive method for monitoring a person's oxygen saturation, which is related to the heart pulse.

3.4   Space and time digitalization

We digitize the measurement data, synchronize it, and build a huge data network from it in space and time. This data network forms the basis of different types of analytics and services.

3.5 Data Management Module

Telehealth systems always include a subject-oriented and time-variant digital platform to access, store, share, and secure patient medical data.

3.6 Artificial Intelligence

Artificial Intelligence also ensures that the correlations of different types of data are synchronized in space and time, building a huge data network. With the help of data network analysis, we can provide a high level of decision support for experts and users. We used the following sub-areas of Artificial Intelligence during the developments: Signal Processing, Control Engineering, Navigation Systems, Data Mining, Expert Systems, Decision Support Systems, Sensor Technology and Fusion, Time Series Analysis, Medical Research, Automatic ECG Analysis, Uncertain Knowledge Management, Machine Learning, Radio Transmission Algorithms, Distributed Algorithms, BigData technology, Face recognition, and other fields.

Each of these technologies is applied in different modules of the Platform in order to produce an integrated service.

Our company has been working for more than 10 years on the development of appropriate high-tech devices and the development of special expert modules for various fields, such as cardiovascular rehabilitation, anthropometry, sports, stress research [7, 8], building control and automation, healthcare, safety and security, industrial parks, horse training.

3.7 Analytic Applications, Reports and Alarms:

The various functionalities are implemented in the form of reports and alerts. Reports can be exported in several different formats (Fig. 4).

4 Functions and Potential Use Cases

Such a platform can be used not only in the field of Telemedicine, but also in institutions, companies, industrial parks, sites, entertainment centers, etc.

Contact Tracing			
Harry Wilson	ID	SUM Time (hour)	
2020.05.12	2020.05.30	80,00	
Person	Personal Contact Time (hour)	Same Trace in 8 hours Time (hour)	Risk (1-100)
Alexander	45,00	12,00	86
Catrin	12,60	23,00	53
James	10,34	10,00	28
Mason	2,04	4,50	9
Oliver	0,64	1,20	3
Lucas	0,02	0,90	1,38

Sanitation state Report				
Date	Time	Foreground	Office1	Office2
2020.05.12	6:12:00	20%	12%	24%
2020.05.12	8:12:00	32%	38%	79%
2020.05.12	10:12:00	88%	88%	97%
2020.05.12	12:12:00	88%	14%	14%
2020.05.12	14:12:00	10%	49%	77%
2020.05.12	16:12:00	35%	87%	94%

Capacity Report				
Date	Time	Foreground	Office1	Office2
2020.05.12	10:12:00	23	98	92
2020.05.12	12:12:00	87	67	39
2020.05.12	16:12:00	12	59	85

Warnings				
Date	Time	Foreground	Office1	Office2
2020.05.12	10:12:00		SANITATE	SANITATE
2020.05.12	12:12:00	SANITATE		
2020.05.12	16:12:00			SANITATE

6ft of separation warnings				
Date	Time	Person1	Person2	Intervall
2020.05.12	6:04:00	Alexander (ID1)	Robert (ID2)	0:04:04
2020.05.12	8:12:00	Catrin (ID1)	Michael (ID2)	0:24:02
2020.05.12	9:14:00	Catrin (ID1)	Robert (ID2)	0:09:03
2020.05.12	10:36:00	Alexander (ID1)	Thomas (ID2)	0:04:14
2020.05.12	16:04:00	Oliver (ID1)	Noah (ID2)	0:54:10
2020.05.12	16:04:00	Alexander (ID1)	Harry (ID2)	0:03:03

6ft of separation SUMMARY			
Date	Name	ID	Number of warnings
2020.05.13	Catrin	234	2
2020.05.13	Harry	123	74
2020.05.13	Oliver	345	30

Fig. 4 Reports

4.1 COVID-19's Protection Specialized Functions

- tracking: enforce social distancing, perform contact tracing
- safety and health: high contact areas are being routinely sanitized, send an immediate alert if a symptom of an infection occurs
- controlling: risk are isolated faster, keep under control the spread of the virus, ban on prohibited areas
- optimization: risk forecasting, workflow optimization, the scope of the impact is reduced
- cost-effectiveness: testing and other costs are minimized, minimized revenue loss
- legality and transparency: promoting and monitoring compliance with current safety standards
- employee satisfaction: employees feel safe
- facilities capacity management: buildings, rooms capacity and time scheduling management based on the position and tracking information.

Other use cases and functions that support the operation of organizations: control the day-to-day tasks; improvement of work conditions and workflows; track your assets, people and actions; real time location, condition and timing; accuracy and speed of events occurring throughout the operation; various sensory modules for full-spectrum observation of persons and objects; better and more efficient operation at organizational level; risk reduction; reduce operating costs; control, manageability and optimization of processes; transparency; improving work protection; more efficient labor utilization; check the intensity, quality of work; and optimization of the logistic.

5 Conclusions

A special system has been presented that provides real answers to the problems posed by the COVID-19 situation. We have developed a complex solution that includes social distance control [1], effective contact detection [2], continuous monitoring of people's health status, and is based on digitizing and monitoring spaces, time, motion and physiological parameters with special measurement devices, and artificial intelligence.

The basic idea of building home care systems is not new [3, 6], but their real-life usability depends to a large extent on whether integrated measuring devices are really capable of fulfilling their function, i.e. measuring reliable, accurate, diverse data that can really be used by professionals. It depends on whether thousands of devices can operate simultaneously even within a building, and time-shared radio protocols are capable of operating a large number of measuring tags simultaneously. Another important factor is whether signal processing algorithms are capable of automatically analyzing signals. Also an important challenge is whether expert systems have been developed to an appropriate professional level, which combine the knowledge base of a specific field, and whether artificial intelligence algorithms are able to provide a high level of decision support.

Our COVID-19's Platform provides an effective response to these challenges in terms of both hardware and software, and expert systems.

References

1. Lewnard JA, Lo NC (2020) Scientific and ethical basis for social-distancing interventions against COVID-19. Lancet Infect Dis 20(6):631–633. https://doi.org/10.1016/S1473-3099(20)30190-0
2. Dubov A, Shoptaw S (2020) The value and ethics of using technology to contain the COVID-19 Epidemic. Am J Bioethics. https://doi.org/10.1080/15265161.2020.1764136
3. Lokken TG et al (2020) Overview for implementation of telemedicine services in a large integrated multispecialty health care system. Telemed e-Health 26(4)
4. Mendoza-Silva GM, Torres-Sospedra J, Huerta J (2019) A meta-review of indoor positioning systems. Sensors 19(20):4507
5. Jennifer Y, Biswanath M, Dipak G (2008) Wireless sensor network survey. Comput Netw 52(12):2292–2330
6. Kocur D, Švecová M, Zetik R (2019) Basic signal processing principles for monitoring of persons using UWB sensors. An Overview. Acta Electrotechnica et Informatica 19(2):9–15. doi: https://doi.org/10.15546/aeei-2019-0009
7. Móra Á, Szilágyi GY et al (2019) Differences of autonomic nervous system regulation at physical and psychological stress. In: ACSS 2019, Asia-Singapore conference on sport science. ISBN:978–981–11–6605
8. Ligetvári R, Szilágyi GY et al (2019) Increased levels of plasma endothelin-1 (ET-1) in response to acute extreme physical but not to mental stress with preserved left ventricular function in male Hungarian athletes. In: 24th annual congress of the european college of sport science

Optimizing IoT Networks Through Combined Estimations of Resource Allocation and Computation Offloading

Saibal Ghosh and Dharma P. Agrawal

Abstract The low energy requirements of devices comprising the Internet of Things have made them ubiquitous. However, it is also a huge bottleneck in real world deployments. Computational offloading can improve network lifetimes wherein these devices move heavy computation away to the edge and the cloud. However, this also increases power consumption and network latencies. Optimum performance can be achieved by intelligently deciding if the computation should be offloaded or performed locally—a task that is inherently difficult given the complexity of the systems. In this work, we formulate a distributed network optimization problem to achieve a balance between performance and energy consumption. Our simulation results show substantial improvements in network latency and power consumption.

Keywords Internet of things · Computation offloading · Edge computing · Energy efficiency

1 Introduction

The rapid growth of the Internet of Things (IoT) in the past decade has been fuelled by improvements in computing capabilities and battery technology. However, this widespread usage has also meant that these devices are being used in scenarios that were not originally envisaged resulting in reduced performance and expected lifetimes. While traditional cloud computing can help to alleviate some of these issues, extending the cloud to the edge can improve the perceived latency and lack of computing capability immensely.

S. Ghosh (✉) · D. P. Agrawal
University of Cincinnati, Cincinnati, OH 45221, USA
e-mail: ghoshsl@mail.uc.edu

D. P. Agrawal
e-mail: agrawadp@ucmail.uc.edu

Current deployments of IoT devices include a number of edge computing nodes to accelerate computation. These nodes are typically more powerful than individual IoT devices but lack the full fledged computation capability of a cloud. For efficiency, these nodes can also dynamically modify their resources based on the computational load [1]. A hierarchy of nodes are generally employed based on their capabilities to further improve the efficiency of the whole system. As the computational load increases in individual IoT devices, they can make a decision to offload some of the computation to edge devices. Each edge device can in turn offload to the next higher layer based on their current load until it reaches the cloud. The crux of this mechanism is to ensure that an individual node is never inundated with too much computation that might result in longer wait times and faster depletion of energy. However, computation offloading also introduces latency and results in more energy utilization. In order to ensure optimal utilization of system resources, it is important to decide thresholds and resource allocation for offloading based on the current computation and network state and the overall computation to be performed.

The nature of network traffic between IoT devices and the edge is inherently complex and depends on a variety of factors including the current system load, the wireless channel capacity and overall energy consumption. If IoT devices merely offload all the computation to the edge nodes, it would increase the energy consumption exponentially and greatly increase latencies on the wireless channel. However, if these devices perform all computations locally, the total time for computation would also increase exponentially. Modelling our system would therefore involve intelligent handling of local and offloaded computation taking into account the overhead from the offloading decision making process. The offloading decision is based on a continuous evaluation of the prevalent network conditions and the nature of the computation.

In this work, we have modeled the problem of computation offloading and resource allocation as a network optimization problem with the goal of minimizing power consumption across the IoT devices and the edge while ensuring added latencies incurred from offloading are kept to a minimum.

The rest of the chapter is organized as follows. Section 2 describes the related research and how our work differs from the existing work in this direction. We describe our system architecture and define the problem formally in Sect. 3. Section 4 describes our simulation setup while we analyze the results in Sect. 5. Section 6 concludes the paper and describes our goals for the future.

2   Related Work

Traditional power saving in IoT networks offloads computation to the nodes at the edge. This method has improved power efficiency in smartphones wherein they just act as portals while the actual computation is performed at the data center [2, 3]. Average task duration and response times under limited energy constraints can be minimized by utilizing the alternating direction method of multipliers [4, 5]. A greedy heuristic scheme for multi hop offloading with offloading path selection was studied in [6]. A study on the social relationships in IoT users was used to develop a socially aware offloading scheme through a game theoretic approach in [7]. A study on the joint minimization of latency and energy consumption on IoT networks was performed in [8]. The study formulated the problem as a continuous time Markov decision process and a solution was developed using dynamic programming.

A dynamic offloading scheme was developed for energy harvesting IoT devices as part of [9]. An intelligent mechanism for placement of IoT nodes that takes into account resource availability was formulated in [10]. An adaptive offloading scheme minimizing energy consumption in transmission with a guaranteed queuing latency was developed in [11]. A socially aware network resource allocation scheme for device to device communications was developed in [12]. Chen et al. studied a dynamic service migration mechanism for cognitive computing at the edge in [13]. An optimal strategy for offloading computation in energy harvesting devices was developed in [14]. This study was based on Lyapunov optimization techniques and employed the Vickrey-Clarke-Groves auction method to determine rewards and improve the system.

Most of these works assume that statistical information on computation tasks are readily available. However, this is inherently difficult in real world deployments [15]. Moreover, these works assume a static dual level network wherein IoT devices send and receive computation to/from the cloud. However, as we observe in this work, increasing the number of tiers increases the efficiency albeit with a finite increase in the network complexity and traffic. Our simulation results attest to this observation and prove that the overall efficiency can be improved even with the added overhead of maintaining a multi-tier edge network.

3 System Architecture and Problem Formulation

We consider a multi-tiered IoT network comprising of a number of devices, a number of edge computing nodes and the cloud. The edge layer bridges the IoT devices and the computing nodes over wireless links. Each IoT device can connect to at least one of these edge nodes. The node that the device chooses to offload its data is evaluated based on network load and latencies. Each edge node can connect to at least of the edge nodes that are one level higher. If the computation exceeds the capacity of the edge nodes at the highest tier, they offload to the cloud. The highest level edge nodes access the cloud over high capacity wired links. Our work models the energy consumption and perceived latencies in the edge node tiers. The energy consumption is comprised of power consumed in processing and transmission. The perceived latency is a combination of the queuing latency, processing latency and the transmission latency. For the purposes of our simulation we consider each work unit to complete processing within its allotted time slot. We would now briefly describe the mechanism to compute these factors in the following sections.

3.1 CPU Workload

The exponential moving average metric is used to compute the average CPU utilization [16]. For a given time period t, the CPU utilization C_t is given by:

$$C_t = \alpha \cdot c_t + (1 - \alpha) \cdot C_{t-1} \tag{1}$$

where, $\alpha = 2/(N + 1)$ and c_t is the instantaneous CPU utilization at time t, α is the weighted decreasing coefficient, and N denotes the total time period.

3.2 Channel Traffic and Bandwidth

The exponential moving average metric is also used to compute the instantaneous traffic before making an computation offloading decision. For a given time period t, the bandwidth B_t is given by:

$$B_t = \beta \cdot b_t + (1 - \beta) \cdot B_{t-1} \tag{2}$$

where, $\beta = 2/(N + 1)$ and b_t is the instantaneous bandwidth at time t, β is the weighted decreasing coefficient and N denotes the total time period.

3.3 Modeling the Access Node

We consider that during time slot t, there is a finite amount of computation $C_i(t)$ generated from IoT devices that need to be processed at the access node i. The task arrival rates are different for each access node. Each access node employs time series forecasting methods to estimate the arrival rate based on the previous work load and generates an estimate window W_i [17].

We consider each access node to maintain four buffers:

1. Input buffer for incoming computations $C_{i,-1}(t)$
2. Estimation buffer for estimated computations $C_{i,0}(t), ..., C_{i,W_i-1}(t)$
3. Processing buffer for local computations $P_i^{(a,p)}(t)$ and
4. Offloading buffer for computations to be offloaded $O_i^{(a,o)}(t)$

At any time slot t, the estimation buffer $C_{i,w}(t)(0 \leq w \leq W_i - 1)$ stores the computation that is expected to arrive in the time slot $(t + w)$. Computations that actually arrive at access node i is stored in the input buffer $C_{i,-1}(t)$, before they are moved to the processing or offloading buffer. Computations in the processing buffer $P_i^{(a,p)}(t)$ are processed by the access node i locally, while those in the offloading buffer $O_i^{(a,o)}(t)$ would be offloaded to one of the computing nodes.

In each time slot t, the access node i can move estimated workloads into the estimation buffer. We define $\mu_{i,w}(t)$ as the amount of computational workload from $C_{i,w}(t)$, for $w \in \{-1, 0, ..., W_i - 1\}$, the amount of computation assigned to the processing buffer as $\rho_i^{(a,p)}(t)$ and that to the offloading buffer as $\rho_i^{(a,o)}(t)$. Thus, we have:

$$0 \leq \rho_i^{(a,\beta)}(t) \leq \rho_{i,max}^{(a,\beta)}(t) \forall \beta \in \{p, o\} \tag{3}$$

where, $\rho_{i,max}^{a,\beta}$ is ≥ 0. Therefore, we have:

$$\sum_{w=-1}^{W_i-1} \mu_{i,w}(t) = \rho_i^{(a,p)}(t) + \rho_i^{(a,o)}(t) \tag{4}$$

In each time slot t, computations in the offloading buffer $O_i^{(a,o)}(t)$ would be transferred to the set of computing nodes M_i connected to the access node. The transmission capacity is determined from the transmission power $(p_{i,j}(t))_{j \in M_i}$, where $p_{i,j}(t)$ is the power required to transmit from the access node i to the computing node j.

Therefore, we have:

$$p_{i,j}(t) \geq 0, \forall i \in N, j \in M_i, t \tag{5}$$

and,

$$\sum_{j \in M_i} p_{i,j}(t) \leq p_{i,max}, \forall i \in N, t \tag{6}$$

The transmission capacity from access node i to computing node j is [18]:

$$U_{i,j}(t) \triangleq \hat{U}_{i,j}(p_{i,j}(t)) = \tau_0 B \log_2 \left(1 + \frac{p_{i,j}(t)H_{i,j}(t)}{N_0 B}\right) \tag{7}$$

where, $U_{i,j}(t)$ is the total transmission capacity in time slot t, τ_0 is the length of each time slot, B is the channel bandwidth, $H_{i,j}(t)$ is the wireless channel gain between the access node i and the computing node j and N_0 is the power spectral density of the additive white Gaussian noise. In this work, we ignore the intra-node wireless interference.

Modulating the transmission power allows us to control the amount of computation offloaded from access node i to computing node j. Therefore, changes to the offloading buffer is governed by:

$$O_i^{a,o}(t+1) \leq (O_i^{a,o}(t) - \sum_{j \in M_i} U_{i,j}(t)) + \rho_i^{a,o}(t) \tag{8}$$

where, $\sum_{j \in M_i} U_{i,j}(t)$ is the total transmission capacity at time slot t. We ignore the transmission latency between the access node i and the computing node j when compared to the length of the time slot t.

3.4 Modeling the Computing Node

We consider each computing node $j \in M$ to maintain three buffers:

1. Input buffer for incoming computations $I_j^{(c,i)}(t)$
2. Processing buffer for local computations $P_j^{(c,a)}(t)$ and
3. Offloading buffer for computations to be offloaded $O_j^{c,o}(t)$

The computational workloads arriving at computing node j consist of workloads that have been offloaded from the set of access nodes N_j. Denoting computation for the processing buffer as $\rho_j^{(c,i)}(t)$ and the offloading buffer as $\rho_j^{(c,o)}t$:

$$0 \leq \rho_j^{c,\beta}(t) \leq \rho_{j,max}^{(c,\beta)}, \forall \beta \in \{p, o\} \tag{9}$$

where $\rho_{j,max}^{(c,\beta)} \geq 0$. Therefore, we have:

$$O_j^{c,i}(t+1) \leq (O_i^{c,i}(t) - (\rho_j^{c,p}(t) + \rho_j^{c,o}(t)) + \sum_{j \in N_j} U_{i,j}(t)) \tag{10}$$

For each computing node, $j \in M$, its offloading buffer $O_j^{c,o}(t)$ contains the workload that is destined for the cloud. We consider $G_j(t)$ as the capacity of the wired link from the computing node j to the cloud during time slot t. Therefore, the capacity from the computing node j to the cloud is [18]:

$$O_j^{c,o}(t+1) \leq (O_j^{c,o}(t) - G_j(t)) + \rho_j^{c,o}(t) \tag{11}$$

The workload offloaded to the cloud is bounded by $min\{O_j^{c,o}, G_j(t)\}$.

3.5 Energy Expenditure

The energy $E(t)$ consumed by the system in time slot t is a combination of the energy expended in processing and transmission. Considering a CPU running at frequency f, the power consumption is $\tau_0 \sigma f^3$ where σ is specific to the CPU hardware [19].

Therefore, we have:

$$E(t) \triangleq \hat{E}(f(t), p(t)) = \sum_{i \in N} \tau_0 \sigma (f_i^{(e)}(t))^3 + \sum_{j \in M} \tau_0 \sigma (f_j^{(c)}(t))^3 + \sum_{i \in N} \sum_{j \in M_i} \tau_0 p_{i,j}(t) \tag{12}$$

where, $f(t) \triangleq ((f_i^{(a)})(t))_{i \in N}, ((f_j^{(c)})(t))_{j \in M}$ is the set of all CPU frequencies in the system, and $p(t) \triangleq (p)i(t))_{i \in N}$. $p_i(t) = (p_{i,j}(t))_{j \in M}$ is the power allocated for transmission in the access node i.

3.6 *Problem Formulation*

Having defined our system parameters we can now define our problem formally. The overall power consumption $\hat{P}$ is defined as:

$$\hat{P} \triangleq \lim_{T \to \infty} \sup \frac{1}{T} \sum_{t=0}^{T-1} E\{P(t)\} \tag{13}$$

and the overall buffer backlog as:

$$\hat{B} \triangleq \lim_{T \to \infty} \sup \frac{1}{T} \sum_{t=0}^{T-1} \sum_{\beta \in \{a,p,o\}} (\sum_{i \in N} E\{B_i^{(a,\beta)}(t)\} \sum_{j \in M} E\{B_j^{(c,\beta)}(t)\}) \tag{14}$$

Our objective is to minimize power consumption $\hat{P}$ while ensuring that all computations complete within a reasonable time. We use a variation of the Lyapunov optimization techniques to split the problem into sub problems over time slots [20–22]. We are omitting the details of this process due to space constraints.

In each time slot t, each node $k \in N \cup M$ determines the computation to be performed locally and offloaded as $\rho_k^{(\alpha,a)}(t)$ and $\rho_k^{(\alpha,o)}(t)$ respectively where $\alpha = \{a, c\}$ is dependent on the hardware. The decision to offload is determined by:

$$\min_{0 \le \rho_k^{(\alpha,\beta)} \le \rho_{k,max}^{(\alpha,\beta)}} (O_k^{(\alpha,\beta)}(t) - O_k^{(\alpha,a)}(t)) \rho_k^{(\alpha,\beta)}$$

wherein, $\beta = \{a, o\}$, and the optimal solution is given as:

$$\rho_k^{(\alpha,\beta)(t)} = \begin{cases} \rho_{k,max}^{(\alpha,\beta)}, & if O_k^{(\alpha,\beta)}(t) < O_k^{(\alpha,a)}(t) \\ 0, & otherwise \end{cases} \tag{15}$$

4 Simulation Setup

We have used a WSNet-based simulator written in C/Modern C++ [23]. We have extended and improved the simulator in the the spectrum usage and orthogonal models to reduce wireless interference. The *spectrum usage* model represents the time-frequency dependent aspects of communication. It can be used to exploit the available spectrum in terms of spectral resources rather than the traditional logical channel approach. This feature allows us to support the heterogeneity in the Physical layer arising from the different IoT devices.

Table 1 Simulation parameters

Parameter	Value
Input data size	0–1500 KB
Channel bandwidth	0- 500 KB/s
CPU utilization	10–70%
Latency threshold	10–30 s

The simulation parameters are in Table 1. Our study determined if our mechanism correctly decided to offload by evaluating if an offload would provide tangible benefits over local computation. An explanation and discussion of our results follows from our analysis of these parameters is presented in the next section.

5 Analysis and Discussion

We have compared the variation of input data size, channel bandwidth, CPU utilization and the latency threshold with respect to the overall energy consumption and overall execution time.

Figure 1 shows more energy is consumed for local computations, as compared to offloading and the cost increases with the size of the input data. This is because

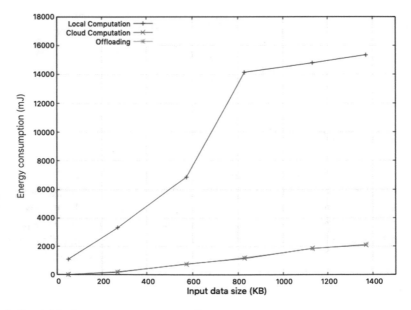

Fig. 1 Input data size versus energy consumption

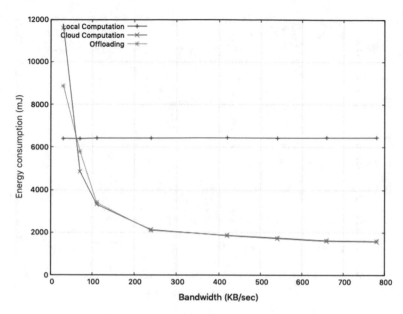

Fig. 2 Channel bandwidth versus energy consumption

the energy expended in transmission is less than computation and the local energy consumption increases faster than on the cloud. Figure 2 shows computations use less energy locally when the bandwidth is less but changes as bandwidth requirements increase. This is because, while the transmission power remains constant, the effect is more pronounced in lower bandwidth ranges.

Figure 3 shows that the utilization of CPU is less on the cloud as compared to local execution since weaker IoT devices take longer and expend more energy to compute the same amount of work. The latency threshold data in Fig. 4 shows it is more economical to run the computations locally when the latency threshold is low since extra delays in transmission increases the overall latency.

Figure 5 shows that the execution time grows faster with increasing data sizes when the computation is performed locally. This is due to the weaker CPUs in the IoT nodes. Figure 6 shows that as the bandwidth increases the execution time decreases for offloading. This is because the transmission overhead is higher in lower bandwidth scenarios. Figure 7 shows that CPU utilization has little impact on the overall execution time when the computation is offloaded because of powerful hardware. The latency threshold data in Fig. 8 shows that offloading the computation reduces the execution time once the threshold exceeds a specified value and we validate this with our simulations.

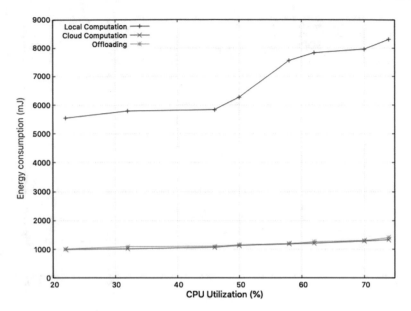

Fig. 3 CPU utilization versus energy consumption

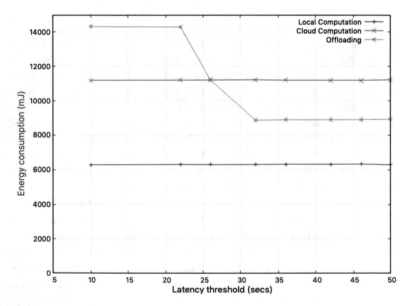

Fig. 4 Latency threshold versus energy consumption

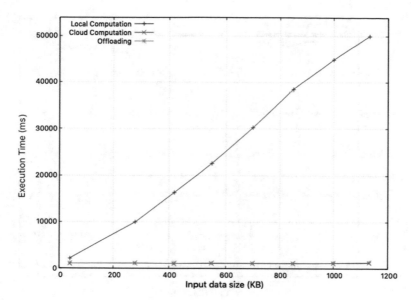

Fig. 5 Input data set size versus execution time

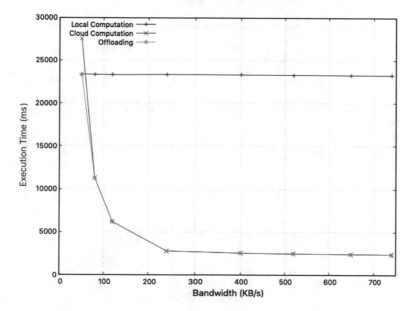

Fig. 6 Channel bandwidth versus execution time

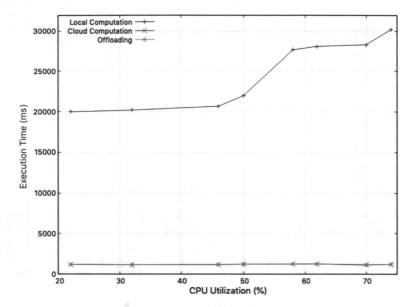

Fig. 7 Average CPU utilization versus execution time

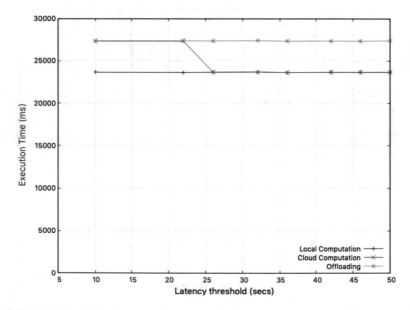

Fig. 8 Latency threshold versus execution time

6 Conclusion and Future Work

In this work we have explored the problem of computation offloading and resource allocation in IoT networks with multiple intermediate edge levels. We formulated an optimization problem and showed that by estimating computation requirements we can reduce overall power consumption and latency. We are currently exploring estimation models with machine learning techniques that can dynamically adapt, improve estimations and lead to better results.

References

1. Yi S, Hao Z, Qin Z, Li Q (2015) Fog computing: platform and applications. In: Proceedings of IEEE HotWeb
2. Cuervo E, Balasubramanian A, Cho DK, Wolman A, Saroiu S, Chandra R, Bahl P (2010) MAUI: making smartphones last longer with code offload. ACM MobiSys
3. Kumar K, Lu YH (2010) Cloud computing for mobile users: can offloading computation save energy? Computer 43(4):51–56
4. Xiao Y, Krunz M (2017) QoE and power efficiency tradeoff for fog computing networks with fog node cooperation. In: Proceedings of IEEE INFOCOM
5. Wang Y, Tao X, Zhang X, Zhang P, Hou YT (2019) Cooperative task offloading in three-tier mobile computing networks: an ADMM framework. IEEE Trans Veh Technol 68(3):2763–2776
6. Misra S, Saha N (2019) Detour: dynamic task offloading in software defined fog for IoT applications. IEEE J Sel Areas Commun 37(5):1159–1166
7. Liu L, Chang Z, Guo X (2018) Socially-aware dynamic computation offloading scheme for fog computing system with energy harvesting devices. IEEE Internet Things J 5(3):1869–1879
8. Lei L, Xu H, Xiong X, Zheng K, Xiang W (2019) Joint computation offloading and multi-user scheduling using approximate dynamic programming in NB-IoT edge computing system. IEEE Internet Things J 6(3):5345–5362
9. Mao Y, Zhang J, Song S, Letaief KB (2016) Power-delay tradeoff in multi-user mobile-edge computing systems. In: Proceedings of IEEE GLOBECOM
10. Taneja M, Davy A (2017) Resource aware placement of IoT application modules in fog-cloud computing paradigm. In: Proceedings of IFIP/IEEE IM
11. Chen Y, Zhang N, Zhang Y, Chen X, Wu W, Shen XS (2019) Energy efficient dynamic offloading in mobile edge computing for Internet of Things. IEEE Trans Cloud Comput
12. Gao Y, Tang W, Wu M, Yang P, Dan L (2019) Dynamic social-aware computation offloading for low-latency communications in IoT. IEEE Internet Things J
13. Chen M, Li W, Fortino G, Hao Y, Hu L, Humar I (2019) A dynamic service migration mechanism in edge cognitive computing. ACM Trans Internet Technol 19(2):30
14. Zhang D, Tan L, Ren J, Awad MK, Zhang S, Zhang Y, Wan P-J (2019) Near-optimal and truthful online auction for computation offloading in green edge-computing systems. IEEE Trans Mobile Comput
15. Zhang D, Chen Z, Cai LX, Zhou H, Duan S, Ren J, Shen X, Zhang Y (2017) Resource allocation for green cloud radio access networks with hybrid energy supplies. IEEE Trans Veh Technol 67(2):1684–1697
16. Burgstahler L, Neubauer M (2002) New modifications of the exponential moving average algorithm for bandwidth estimation. In: Proceedings of the 15th ITC specialist seminar (2002)
17. Ahmed NK, Atiya AF, Gayar NE, El-Shishiny H (2010) An empirical comparison of machine learning models for time series forecasting. Econ Rev 29(5–6):594–621

18. Gallager RG (2008) Principles of digital communication. Cambridge University Press, Cambridge
19. Kim Y, Kwak J, Chong S (2018) Dual-side optimization for cost-delay tradeoff in mobile edge computing. IEEE Trans Veh Technol 67(2):1765–1781
20. Huang L, Zhang S, Chen M, Liu X (2016) When backpressure meets predictive scheduling. IEEE/ACM Trans Network 24(4):2237–2250
21. Neely MJ (2010) Stochastic network optimization with application to communication and queueing systems. Synthesis Lect Commun Networks 3(1):1–211
22. Leon-Garcia A (2017) Probability, statistics, and random processes for electrical engineering, 3rd ed. Pearson Education
23. Center for Research and Specialized Technology in micro and nanotechnologies, WSNet - An event-driven accurate and realistic network simulator written in C/C++ for wireless networks on a large scale. 2020 [Online]. Available: https://github.com/CEA-Leti/wsnet

A Decision Support Framework to Enhance Performance in Resource Constrained Devices

Saibal Ghosh and Dharma P. Agrawal

Abstract The last few decades have seen an exponential increase in capabilities of smaller computing devices with an equally substantial decrease in their form factors. The first wave was ubiquitous mobile connectivity which paved the way for the widespread adoption of the Internet of Things (IoT) paradigm and its later incarnation: The Internet of Everything (IoE). However, many of these devices may not be able to fulfill all the requirements of applications that are designed to run on them. In this work, we explore mechanisms to optimize computing offloading decisions while minimizing transmission power to create a decision framework. Our simulations show that the framework can achieve substantial improvements in computing performance and power consumption.

Keywords Mobile wireless networks · Mobile user experience · Distributed resource allocation · Distributed computing · Internet of things

1 Introduction

As more and more users have switched to smartphones, the demand for rich mobile computing experiences have skyrocketed. New mobile applications such as natural language processing, facial recognition, virtual and augmented reality and massively multiplayer online gaming have emerged [1–3]. The power of ubiquitous connectivity has even extended to smart vehicles helping in early fault detection and remote diagnosis, requiring minimal human interaction and helping to keep our roads safer [4, 5]. For such experiences to be feasible, engineers have to keep the device's power and form factors to an acceptable level while allowing enough processing power to create immersive experiences for effective user engagement [6].

S. Ghosh (✉) · D. P. Agrawal
University of Cincinnati, Cincinnati, OH 45221, USA
e-mail: ghoshsl@mail.uc.edu

D. P. Agrawal
e-mail: agrawadp@ucmail.uc.edu

© The Author(s), under exclusive license to Springer Nature Singapore Pte Ltd. 2021
H. Kim et al. (eds.), *Information Science and Applications*, Lecture Notes
in Electrical Engineering 739, https://doi.org/10.1007/978-981-33-6385-4_32

A solution to this problem is mobile edge and cloud computing. Traditional Cloud infrastructure providers such as Amazon Web Services, Microsoft Azure and Google Cloud can provide the resources to run these demanding applications. However, moving computation away to the edge and cloud introduces unexpected latencies that impacts performance. More powerful IoT devices can act as a bridge between traditionally inaccessible areas and the cloud [7]. We propose a tiered approach wherein parts of the computation are moved to the edge for intermediate computing while the cloud handles tasks that require more computing power[8–12]. The edge and the cloud differ in computing power and network latencies but a balance can be achieved to provide the best of efficiency and user experiences.

In this work, we study the offloading strategies for mobile devices and find a middle ground, leveraging the edge and the cloud. Our simulations show that an informed decision on computation offloading can be achieved from observing the size of the payload data and computing load. By leveraging the combined power of the edge and the cloud, we can obtain a framework for an optimal offloading strategy that efficiently allocates workloads and minimizes power expenditure.

We describe the problem of distributing computation in the next section and our simulation model in Sect. 3. We objectively formulate the problem of offloading computation decisions in Sect. 4 and discuss the simulations and results in Sect. 5. We finally conclude with our observations in Sect. 6.

2   Effectively Distributing Computation

An IoT network distinguishes itself from traditional computer networks in the sheer number of nodes. Efficient utilization of the system requires a majority of these nodes to perform computations. However, this also results in increased traffic for housekeeping. Ideally, the computation should be distributed to the maximum possible extent while minimizing traffic. We split the task into atomic units that can be independently computed at a node. Each group of nodes taking part in a larger computation task is led by a leader node. We use a lightweight implementation of Paxos to maintain consensus in the network [13].

Considering a group of nodes denoted by N and a group of leaders L, we construct a bipartite graph denoted as $G = (N \cup L, E)$, where E is the set of edges between nodes and leaders. An edge between $n_i \in N$ and $l_j \in L$ indicates the presence of a communication channel between the two. Each node has one leader while each leader is responsible for one or more subordinate nodes. We consider t_i as the task to be executed on node n_i and that each node is able to connect to other nodes either as a group or through leaders that communicate between groups. In order to optimize for network traffic we have:

$$min \, {}^{max}_{j \in L} \sum_{i=1}^{|N|} t_i \, x_{ij}$$

wherein,

$$\sum_{j=1}^{|L|} x_{ij} = 1, i = 1, \ldots, |N|$$

$$x_{ij} \in 0, 1$$

(1)

x_{ij} is 1, if the node n_i is part of the group that has l_j as the leader and 0 otherwise.

3 Modelling the System

We consider a system consisting of a large number of low powered devices clustered in small groups with leaders. The leader coordinates task migrations and groups tasks to be dispatched to the edge or cloud. All tasks assigned to an individual node are atomic. However, for the overall computation, a number of these smaller sub-tasks may need to be computed in parallel. Also, for efficiency, some tasks are grouped together and computed at the edge or dispatched to the cloud. In general, we consider the following scenarios:

1. Tasks that can be computed individually at each node.
2. Tasks that need to be grouped and computed at the edge.
3. Tasks that require intensive computation and are dispatched to the cloud.

We believe that this model accurately models a modern, large sized, real world mobile network. Most computational tasks have a large number of low intensity tasks and a minute number of computationally expensive tasks. Moreover, real world computation demands that some sub-tasks must follow an order of computation. Our simulation considers the absolute order of computations for an atomic task in each node to determine the overall order of computation. Building the graph for computations helps us to formulate the decision support framework to assist in real world IoT setups and deployments in the future.

Figure 1 shows a schematic diagram of the network with mobile devices, base stations, edge computing devices and the cloud. Our network is comprised of a number of macro base stations connected to the cloud over very high capacity backbone links. We have considered a tiered deployment of base stations wherein each macro base station can also be connected to a number of smaller base stations. Each macro base station has a number of Mobile Edge Computing (MEC) servers. The smaller base stations are connected to the macro base stations over high capacity backbone or wireless links. Each smaller base station is connected to one or more devices over wireless links. Leader election strategies are applied to elect the base station and macro base station leaders based on the nature of computation and the prevalent network conditions.

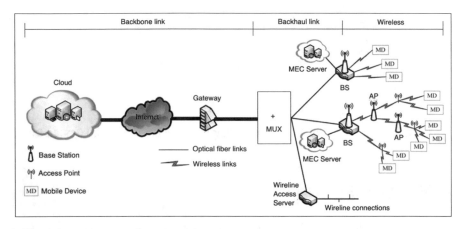

Fig. 1 The tiered network architecture

We denote the set of macro (S_m) and small base stations (S_s) as a nested list of resources. Each macro base station in the list can be the leader of one or more smaller base stations. We denote this as:

$$S_m = S_{m_0}, S_{m_1}, S_{m_2}, \ldots, S_{m_n}$$

where,

$$S_{m_i} = S_{s_0}, S_{s_1}, S_{s_2}, \ldots, S_{s_n}$$

The macro and small base stations operate in non-overlapping frequency bands. The spectrum is divided into C channels where $C = 1, 2, \ldots, C$, and each channel is orthogonal to others to minimize interference. The bandwidth for each channel in the base stations is ω_b and we have $\omega_1 = \omega_2 = \ldots = \omega_b \neq \omega_m$. Let us consider the set of all nodes as $N = 1, 2, \ldots, N$. Each node has an assigned task T_n, where T_n is defined by the tuple $T_n = R_n, P_n, d_n^{max}$, where R_n is the required resource for performing the computation, P_n is the size of the input and d_n^{max} represents the maximum allowed computational delay.

For each task, we choose between local computation, group and/or offload to the network edge or offload to the cloud. The offload decision set is:

$$\Omega = -O, -1, -O, \ldots, -1, 0, 1, \ldots, O, O + 1$$

and the computational offload strategy is:

$$Y = y_n | y_n = i, i \in \Omega, n \in N \tag{2}$$

where, $y_n = 0$ for local computation at n, $y_n = m$, $(m \in S_{s_i})$ for offloading at an edge node through a small base station s, $y_n = -m$, $(m \in S_{s_i})$ for offloading to the cloud through a small base station s, $y_n = O + 1$ for offloading at an edge node through a macro base station and $y_n = -O - 1$ for offloading at the cloud through a macro base station.

3.1 Modelling Local Node Computation

Let us consider u_n^0 as a device's computing capacity, R_n as the computation that needs to be performed, n and k to be the energy capacity of the device [14]. Therefore, the time taken to process task T_n locally is given by:

$$d_0^m = \frac{R_n}{u_0^n} \tag{3}$$

and the energy consumption for the node as [14, 15]

$$c_0^n = k(u_n^0)^2 R_n \tag{4}$$

3.2 Modelling Edge Node Computation

This scenario requires I_n data to be transmitted over links to the small base stations or to the macro base station s, $s \in S_m$ to be sent over to the edge. Let us consider the transmission rate of a device n as:

$$r_n^m = n_m \omega_m \log_2 (\Delta_n p_{nm}^c + 1) \tag{5}$$

where, p_{nm}^k is the transmitting power of the device on channel c and,

$$\Delta_n = \frac{g_n^m}{N_o + I(O < |y_n| < S_m) \sum_{n' \neq n}^N \sum_{m \neq m'}^{S_m} p_{n'm'}^c g_{n'}^{m'}}$$

where N_o is the additive white Gaussian noise, g_n^m is the power gain from node n to base station m and $\sum_{n' \neq n}^N \sum_{m \neq m'}^{S_m} p_{n'm'}^c g_{n'}^{m'}$ is the interference from other edge nodes to the node n transmitting to device m on the same sub-channel, $I(x)$ is 1 if x is $true$ and 0 otherwise, n_m is the number of orthogonal sub-channels of the node n assigned by the leader node. For simplicity, we assume that:

$$C \geq \sum_{m=1}^{S_m} \sum_{i=1}^N I(|y_i| = m) \tag{6}$$

which gives us:

$$n_{S_m} = \frac{c}{\sum_{m=1}^{S_m} \sum_{i=1}^{N} I(|y_i| = m)} (m \in S_m) \tag{7}$$

and,

$$n_{S_{m+1}} = \frac{C}{\sum_{i=1}^{N} I(|y_i| = S_m + 1)} \tag{8}$$

using the uniform zero frequency reuse method [16]. Since the input data volume exceeds the output data, the time required to receive the computation is ignored.

Now, the delay and energy consumption at the leader nodes are:

$$t_n^u = I(1 \leq y_n \leq O)t_{n,m}^u + I(y_n = O + 1)t_{n,O+1}^u \tag{9}$$

and,

$$e_n^u = I(1 \leq y_n \leq O)e_{n,m}^u + I(y_n = O + 1)e_{n,O+1}^u \tag{10}$$

where, $t_{n,s}^u = \frac{R_n}{u_{n,s}^u + \frac{P_n}{r_n^s} + P_n t_b}$ is the delay in moving the task from the node to the leader through a small base station s, $t_{n,O+1}^u = \frac{R_n}{u_{n,O+1}^u + \frac{P_n}{r_n^{O+1}}}$ is from the node to the leader through a macro base station, u_n^u is the computing capacity of the node n, t_b is the transmission delay for the backbone links for a unit of data, $e_{n,s}^u = e_n^s + p_n^c + \left(\frac{R_n}{u_{n,s}^u} + P_n t_b\right) + \frac{p_{n,s}^t P_n}{r_n^s}$ is the energy for moving the task from the node to the leader through a small base station s, $e_{n,O+1}^u = e_n^s + \left(\frac{p_n^c R_n}{u_{n,O+1}^u}\right) + \left(\frac{p_{n,O+1}^t P_n}{r_n^{O+1}}\right)$ is the energy for moving the task from the node to the edge through a macro base station, e_n^s represents the energy consumed for scanning available base stations, p_n^c is the idle power draw, $p_{n,m}^t (m \in S_m)$ is the total transmission power n and $p_{n,m}^t = n_m p_{n,m}^k$.

Therefore, the cost of migrating the computation to the edge nodes is:

$$H_n^u = \zeta^u + u_{n,m}^u + \eta r_n^m, (m \in S_m) \tag{11}$$

where, ζ^u is the unit cost of computation at the edge, η is the unit cost of transmission for the base station m. Thus, the total cost of moving the computation to the leader nodes is:

$$c_n^u = \beta_n e_n^u + \alpha_n H_n^u \tag{12}$$

where, β_n and α_n are factors that govern the energy consumption and cost.

3.3 Modelling Cloud Computation

We assume that the cloud always has enough computing capacity and that any task arriving at the cloud will always proceed to completion[17].

The computational delay and energy consumption for the cloud node n are:

$$t_n^d = I(-O + 1 \leq y_n \leq -1)t_{n,m}^d + I(y_n \leq -O)t_{n,O+1}^d \tag{13}$$

and,

$$e_n^d = I(-O \leq y_n \leq -1)e_{n,m}^d + I(y_n \leq -O)t_{n,O+1}^d \tag{14}$$

where, $t_{n,O+1}^d = \frac{R_n}{t_{n,O+1}^d + \frac{P_n}{r_n^{O+1}}} + P_n \tau$ is the delay for moving the task from the leader
to the cloud through a macro base station, $t_{n,s}^d = \frac{R_n}{l_{n,s}^d + \frac{P_n}{r_n^s}} + P_n t_b + P_n \tau$ is from
the node n to the cloud through a small base station s, τ is the delay from a
macro base station to the cloud for a unit amount of data transferred, l_n^d is the
computing capacity of the node n, t_b is the delay for the backbone links for a
unit of data, $e_{n,O+1}^d = e_n^s + p_n^d \left(\frac{R_n}{l_{n,O+1}^d} + P_n \tau \right) + \frac{p_{n,O+1}^t P_n}{r_n^{O+1}}$ is the energy consumed
in moving the task from the node n to the cloud through a macro base station,
$e_{n,s}^d = e_n^s + p_n^c \left(\frac{R_n}{l_{n,s}^d} + P_n t_b + P_n \tau \right) + \frac{p_{n,s}^t P_n}{r_n^s}$ is the energy consumed for moving the
task from the node n to the cloud node through a small base station s, e_n^s is the
energy consumed for scanning available base stations, p_n^c is the idle power draw,
$p_{(n,s)}^t (m \in S_s)$ is the total transmission power of the node n and $p_{(n,s)}^t = n_m p_{(n,s)}^k$.
    Therefore, the total cost of migrating the computation to the cloud is:

$$c_n^d = \beta_n e_n^d + \alpha_n H_n^d \tag{15}$$

where,

$$H_m^d = \zeta^d l(n, m)^d + \eta r_n^m, (m \in S_m)$$

and ζ^d is the unit computation cost in the cloud.

4   Problem Formulation

The overall cost of computation can be minimized by optimizing the following:

- Migrating data and tasks to the edge nodes and the cloud.
- Resource allocation for the overall computation.
- Power utilization for data transfers.

    These can be expressed as the following mixed-integer programming problem:

$$\min_{p, Y, u} \sum_{m=-O-1}^{O+1} \sum_{n=1}^{N} I(Y_n = m)c_n \tag{16}$$

with the following six constraints:

$$C1 : u_n^0 \geq 0, \; n \in N,$$

$$C2 : 0 \leq p_{n,m}^t \leq p_n^{max}, \; n \in N,$$

$$C3 : u_{n,m}^d \geq 0, \; m \in S_m, n \in N,$$

$$C4 : \sum_{m=1}^{O+1} \sum_{n=1}^{N} u_{n,m}^u \leq U, \; m \in S_m,$$

$$C5 : \sum_{m=-O-1}^{O+1} I(y_n = m) = 1, \; m \in \Omega,$$

$$C6 : I(y_n = m)(t_n^l + t_n^d + t_n^0) \leq t_n^{max}, \; m \in \Omega$$

where, $C1$ denotes that each device has a finite, non-zero computing capacity, $C2$ denotes the range of the transmission power, $C3$ implies that the computing resources are tangible, $C4$ puts a constraint that the computing resources of individual nodes are less than the computing resources at the edge, $C5$ guarantees that the decision to move tasks from a particular device is deterministic, $C6$ enforces that the computation proceeds to completion within a specified deadline and

$$c_n = \begin{cases} c_n^0, & y_n = 0 \\ c_n^l, & y_n > 0 \\ c_n^d, & y_n < 0 \end{cases}$$

5 Simulation

We compared performance based on local computation, offloading to the edge and the cloud. The crux of the problem lies in efficiently making the decision to move computation away from the local nodes to the edge or to the cloud. We consider an iterative approach for arriving at a solution to this problem. We apply the Ford-Fulkerson algorithm repeatedly to compute the efficacy of the migration between the low powered nodes and the edge nodes and the cloud [18]. At each iteration, each device in the network has the opportunity to compute and update its task offloading decision based on the previous iteration. Each node computes the efficacy of migration based on the strategy described in Sect. 2 and performs the migration accordingly. Therefore, at each iteration, each device competes to choose the optimal offloading strategy that minimizes the respective cost. Thus, the device with the maximal reduction in the computation cost wins and is able to update its computation offloading decision. This iterative process continues until no changes to the overall cost are detected with the changes to the offloading decision strategy and the process terminates.

6 Discussion and Conclusion

Figure 2 shows the averaged computation cost with respect to the input data size. The computing cost decreases with each iteration until the algorithm converges and we obtain an optimum resource allocation. The system always converges even with increased payload sizes. This is because, our algorithm actively redistributes the computation and more intensive tasks are ultimately moved to more powerful hardware at the edge or in the cloud. Hence, although the overall system cost is more for a larger payload, we do reach an optimum utilization of available resources after a few iterations. Figure 3 shows a comparison of the averaged computational cost for the three schemes with respect to the available computational resources at the edge. The computation cost decreases as the capacity at the edge increases.

However, we observe that at the start of the simulation, when resources at the edge are low, offloading to the cloud outperforms offloading to the edge. This observation is explained from the fact that at low edge node densities, it is more efficient to offload the entire computation directly to the cloud rather than wait for more edge nodes to become available which could slow down and increase the overall cost. As more edge nodes become available, the average system cost converges and becomes constant. This is because for each task there is a finite limit for fragmentation. Once the system reaches this limit, the throughput becomes constant and cannot be improved further even by adding more nodes.

Figure 4 shows the variation of the average system cost with respect to the various offloading schemes and the changes in the computation cost ζ. For low CPU

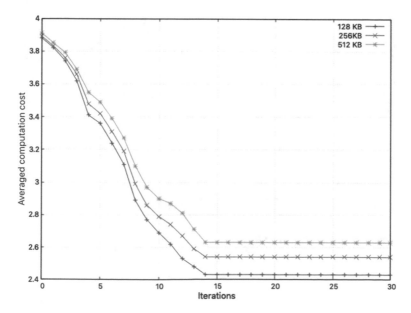

Fig. 2 Averaged computation cost versus payload size

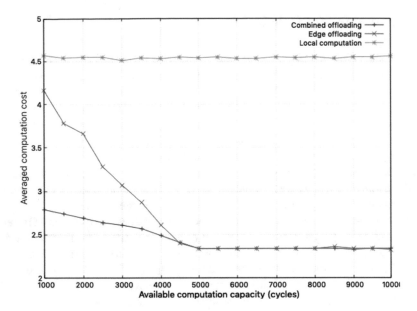

Fig. 3 Averaged computation cost versus edge capacity

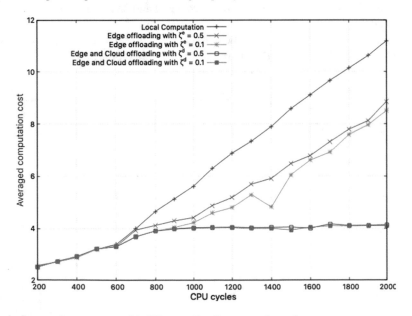

Fig. 4 Averaged system cost with different offloading strategies and ζ

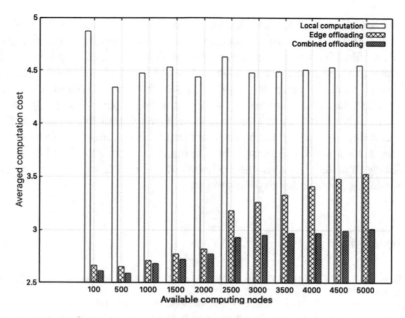

Fig. 5 Variation of the averaged system cost with the number of nodes

requirements, the average system cost is almost the same for all schemes. This is because, each node can perform low intensity tasks without having to offload. As such, the whole system operates in local computation mode irrespective of the chosen offloading scheme. However, as the required CPU resources increases, we see that the offloading schemes start performing better than purely local computation. Once the required CPU cycles exceed the gigahertz level, the combined edge and cloud computing offloading scheme performs even better as the increased CPU requirements are now met by the edge and the cloud improving the efficiency of the devices.

Since the combined strategy effectively offloads the computation to the cloud due to lack of available CPU cycles at the edge, a substantial improvement is achieved over the fixed offloading strategy and the edge, albeit with a small network latency penalty. As the capacity of the edge is increased, the overall system cost for the two strategies tend to converge as most of the computational requirements can be handled by the edge itself. Therefore, once the system reaches an optimum utilization at the edge, the overall cost of the computation becomes constant.

The variation of the average system cost with the number of available nodes for all the computational strategies is shown in Fig. 5. Both edge and combined cloud and edge computing perform better than the local computation. However, the combined strategy performs better and is able to reduce the overall system cost above and beyond what can be achieved from only the edge. This is more evident when the overall computational capability of the edge is small.

The fact that offloading and efficient resource allocation can greatly reduce the overall cost of computation over strictly local computation is evident from our results wherein both the offloading strategies consistently outperform the strictly local strategy. However, as the number of devices increases, we observe that the combined edge and cloud offloading strategy tends to outperform the edge offloading strategy. This is attributed to the fact that even with an increase in the number of edge nodes, they still cannot match the performance of the cloud where some computational tasks maybe offloaded by the combined strategy. For the purposes of this study we have assumed that the cloud can always compute the overall task to completion. In the real world there is always a finite possibility that the cloud may run out of capacity and may cause a bottleneck in the overall computation as the local and edge nodes are already saturated. However, the likelihood of that happening is extremely small and we do not consider that as part of our study.

References

1. Kumar K, Lu Y (2010) Cloud computing for mobile users: can offloading computation save energy? IEEE Comput 43(4):51–56
2. Soyata T, Muraleedharan R, Funai C, Kwon M, Heinzelman W (2012) Cloud-vision: real-time face recognition using a mobile-cloudlet-cloud acceleration architecture. In: Proceedings of the IEEE ISCC 2012, pp 59–66
3. Cohen J (2008) Embedded speech recognition applications in mobile phones: status, trends, and challenges. In: Proceedings of the IEEE ICASSP 2008, pp 5352–5355
4. Zhao J, Chen Y, Gong Y (2016) Study of connectivity probability of vehicle-to-vehicle and vehicle-to-infrastructure communication systems. In: Proceedings of the 2016 IEEE 23rd vehicular technology conference (VTC Spring), 2016, pp 1–4
5. Zhang Y, Yu R, Nekovee M, Liu Y, Xie S, Gjessing S (2012) Cognitive machine-to-machine communications: visions and potentials for the smart grid. IEEE Network 26(3):6–13
6. Cuervo E et al (2010) MAUI: making smartphones last longer with code offload. In: Proceedings of the 8th international conference on mobile systems, applications, and services, 2010, pp 49–62
7. Aazam M, Huh E-N (2016) Fog computing: the cloud-IoT/IoE middleware paradigm. IEEE Potentials 35(3):40–44
8. Li H et al (2016) Mobile-edge computing progress and challenges. In: 4th IEEE conference on mobile cloud computing services and engineering, Mar-Apr 2016
9. Drolia U et al (2013) The case for mobile edge-clouds. In: Proceedings of the 10th IEEE international conference on ubiquitous intelligence and computing, 2013, pp 209–215
10. Ericsson, The telecom cloud opportunity. 2012 [Online]. Available: http://www.ericsson.com/res/site_AU/docs/2012/ericsson_telecom_cloud_discussion_paper.pdf
11. Barbarossa S, Sardellitti S, Lorenzo PD (2013) Joint allocation of computation and communication resources in multiuser mobile cloud computing. Proceedings of the IEEE workshop SPAWC 2013, pp 26–30
12. Satyanarayanan M, Bahl P, Caceres R, Davies N (2009) The case for VM-based cloudlets in mobile computing. IEEE Pervasive Comput 8(4):14–23
13. Lamport L (1998) The part-time parliament. ACM Trans Comput Syst 16:133–169
14. Ma X, Zhang S, Li W, Zhang P, Lin C, Shen X (2017) Cost-effective workload scheduling in cloud assisted mobile edge computing. Proceedings of the IEEE/ACM international symposium on quality of service (IWQoS) 2017, pp 1–10
15. Zhang J, Xia W, Cheng Z, Zou Q, Huang B, Shen F, Yan F, Shen L (2017) An evolutionary game for joint wireless and cloud resource allocation in mobile edge computing. In: Proceedings of the 2017 9th international conference on wireless communications and signal processing (WSCP), 2017, pp 1–6

16. Zhang J, Xia W, Yan F, Shen L (2018) Joint computation offloading and resource allocation optimization in heterogeneous networks with mobile edge computing. IEEE Access 6:19324–19337
17. Nishtala R et al (2013) Scaling memcache at facebook. In: Proceedings of the 10th USENIX conference on networked systems design and implementation, Apr 2013, pp 385–398
18. Ford LR Jr, Fulkerson DR (2015) Flows in networks. Princeton University Press, Princeton

Proactive Personalized Primary Care Information System (P³CIS)

V. Lakshmi Narasimhan

Abstract Personalized primary care is becoming very important as individualized medical datasets are being collected on a vast scale by many systems and organizations around the world. However, allegations are there that current primary care treatments and procedures therein are not optimized for all people, but only to certain type of people. This paper concerns the design of an Internet of Things (IoT) based personal information system that can facilitate personalized primary care. In addition, the system also offers proactive care in that key features from automatically extracted data from the user is analyzed and, where required, sent to their primary physician. The Proactive Personalized Primary Care Information System (P³CIS) is based on private cloud architecture and hence offers reasonable data security and privacy to user datasets through encryption and decryption procedures. The performance indices indicate that the P³CIS system is practically viable and further offers good value-add to proactive personalized primary care to every individual patient.

Keywords Proactive personal care · IoT · Data security and privacy · Private cloud · Parametric performance model

1 Introduction

Personalized primary care concerns the provision of customized medications and treatment procedures based on individual characteristics such as race, gender, height, weight and the like [1]. The current system of medical care is not personalized and indeed, there are allegations that both treatment procedures and medications have been developed for White men and White women only [2]. Further, allegations are there that such procedures are not appropriate for non-White people. This paper concerns the design of a Proactive Personalized Primary Care Information System (P3CIS), which aims to offer proactive and personalized care to each and every

V. Lakshmi Narasimhan (✉)
Department of Computer Science, University of Botswana, Gaborone, Botswana
e-mail: srikar1008@gmail.com

H. Kim et al. (eds.), *Information Science and Applications*, Lecture Notes
in Electrical Engineering 739, https://doi.org/10.1007/978-981-33-6385-4_33

individual. The system is proactive in that datasets can be analyzed locally and key features can be sent to primary physician on a need basis.

The rest of the paper is organized as follows: Sect. 2 provides an overview of Internet of Things (IoTs), while Sect. 3 describes the nature of proactivity and personalization in primary care. Section 4 presents the architecture of the P3CIS, followed by the need for security and privacy of data and information in the P3CIS system in Sect. 5. The performance of the P3CIS system is evaluated using parametric modeling technique as described in Sect. 6, while the conclusion summarizes the paper and offers pointers for further R&D work in this arena.

2   Internet of Things (IoT) and Sensors

Internet of Things (IoTs) are devices that contain a processor, variety of sensors, wireless modem, energy source and sometimes GPS (Fig. 1) [3, 4]. Sensors are available in IoTs include those for monitoring weight, temperature, pulse rate, heart rate, blood oxygen level, blood sugar level and other details need to monitor an individual. IoTs have an operating system and other associated programs so that most processing requirements are met locally. Networking systems over IoTs allow

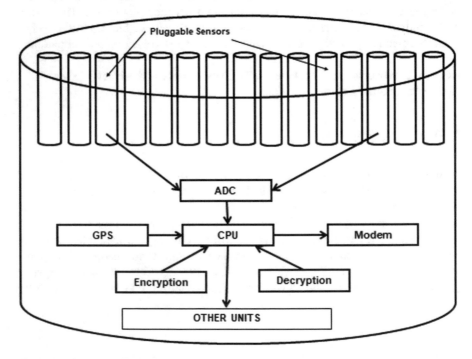

Fig. 1 Configuration of an IoT

them to cooperate with each other so that a complete coordinating system can be built rapidly in order monitor individual patient and offer proactive personal primary care [7, 8].

3   Proactivity and Personalization in Primary Care

Personalization in primary care concerns the provision of individual care attention based on their personal history, their gender, race and their body needs and conditions. Personal care varies from one patient to another and personalized medical care offers better recovery and management from various diseases [5]. Further, the P3CIS system offers proactive care [6] in that key features from individual patients can be automatically extracted and locally analyzed and, when and where required, the data and features can be sent to their primary physician via built-in modem in the IoT device. The IoT based device contains personalized sensors in order to monitor conditions of a user on an on-going basis and record them. A care provider can download data and/or extracted key features using their smart phone and monitor them on a regular basis. The IoT based device is similar to a wrist band and does not intrude with the patient movements or comfort.

4   P³CIS Information Architecture

The P3CIS data architecture is provided in Fig. 2, whose core is ISO Standard Asset Registry [11] and a data warehouse. Various types of data cleansers and handlers (e.g., XML extensions for import and export [9]) are also available in this data architecture. Figure 3 provides the private-cloud based information architecture [10] of the P3CIS system offering a wide variety of services, which include data collection, processing, analytics, besides encryption and decryption. In addition, Query-By-Example (QBE) service is also available, whereby a service provider can search over a given condition of a particular patient—be they signals, data or image. The QBE service thus provides image and signal comparators, besides a variety of data related comparators.

5   Security and Privacy of Data and Information

Security and privacy of medical data are two major issues with regard to medical data. The P3CIS system is private cloud based and hence only a limited number of approved personnel can operate the system. Further, the system offers a limited degree of data encryption and decryption and, in addition, provides data dithering also. As a result, the P3CIS system has a decent control over the two thorny issues of security and privacy of user's medical data.

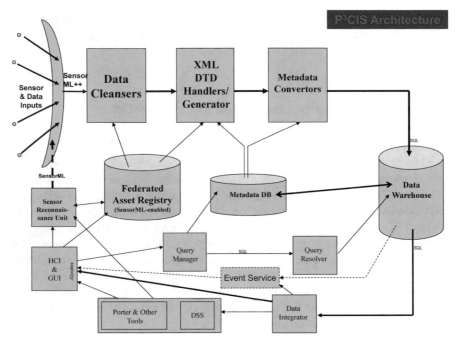

Fig. 2 P3CIS data architecture

6 Parametric Performance Evaluation

A parametric model-based evaluation of the P3CIS Cloud system has been carried out. Table 1 provides typical parameters used for the evaluation of the P3CIS Cloud, which have been obtained after discussions with several experts. Table 2 provides a list of performance indicators and their values, wherein the values are calculated using Relative Cost Unit (RCU) so that depending on the actual cost of individual companies a suitable multiplier can be employed to calculate the corrected values of various performance indices. It is hoped that these indicators will provide the way forward for the advancement of such systems in various marine sensor network R&D centers around the world.

7 Conclusions

This paper describes a wireless sensor network environment for managing marine environment. A CCOM model of the sensor network system has been developed. In addition, a four-layer information management architecture containing modules that collects and cleans data, an integrated asset registry, tools and technologies for information exploitation and output management in order to handle various types

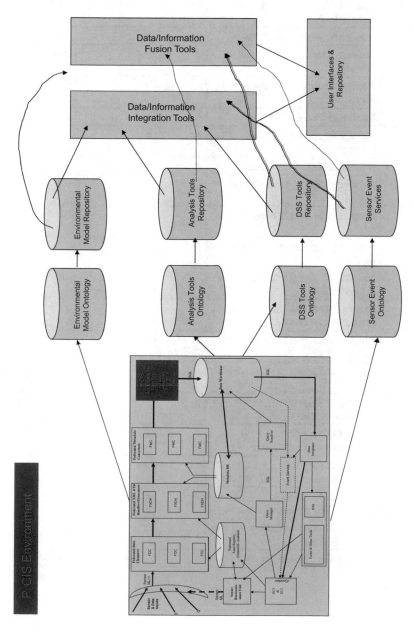

Fig. 3 Private cloud based P3CIS information architecture

Table 1 P^{3}CIS private cloud information systems architecture

S. No.	Explanation	Symbol	Average value	Max. value
	Size of P^{3}CIS PDF file (data or sonar or image)	a	0.001 GB	1 GB
	Number of sensors per person	b	8	15
	Number of persons per field of interest	c	50	200
	Average number of messages per sensor	d	3	5
	Number of gateway nodes	e	5	7
	Average sensor message size per message	f	5	10
	Number of services in P^{3}CIS	g	4	10
	Number of Maintenance calls per day	h	4	6
	Number of *To-Act-On* messages per day	i	5	20
	Number of Upgrade requirements per day	j	4	6
	Number of Internal low-end service calls per day	k	10	18
	Number of Internal medium-end service calls per day	l	5	8
	Number of Internal high-end service calls per day	m	3	5
	Number of Reports to be generated per day	n	40	100
	Number of Knowledge query management per day	p	50	150
	Number of Help line management--simple call per day	q	20	30
	Number of Help line management--medium call per day	r	10	15
	Number of Help line management--complex call per day	s	5	8
	Number of compliance requirements per day (if any)	t	5	10
	Average viewing time per report	u	4 min.	7 min.

(continued)

Table 1 (continued)

S. No.	Explanation	Symbol	Relative cost units
1.	Data storage cost per GB per month	C1	25
2.	Data access cost per GB	C2	2
3.	Internal low-end service cost per service call	C3	1
4.	Internal medium-end service cost per service call	C4	3
5.	Internal high-end service cost per service call	C5	8
6.	Maintenance cost per service call	C6	15
7.	Upgrade cost per service call	C7	10
8.	Encryption cost per file (1 MB)	C8	5
9.	Decryption cost per file (1 MB)	C9	5
10.	Air conditioning costs per day	C10	200
11.	Average Downtown costs for services upgrade per day	C11	400
12.	Compliance Management costs per compliance requirement (if any)	C12	500
13.	Average Report generation cost per report	C13	20
14.	Average Knowledge query management cost per query	C14	2
15.	Help line management per simple call	C15	1
16.	Help line management per medium call	C16	5
17.	Help line management per complex call	C17	10

of customer requirements. An important feature of this architecture is the use of ISO-compliant federated asset registry [11], which ensures all data, metadata and information can be stored and exported in a standardized manner. We are currently working on following up on the MIMOSA Standards [12] and ISO Standards [13] and enhancing them to suit marine information management. The issues to be addressed include, conditional monitoring and facilitation for diagnostics and prognostics of various asset types inside a marine sensor network, including sensors, thrusters and others. The performance of the P3CIS private Cloud has been evaluated using parametric modeling technique and the results indicate that P3CIS private Cloud system can successfully enhance the performance of the scientists and technical people involved in this field; the details are also provided in this paper. We intend to develop operational IT systems, in addition to developing international Standards in these areas.

Table 2 P^{3}CIS private cloud information systems architecture performance indicators

S. No.	Metric name	Symbol	Formula	Typical average value	Max. value
1.	Average Execution time per person	PI-1	(a * b + d * e * g) * h * i	60.16	43,800
2.	Average bandwidth used per day	PI-2	a * b * c * i	2	60,000
3.	Average Downtime Management per person	PI-3	C11 / (b * c)	1	0.14
4.	Average cost of security per person	PI-4	(C8 + C9) * b * e * d	1,200	5,250
5.	Average Ease of Use per person == Average Execution time per person cost + Weighted average service call time + Weighted average Help call time	PI-5	{(a * b + d * e * g) * h * I} + {C3 * k + C4 * 1 + C5 * m} + {C15 * q + C16 * r + C17 * s}	229.16	44,067
6.	Average report generation cost per day	PI-6	C13 * n	800	2,000
7.	Average compliance requirement cost per day (if any)	PI-7	C12 * t	2,500	5,000
8.	Average network usage cost == Average execution time cost per person+ Visit cost + specialty related cost + InfoSec cost + data access and storage cost	PI-8	(a * b * c) + (d * f) + (g * i) + (C8 + C9) * i + (C1 + C2) * b * c * i	54,085.4	1,623,550
9.	Average P3CIS cloud usage cost == PI-18 + Knowledge query cost + upgrade cost + maintenance cost + Aircon cost	PI-9	PI-8 + (C14 * p) + (C7 * j) + (C6 * h) + C10	54,485.4	1,624,200
10.	Average Cost of Ownership per acre	PI-10	PI-9/c	1,089.71	8,121

References

1. Bleijenberg N (2013) Personalized primary care for older people: an evaluation of a multicomponent nurse-led care program: an evaluation of a multicomponent nurse-led care program. Ph.D. thesis, Utrecht University Repository
2. LaVeist TA, Nickerson KJ, Bowie JV (2000) Attitudes about racism, medical mistrust, and satisfaction with care among african american and white cardiac patients. Med Care Res Rev 1
3. Atzori L, Lera A, Morobeto G (2010) The Internet of Things: a survey. Comput Netw 54(15)
4. Mendez DM, Papapanagiotou I, Yang B (2018) Internet of Things: survey on security and privacy. Inf Secur J A Glob Perspect
5. Porter ME, Pabo EA, Lee TH (2013) Redesigning primary care: a strategic vision to improve value by organizing around patients' needs. Health Affairs 32(3)
6. Madushanki AAR, Halgamuge MN, Wirasagoda WAHS, Syed A (2019) Adoption of the Internet of Things (IoT) in agriculture and smart farming towards urban greening: a review. Intl Jol Adv Comput Sci Appl 10(4):11–28
7. Lakshmi Narasimhan V (2020) Precision agricultural management information systems (PAMIS). In: Proceedings IEEE ICIOTCT 2020 conferences, Patna, India
8. Lakshmi Narasimhan V, Arvind A, Bever K (2007) Greenhouse asset management using wireless sensor-actor networks. In: International conference on mobile ubiquitous computing, systems, services and technologies (UBICOMM'07)
9. Khan AR, Othman M, Madani SA, Khan SU (2014) A survey of mobile cloud computing application models. IEEE Commun Surv Tutor 16(1)
10. Johnston A (2009) OpenO&M™ and ISO 15926 Collaborative Deployment. POSC Caesar Members Meeting 20 Oct 2009 Kuala Lumpur, Malaysia
11. MIMOSA Standards, https://www.mimosa.org/. 28 May 2020
12. ISO TC 108 Asset Management Standards, ISO/TC 108/SC 5, Condition monitoring and diagnostics of machine systems. https://www.iso.org/committee/51538.html. 28 May 2020

CADAES – Cooperating Autonomous Drones for Agricultural and Entomological Safety

V. Lakshmi Narasimhan

Abstract Unmanned Ariel Vehicles (UAV), also known as Drones, are being employed in the Agricultural industry widely in many countries—Japan being the lead example. Drones are employed to spray insecticides and also gather imagery of the field. The latter is used for managing and modelling plant growth, disease modelling and even for entomological applications. This paper describes the design of a Drone-underpinned Agricultural systems, which runs over a Cloud environment. It uses 3-D Digital Elevation Models (DEM), besides facilitating access to a variety of publicly available and privately held agricultural databases on diseases, growth modelling and others. The system employs an agent-based approach towards decision support system design. The performance of the cloud-based system is evaluated using parametric modelling technique and, they indicate that the system is well-suited for a variety of Agricultural applications.

Keywords Agricultural drones · Cloud-based drone information architecture · Agent-based co-ordination · Parametric performance modelling · Entomological safety

1 Introduction

In the context of Agriculture, drones are simply Unmanned Ariel Vehicles (UAVs) carrying a number of useful devices such as cameras, GPS, specialized software and hardware for processing, and spray-able resources (see Fig. 1). Drones have successfully been employed in a wide variety of applications such as, law enforcement, fisheries, surveillance, water management, various military applications and others. Indeed, the Association for Unmanned Vehicle Systems International (AVUSI) estimates [1, 2] that "drones will create more than 34,000 manufacturing jobs and 70,000 new jobs in the United States in the next three years, with an economic impact of more than \$13.6 billion, growing to more than 100,000 jobs and \$82 billion by 2025".

V. Lakshmi Narasimhan (✉)
Department of Computer Science, University of Botswana, Gaborone, Botswana
e-mail: srikar1008@gmail.com

© The Author(s), under exclusive license to Springer Nature Singapore Pte Ltd. 2021   367
H. Kim et al. (eds.), *Information Science and Applications*, Lecture Notes
in Electrical Engineering 739, https://doi.org/10.1007/978-981-33-6385-4_34

Fig. 1 Applying drone technology to agriculture and entomology

The AUVSU further calls for the legalization of commercial drones in the US and other Western countries. In Asia, China and Japan are heavy drone users too [3]. In fact, Japanese rice farmers have been using robotic RMAX crop-dusting helicopters for nearly 30 years or more. It is heartening to note that more than 127,000 drones have been sold on eBay since March 2014, generating about $16 million in sales despite the fact that drone use faces considerable hurdles in personal privacy issues tied to the onboard camera and other on-board entities [3].

According to Chad Colby, a UAV specialist, three quarters of the USA's drone fleet will be flown in the business of agriculture [1] and that producers are using drone technology to help identify field and livestock issues early and rectify them before crop loss occurs. This new airborne farming tool will allow producers to aerially view fields as never before with high-resolution Infrared cameras to detect insect, disease, and water threats more effectively than the human eye. However, Federal Aviation Administration in USA still maintains its strict regulations on drone use that potentially handcuff farmer ingenuity and limit its application to food and fiber production.

Matt Wade of senseFly [4], a drone maker based in Switzerland, provides three main drivers for the raise in popularity of drones among farmers; the first driver is technological: agriculture-specific drones are becoming easy enough to operate without any piloting skills. They are also getting relatively low-cost, ranging from a few hundred to a few thousand dollars. The aircraft isn't the whole cost, though. Toby Waine [5], a remote sensing expert at Cranfield University, explains: "As a farmer, you can get a cheap drone and take some nice photos, but to make measurements, you need better quality cameras with near-infrared capability for vegetation monitoring. And you'll need supporting agronomic data—such as soil, water and weather—for calibration of crop models to make informed decisions". The second driver is awareness, as word spreads about the productivity benefits. In the UK, for example, the government even provides grants to farmers who want to purchase a drone. The third driver is the evolving regulatory situation (or is the lack of one?). While just a few years ago a drone flying over the countryside was a rarity, now they're becoming so common that countries increasingly move to more pragmatic regulations to manage commercial use.

Indeed the Vision 2020 document [6] of ICAR aims for, *"computer software developed for creating, updating and for retrieval of data thru' Agricultural Research Information System (ARIS) ... that will emphasise efficiency, sustainability, diversification, post-production management, ... Innovative approaches will have to be adopted to upgrade skills of farmers and for technological empowerment of women engaged in agriculture....".* Furthermore, the Vision 2030 document [7] of ICAR aims to, *"Develop a futuristic human resource development programme in cutting-edge science and technology in order to improve access to information through effective use of Information and Communication Technology......".* These objectives comprehensively met by the design of a system called, CADAES—Coordinating Autonomous Drone Technology for Agricultural and Entomological Safety. The system collects copious amount of historically annotatable datasets and would employ Big Data Analytics [8, 9] over the Cloud computing environment [10] and extensive software systems so that small and medium size farmers can benefit from integrated farm and crop management processes. This design and implementation methodology forms part of the larger Agricultural Internet of Everything (also known as Internet of Things (IoT)) [11, 12] in that all farm related monitoring can now be brought from the field to the farmer-at-home.

The rest of the paper is organized as follows: Sect. 2 surveys related literature on the use of Agricultural drones, while Sect. 3 describes the CADAES environment. Section 4 details the operation of CADAES, followed by the description of the detailed design of the CADAES environment in Sect. 5. The parametric model for evaluating the performance of CADAES is presented in Sect. 6, while the Conclusion summaries the paper and offers pointers for further R&D work in this arena.

2 Related Research

Related research work includes those by the authors of [5], who have contributed to near-Infrared imaging, while authors of papers [13, 14] have dealt with design of drone technologies only. Furthermore, the authors of [5, 15, 16] have dealt with many aspects of drone (and remote sensing) based image processing of agricultural crops. Lastly, the work using Depth-sensing camera [17] will be unique and to date not much filed results have been reported based on such a device. Greenhouse sensors and related automatic robotic manipulators and algorithms have been dealt with in [18, 19]. While many of these papers are related to this research, our proposal contains significant enhancements and novelties that we need to study them in a comprehensive manner and hence this proposal seeking in-depth research funding. Furthermore, this paper envisages value-add to farmers and farm-related product developers thru' actual database generation and a number of predictive tools and technologies. Such a comprehensive approach has not been attempted yet—to the best of the investigators.

3   The CADAES Environment

The Cooperating Autonomous Drones for Agricultural and Entomological Safety (CADAES) environment can coordinate autonomous Drones for the benefit of agricultural and entomological safety. The technology treats drones as independent and coordinating entities so that they can operate on a cooperative basis. Further, a comprehensive 3D Digital Elevation Model (DEM) [20] for the field at large—i.e., in the case of small farmers this implies that the farms surrounding them—is modelled so that they can get extensive view of their agricultural necessities and understand the limitations within which they can operate. Big Data over a cloud computing environment would be used to perform necessary analytics in order to provide a comprehensive decision support system. Specifically, the CADAES environment offers the following features:

- Design of a multi-device small-scale drone using Commercial Off-The Shelf (COTS) products and software in addition to using a commercial COTS drone (so that one can perform comparisons of various nature),
- Development of a suite of near-Infrared image processing software for crop growth modelling and create crop-specific database,
- Provides a suite of near-Infrared image processing software for dynamic disease profiling (including their life-cycle analyses) and create crop-specific database,
- Provides a suite of near-Infrared image processing software for seasonal entomological modelling and create crop-specific database,
- Offers a software system that can identify entomological corridors of activities and create crop-specific database,
- Identifies 3D vectors of spread for diseases and insects.
- Inducts post-disaster management and monitoring processes,
- Facilitates 3-Dimensional Digital Elevation (DEM) model of the land in order to understand water and nutrient usage, and,
- Provides of a cost-benefit model for the usage of the CADTAES-like technology

4   Operation of the CADAES Environment

Modern agricultural drones employ fly-by-wire technology and are fully autonomous and programmed to follow a given trajectory or corridor. They are equipped with tools such as, accelerometers, gyroscopes, a compass, and hardware to avoid obstacles. The autopilot does the flying, computing everything from the take-off and flight to the landing—aiming for maximum coverage of the field, while getting all the necessary data. "Users are never required to manually generate or plot mission paths or build flight plans based on weather conditions. Instead, the aircraft will automatically adapt to its conditions to collect the highest quality data," quotes Earon [21] a drone scientist. The use of modern Information and Communication Technology (ICT) and 3D terrain modelling play an important role. This creates an Internet of Everything (IoE) so that the farmer now becomes empowered by the technology. The datasets derived from these sensors can be analysed on a real-time basis. Further, one could use them in manners so that one can ask several types of what-if questions answered/analysed. The overall evaluation of the effectiveness and operational efficacy of the CADAES environment is performed over through actual on-field experimentation in a real farm – as provided in Sect. 6.

5 Design of the CADAES Environment

This section describes the design of the CADAES environment through the provision of: i) design of a novel drone family, ii) nature of communication and networking mechanisms, iii) underlying agent-based coordination mechanisms, iv) the design of the Drone Operations Management and Information Exploitation Environment (DOMIE[2]) and v) the ways to manage agricultural data safety and security of individual farmers.

5.1 Design of Novel Drone Family

The CADAES environment handles autonomous, wireless, reusable and re-configurable family of drones with several devices as noted in Table 1. These devices

Table 1 Nature of Devices in a Drone

Photogrammetric Devices
• High resolution visual camera
• Near-Infrared camera
• (Hyperspectral camera)
• Depth-Sensing Camera Works in Bright Light and Darkness
Protection Devices
• Dust & Sand protection units
Communication & Coordination Devices
• Agent-based communication devices
Navigation Devices
• GPS unit
• Gyroscope unit

meet the needs of four major functional areas [13, 14], namely, Photogrammetric, Protection, Communication and Coordination and Navigational. These devices are mounted on a single stable platform under the drone. One of the drones carries a hyperspectral camera, as it has been shown in the literature that hyperspectral images can give a better understanding of the crops and disease profile [3, 12]; however, comprehensive work in this area is still not in the public domain. All drones will carry high resolution visual camera, near-Infrared camera and a Depth-Sensing Camera (DSC) Works in Bright Light and Darkness [17]. It has been shown that the DSC can work well through smoke, smog and fog-ridden conditions also. Dust and sand protection will also be available in all drones, besides a GPS and gyroscopic unit. We deployed COTS-based drones such as those offered in [22, 23], but more specialized drones can be designed for specific environments.

5.2 Communication and Networking Mechanisms

Typical drones employ the ZigBee Pro/IEEE 802.15.4 communications protocol [24, 25], whose typical characteristics are provided in Table 2 and its typical communications packet is shown in Fig. 2. The ZigBee standard supports three device types: ZigBee Coordinator, ZigBee Router and ZigBee End Device, with each device type implementing several types of functions, thereby impacting overall cost of the device. ZigBee Pro/IEEE 802.15.4 Packet Communication Format/s is well-documented and relevant software systems are also available. As noted, the ZigBee communication supports point-to-point, point-to-multipoint and peer-to-peer topologies. In addition, self-routing, self-healing and fault-tolerant features are also available in this network.

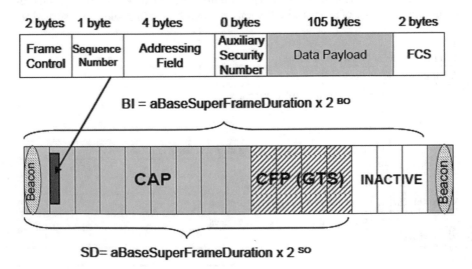

Fig. 2 ZigBee Pro/IEEE 802.15.4 packet communication format

- Works on 2.4 GHz
- Point-to-point, point-to-multipoint and peer-to-peer topologies supported
- Self-routing, self-healing and fault-tolerant
- Based on Atmega128RFA1, SoC
- ZigBee-Pro/IEEE 802.15.4
- On-Chip Antenna
- -100 dBm RX Sensitivity; TX Output Power up to 3.5 dBm
- Ultra Low Power consumption (1.8 to 3.6V) for Rx/Tx : <20 mA
- CPU Active Mode (16MHz): 4.1 mA
- 2.4GHz Transceiver: RX_ON 12.5 mA / TX 14.5 mA (maximum TX output power)
- Deep Sleep Mode: <250nA @ 25°C
- Long-life commercial solid-state (Li/Ni-Ion) Battery and Solar Power

The overall Cloud-based architecture of CADAES, as a drone based agricultural IoE System, is presented in Fig. 3, wherein farmer at the home (or cooperative) can find out about the nature of the soil, plant growth and control nutrient and flows and many other parameters. The architecture is Cloud-based in order to facilitate handling of data sets, various databases and other software systems. Further, this set up is enhanced through cameras placed at strategic places. It is noted that the farmer

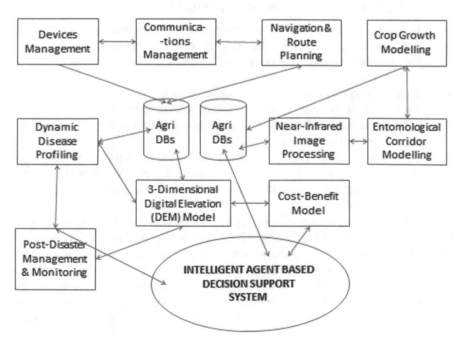

Fig. 3 Overall architecture of the cloud-based CADTAES environment

can communicate with each node via a smart phone and, in addition, appropriate APPs are developed so that on-line engagement with their farms can be effected from anywhere-anytime.

5.3 Agent-Based Coordination Mechanisms

Intelligent agent-based coordination mechanism will be used to manage more than one drone. Specifically, we intend to employ Belief, Desire and Intention (BDI) [21, 22, 27] agents for coordination amongst the drones. BDI framework combines Artificial Intelligence (AI) with cognitive architectures (typically event driven programming environment) in order to pursue long-term objectives through the monitoring of short-term events and follow-up actions. Domain expertise in the handling of cooperating drones is required (cf. more details on cooperative engagement facility of US-DoD [28]) to document standard operating procedures and chains of decision makings. It should be possible to model complex prioritizations and interactions and monitor them as they happen in real-time. Later, if required one should be able to reproduce them—this problem has been shown to be impossible to solve in theory, but in practice a near-optimal solution can be found. Specialized behavioural agents can also be evolved from the BDI architecture. Java language-based JACK is an excellent public domain BDI tool that could be employed for this purpose.

Specifically, the following activities are possible:

i.   Creation of a repertoire of behaviours common to all agents in the agricultural domain—this process of domain tailoring agents is typical in BDI systems,

ii.   Identification protocols for communication/coordination and their reciprocal commitments,

iii. Identification the set of actions and default reactions to events.

iv.   Have an ability to manipulate and operate with common databases, synchronization requirements, and the modularization of engineering hardware-software co-design processes thereof,

v.   Development plan[1] generation tools (including multiple, concurrent and parallel plan generations) that the drones could possibly follow, (we intend to use Petri net-based modelling [29] to check for the correctness of operation of these plans),

vi.   Development a source modelling and usage tool that goes with the plan generator so that resource usage (including fuel usage) can be optimized, and,

vii. Development of advanced concepts—called Capability—in the context of the CADTATES environment based on the works of [30]. Capabilities provide an additional degree of modularization and encapsulation whereby a drone's list of capabilities (viz., its capable list of plans, resources and possible outputs) can be defined in advance so that it can be called up when as and when needed to

---

[1]Plan is a term used to capture all activities associated with a drone such as, navigation, mapping, switching on-and-off of cameras, stability management, etc.

```
capability Foo extends Capability {
    // set of plans
    #uses plan Plan1;
    #uses plan Plan2;

    // knowledge base
    #private  database KnowDB db1;
    #exported database KnowDB db2;

    // events
    #posts event efoo1;
    #posts external event efooex1;
    #handles event efoo2;
    #handles external event efooex2;

    // message events: inter-agent communication
    #sends event messfoo;

    // included capabilities
    #has capability Bar sonOfFoo;
}
```

Fig. 4 Formal definition of the concept of capability (adapted from [30])

solve a complex problem. An example for the formal definition of a capability is provided in Fig. 4.

5.4 Designing a Drone Operations Management and Information Exploitation Environment (DOMIE^2)

Drone Operations Management and Information Exploitation Environment (DOMIE2) is essential so that a farmer can monitor and manage their farm in the best possible manner. The DOMIE2 provides a list of functionalities that include, but not limited to, the following:

- A suite of near-Infrared image processing software for crop growth modelling and create crop-specific database. These information will be provided visually and a pre-constructed narrative will appear in English and prominent local languages that describes the information in farmer-friendly terms.
- A suite of near-Infrared image processing software for dynamic disease profiling (including their life-cycle analyses) and create crop-specific database. Farmer-friendly pre-constructed narrative will appear in English and prominent local languages.

- A suite of near-Infrared image processing software for seasonal entomological modelling and create crop-specific database. Farmer-friendly pre-constructed narrative will appear in English and prominent local languages.
- A software system that can identify entomological corridors of activities and create crop-specific database. Farmer-friendly pre-constructed narrative will appear in English and prominent local languages; this information will also be available visually over a commercial Geospatial Information Management System (GIMS) such as ArcView [31].
- The system will identify 3D vectors of spread for diseases and insects; these information will be made available visually over GIMS.
- Monitor and manage—through appropriate crop modelling software—yield prediction from three weeks of sowing to just prior harvesting; this in turn can be linked to market pricing so that effective harvesting time prediction can be achieved).
- The system will provide post-disaster management (e.g., flood, cyclone, wind, tornado, etc.), and follow-up monitoring processes; this sub-system will spawn the latest disaster management datasets—post disaster.
- The system will provide 3-Dimensional Digital Elevation (DEM) model of the land in order to understand water and nutrient usage, and,
- A cost-benefit model for the usage of the CADTAES-like technology would be available so that a farmer can play with if-then-else conditions on the potential usage of this technology and its benefits.

5.5 Managing Agricultural Data Safety and Security of Individual Farmers

Data security and their safety aspects are key issues in the digital world and they are so in the Agricultural digital world also. While the CADTAES environment would digitally and automatically collect a number of datasets—including those on individual farmers and their farmlands—protecting the privacy of individual farmers' collections are vitally important. It is also noted that the collated statistical information of the collective farmers would be useful to all farmers and hence the cumulative or composite results would be made available in a statistically independent manner.

6 Parametric Performance Analysis of CADAES

A parametric model-based evaluation of the CADAES Cloud system has been carried out. Table 2 provides typical parameters used for the evaluation of the CADAES Cloud, which have been obtained after discussions with several experts. Table 3 provides a list of performance indicators and their values, wherein the values are calculated using Relative Cost Unit (RCU) so that depending on the actual cost of

Tables 2 Parameters for evaluating CADAES-cloud information systems architecture

S. No.	Explanation	Symbol	Average value	Max. value
1.	Size of messages file	a	0.5 MB	1 MB
2.	Number of communication per sensor/IoT per day	b	3	5
3.	Number of sensor/IoT per acre	c	20	35
4.	Number of Follow-up visits per patient	d	3	5
5.	Number of router nodes per acre	e	2	4
6.	Number of Gateways per acre	f	1	3
7.	Number of Sub-Specialties message weightage	g	4	6
8.	Number of Maintenance calls per day	h	4	6
9.	Number of Images per acre	i	5	7
10.	Number of Upgrade requirements per day	j	4	6
11.	Number of Internal low-end service calls per day	k	10	18
12.	Number of Internal medium-end service calls per day	l	5	8
13.	Number of Internal high-end service calls per day	m	3	5
14.	Number of Reports to be generated per day	n	40	60
15.	Number of Knowledge query management per day	p	50	70
16.	Number of Help line management–simple call per day	q	20	30
17.	Number of Help line management–medium call per day	r	10	15
18.	Number of Help line management–complex call per day	s	5	8
19.	Number of Compliance requirements per day	t	1	3
20.	Average viewing time per image	u	4 min.	7 min.

S. No.	Explanation	Symbol	Relative cost units (RCU)
1.	Data storage cost per GB per month	C1	25
2.	Data access cost per GB	C2	2
3.	Internal low-end service cost per service call	C3	1
4.	Internal medium-end service cost per service call	C4	3
5.	Internal high-end service cost per service call	C5	8

(continued)

Tables 2 (continued)

S. No.	Explanation	Symbol	Relative cost units (RCU)
6.	Maintenance cost per service call	C6	15
7.	Upgrade cost per service call	C7	10
8.	Encryption cost per file (0.5 MB)	C8	5
9.	Decryption cost per file (0.5 MB)	C9	5
10.	Air conditioning costs per day	C10	200
11.	Average Downtown costs for services upgrade per day	C11	400
12.	Compliance Management costs per compliance requirement	C12	500
13.	Average Report generation cost per report	C13	10
14.	Average Knowledge query management cost per query	C14	2
15.	Help line management per simple call	C15	1
16.	Help line management per medium call	C16	5
17	Help line management per complex call	C17	10

individual companies a suitable multiplier can be employed to calculate the corrected values of various performance indices. It is hoped that these indicators will provide the way forward for the advancement of such systems in various marine sensor network R&D centers around the world.

7   Conclusions

Drones are currently being employed in a variety of agricultural applications in a number of countries—Japan being the lead example in this arena. This paper describes a Drone based agricultural information system, called CADAES, which collects a variety of data and images using a variety of cameras over a wide spectrum range. The CADAES systems runs over a cloud-based environment and uses DEM models for photogrammetric enhancement of the images. The system employs agent-based data management for storage and retrieval. The system has access to a wide variety of publicly available and privately held databases on such aspects as, growth modelling, disease profiling and entomological considerations for a variety of crop types. The performance of the cloud-based CADAES has been evaluated using parametric modelling technique and, the performance indices indicate that the system is well-suited for a number of agricultural applications. The system is an evolving one and hence has a number of challenges to meet, which include but not limited to:

Table 3 CADAES cloud performance indicators

S. No.	Metric name	Symbol	Formula	Typical average value	Max. value
1.	Average Execution time on data per sensor/IoT	PI-1	(a * b + d * e * g) * h * i	720	25,200
2.	Average bandwidth used per day	PI-2	a * b * c * i	150	1,225
3.	Average Downtime Management per patient	PI-3	C11/ (b * c)	6.667	2.2857
4.	Average Cost of Security per acre	PI-4	(C8 + C9) * b * e * d	450	1,750
5.	Average Ease of Use per acre == Average sensor/IoT Execution time on data per acre + Weighted average service call time + Weighted average Help call time	PI-5	{(a * b + d * e * g) * h * I} + {C3 * k + C4 * l + C5 * m} + {C15 * q + C16 * r + C17 * s}	1,419	10,320
6.	Average report generation cost per day	PI-6	C13 * n	400	600
7.	Average compliance requirement cost per day	PI-7	C12 * t	500	1,500
8.	Average network usage cost == Average Execution time on data per sensor/IoT cost + Visit cost + Specialty related cost + InfoSec cost + Data access & storage cost	PI-8	(a * b * c) + (d * f) + (g * i) + (C8 + C9) * i + (C1 + C2) * b * c * i	8,215	33,397
9.	Average CADAES cloud usage cost == PI-18 + Knowledge query cost + upgrade cost + maintenance cost + Aircon cost	PI-9	PI-8 + (C14 * p) + (C7 * j) + (C6 * h) + C10	8,615	33,887
10.	Average Cost of Ownership per acre	PI-10	PI-9/c	430.75	968.2

(i) data normalization for many zones, (ii) data dictionary design, (iii) overall yield predictions and (iv) detailed entomological modelling along with wind patterns and other environmental considerations; these form future R&D in this arena.

References

1. Moskvitch K, Agricultural drones: the new farmers' market, 21st Aug 2015: http://eandt.the iet.org/magazine/2015/07/farming-drones.cfm
2. Association for Unmanned Vehicle Systems International (AVUSI): http://www.auvsi.org/home
3. FarmPressBlogs—dated: several
4. senseFly, 25th Aug 2015: https://www.youtube.com/user/senseFly
5. Jiménez-Donaire V, Kuang B, Waine T, Mounem Mouazen A, Optimising calibration of on-line visible (vis) and near infrared (NIR) sensor for measurement of key soil properties in vegetable crop fields. EGU general assembly conference abstracts 16, 927
6. "Vision 2020", ICAR, (20th Aug 2015): http://www.icar.org.in/en/node/9638
7. "Vision 2030", ICAR, (20th Aug 2015): http://www.icar.org.in/en/node/9638
8. Big Data, 26 Aug 2015: https://en.wikipedia.org/wiki/Big_data
9. UN GLobal Pulse (2012) In: Letouzé E (ed) Big data for development: opportunities and challenges. United Nations, New York. Retrieved from http://www.unglobalpulse.org/projects/BigDataforDevelopment
10. "Cloud Computing", 26 Aug 2015: https://en.wikipedia.org/wiki/Cloud_computing
11. "Internet of Things", (26th Aug 2015): https://en.wikipedia.org/wiki/Internet_of_Things
12. Perera C, Liu H, Jayawardena S, The emerging internet of things marketplace from an industrial perspective: a survey. IEEE Trans Emerging Topics Comput https://doi.org/10.1109/TETC.2015.23900341. Retrieved 1 February 2015
13. Barnes G, Volkmann W, Sherko R, Kelm K, Drones for peace: part 1 of 2 design and testing of a UAV-based cadastral surveying and mapping methodology, (26th Aug 2015): http://www.fao.org/fileadmin/user_upload/nr/land_tenure/UAS_paper_1_01.pdf
14. da Silva FB, Scott SD, Cummings ML, Design methodology for unmanned aerial vehicle (UAV) team coordination. MIT Technical Report, (26th Aug 2015): http://web.mit.edu/aeroastro/labs/halab/papers/HAL2007-05.pdf
15. Mouazen AM, Alhwaimel SA, Kuang B, Waine TW Fusion of data from multiple soil sensors for the delineation of water holding capacity zones. Precision Agricul 13:745–751
16. Waine TW, Simms DM, Taylor JC, Juniper GR Towards improving the accuracy of opium yield estimates with remote sensing. Int J Remote Sensing 35(16):6292–6309
17. Depth-Sensing Camera Works in Bright Light and Darkness, 26 Aug 2015: http://www.techbriefs.com/component/content/article/1198-ntb/news/news/22832-depth-sensing-camera-works-in-bright-light-and-darkness
18. Lakshmi Prathibha PT, Seetharaman N, Ramesh S, Babu N Lakshmi Narasimhan V (2013) Kinematic analysis and control of a multiaxis manipulator for possible application to land mine detection over an unmanned tracked vehicle. In: Proceedings of the 28th Indian Engineering Congress. Chennai, India
19. Lakshmi Narasimhan V, Arvind A, Bever K (2007) Greenhouse asset management using wireless sensor-actor networks. In: IEEE international conference on mobile ubiquitous computing, systems, services and technologies (IEEE-UBICOMM'07), 4–9 Nov 2007, Papeete, French Polynesia (Tahiti)
20. "Digital Elevation Model (DEM)", (26th Aug 2015): https://en.wikipedia.org/wiki/Digital_elevation_model
21. Earon, Founder of Precision Hawk, (26th Aug 2015): http://dronelife.com/2015/04/06/qa-weith-precisionhawk-founder-dr-ernest-earon/

22. "Sony Aerosense Drone", (26th Aug 2015): http://www.deccanchronicle.com/150826/techno logy-gadgets/article/sony-unveils-aerosenses-drone-prototypes
23. "Top 5 Drones For Commercial And Consumer Applications", (26th Aug 2015): http://www. bidnessetc.com/business/top-5-drones-for-commercial-and-consumer-applications/
24. Keshtgari M, Deljoo A (2013) A wireless sensor network for precision agriculture based on ZigBee technology. Wireless Sen Netw 4:25–30
25. Lu G, Krishnamachari B, Raghavendra CS (2004) Adaptive energy-efficient and low-latency MAC for data gathering in sensor networks. In: Proceedings of WMAN. Institute fur Medien Informatik, Ulm, pp 2440–2443
26. Busetta P, Ronnquist R, Hodgson A, Lucas A (1999) JACK intelligent agents—components for intelligent agents in Java. AgentLink News Lett. Also at: http://www.agent-software.com
27. Agent Oriented Software Pty Ltd (2012) JACK intelligent agents user guide. (26th Aug 2015): http://www.aosgrp.com/documentation/jack/Agent_Manual.pdf
28. Cooperative Engagement Capability (1995) (of US DoD), 26 Aug 2015: http://www.jhuapl. edu/techdigest/td/td1604/APLteam.pdf
29. Peterson JL (1982) Petri nets for modelling of systems—theory and practice. Prentice-Hall
30. Graham JR, Decker KS (2000) Towards a distributed, environment-centered agent framework. In: Intelligent Agents VI—Proceedings of sixth international workshop on agent theories, architectures and langauges (ATAL'99). LNCS, Berlin
31. ArcView Geospatial Information System, 26th Aug 2015: https://en.wikipedia.org/wiki/Arc View
32. Dikaiakos MD, Katsaros D, Mehra P, Pallis G (2009) Cloud computing: distributed internet computing for IT and scientific research. Internet Comput IEEE 13:5

Evaluation of NoSQL Databases Features and Capabilities for Smart City Data Lake Management

Nurhadi, **Rabiah Binti Abdul Kadir**, and **Ely Salwana Binti Mat Surin**

Abstract A data lake means there's an immense data resource or repository. Data lake stores enormous data and uses advanced analytics to pair data from various sources with different types of structured, semi-structured and un-structured information. The lifeblood of a smart city is data. Effective data management is not limited to data collection and storage, but must also involve shared and combined data so that it can be accessed, analyzed and used across agencies, within organizations, and even across the society. NoSQL is a form of database that is becoming increasingly common among web firms. NoSQL databases are non-tabular and store data rather than relational tables in a different way. NoSQL databases come in a variety of forms, mainly document, key-value, wide-column, and graph based on their data model. NoSQL offers easier scalability, better performance compared to conventional relational databases, and consists of many data types, such as document, key-value, wide-column, and graph. This work studies NoSQL database features and capabilities by considering four indicators, namely performance, scalability, accuracy and complexity, in order to measure the compatibility of NoSQL databases with multiple data types. The result of the experiment reveals that when accommodating massive data volumes, MongoDB is the most stable NoSQL database.

Keywords Data lake · NoSQL · Smart city database

Nurhadi · R. B. A. Kadir (✉) · E. S. B. M. Surin
Institute of IR4.0, Universiti Kebangsaan Malaysia, 43600 Bangi, Selangor, Malaysia
e-mail: rabiahivi@ukm.edu.my

Nurhadi
e-mail: p91334@siswa.ukm.edu.my

E. S. B. M. Surin
e-mail: elysalwana@ukm.edu.my

1   Introduction

Smart City uses data lake technology to store data in enormous capacity and there are many types of databases. One of the main and key features of big data technologies is NoSQL database. NoSQL database is able to store structured, semi-structured and unstructured data [1–3] regardless of type, format with 4Vs features: "volume, velocity, variety, veracity" [8–11] which is consisting of several model of data such as document, key-value, wide-column, and graph [4–9].

Data lake provides a scalable framework for storing large amounts of data and generating analytics that can assist multiple stakeholders in making effective decisions and developing new markets [10–13]. However, data lake still has many problems and drawbacks, one of the main issue is the use of trigger functions within ACID (Atomicity, Consistency, Isolation and Durability) to process complex online transactions and text statements [14–16]. This paper focuses on the evaluation of NoSQL database execution of data lake storage to support the use of trigger functions in managing various transactions. Four features and capabilities were evaluated in this study, namely, performance, scalability, accuracy and complexity to measure the execution of the selected NoSQL databases product i.e. MongoDB, Cassandra, Redis and Neo4j.

- **Performance**—The performance measurements in this study include several operations consisting of; select, enter, update, delete [9, 17]. Database performance can be defined as optimizing the use of resources in carrying out operations to in-crease throughput and minimizing errors, allowing as much workload as possible to be processed.
- **Scalability**—The scalability measurement in this study consists of several operations which include; data storage (write), retrieval (read), data sharding, data chacing, cluster management [17, 18].
- **Accuracy**—The accuracy measurements in this study consisted of several operations which included; import data, export data, load data. Accuracy access data is a component of data quality and refers to whether the value of data stored for an object is the correct value [19].
- **Complexity**—The complexity consisting of operations; query, function, variety. for query operations and functions used in this study include; group by, order by, select distinct, aliases, create primary [20].

The rest of the paper is organized as follows: Sect. 2 briefly discusses related work on comparison of NoSQL database features and capabilities followed by the detail description of methodology of comparison in Sect. 3. Section 4 discuss the result of the experiment. Finally, the conclusion and propose relevant expansion suggestions will be described in Sect. 5.

2 Related Work

In order to collect data, a smart city uses distinct types of electronic Internet of Things (IoT) sensors. These information and communication systems are integrated in digital technology throughout all city functions. It is a term that incorporates several ICT solutions in a safe way to manage the assets of a community, including transportation systems, waste management, water management, protection systems, information systems of municipal departments and other community services, as well as data management. With that, data lakes are a perfect place to store large amounts of data redundantly scale and store it. The concept is to connect, store and analyze the various very heterogeneous data sources, and by using NoSQL databases, data must be systematized, organized and modified for further use.

In the case of big data and real-time web applications, NoSQL databases are progressively used. NoSQL databases are particularly useful for working with vast sets of distributed data and are compliant with smart city data collection functionality. For relational data bases (RDBMS) to tackle on their own, this data explosion is proving to be too big and too complex. NoSQL databases are not constrained by the confines of a fixed schema model, unlike relational databases. NoSQL databases implement Schema on Read instead of applying Schema on Write. This makes NoSQL databases especially appropriate for the high-volume, high-variety online applications of today.

Currently, data model is the most important feature in selecting the appropriate NoSQL databases. Though, there are studies conducted by several researchers regarding the comparison of NoSQL databases based on features and capabilities such as performance, integration, and security [9, 21–23]. Meanwhile, other studies support the comparison of performance [17], integrity, cloud service criteria [8, 24], and frameworks [25]. However, those studies do not discuss the accuracy of data access and scalability, which are important features in evaluating of NoSQL databases capabilities for the purpose to select the appropriate NoSQL databases in supporting the trigger function.

Therefore, in this study we conducted a comparison of the NoSQL databases based on four features and capabilities namely; performance, scalability, accurate and complexity. In this study, we compared four NoSQL databases product such as MongoDB, Cassandra, Redis and Neo4j that used for data model document, wide-column, key-value, and graph respectively.

3 Research Evaluation Method

This section presents the research method in evaluating of NoSQL databases product for Smart City Data Lake Management based on features and capabilities. In this study, the research method was implemented based on the framework as shown in Fig. 1.

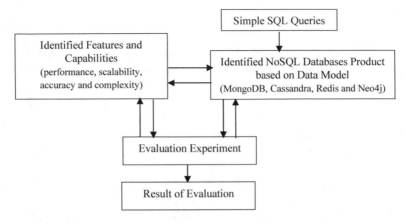

Fig. 1 Framework of research evaluation

In this experiment, 10 simple SQL queries have been tested with the combination of functions and operations such as *SELECT, INSERT, UPDATE, DELETE, IMPORT DATA, EXPORT DATA, LOAD DATA ORDER BY, GROUP BY* etc. as shown in Table 1. Each SQL query was implemented by the selected of NoSQL databases product, which is MongoDB, Cassandra, Redis and Neo4j in executing different data model: document, key-value, wide-column, and graph. The identified features and capabilities of NoSQL databases were measured and compared. The scope of the study is covering four features and capabilities as below:

1. Performance
   Capability to find, analyze and then resolve various database congestion that can impact application response times or hinder application performance.
2. Scalability
   Capability of a system to handle a growing amount of work or potential to perform more total work in the same elapsed time when processing power is expanded to accommodate growth.
3. Accuracy:
   Capability to represent the right data in a form that is consistent and unambiguous and most relevant to historical records stored on computer-accessible digital media.
4. Complexity:
   Capability of the query in evaluating the function and size of the expression.

4   Results and Analysis

The following outcomes of average scores using functions and operations available in NoSQL were obtained from the experiment, as shown in Table 1.

Table 1 Average NoSQL response score

No.	Subject and operation	Document base (MongoDB)	Key-value store (Redis)	Graph store (Neo4j)	Wide column store (Cassandra)
1	Select	95	96	90	95
2	Insert	96	96	88	96
3	Update	96	94	90	94
4	Delete	94	98	92	95
5	Import data	85	79	64	71
6	Export data	84	65	65	70
7	Load data	80	80	66	69
8	Data storage (write)	90	97	86	93
9	Retrieval (read)	90	94	85	96
10	Data sharding	89	93	84	95
11	Data chacing	90	98	85	95
12	Cluster management	91	93	85	94
13	Query (order by)	71	48	72	63
14	Query (group by)	73	50	74	65
15	Function (select distinct)	69	71	72	76
16	Function (aliases/as)	70	73	69	75
17	Function (create primary key)	68	72	69	77
18	Variety	79	60	62	69

Based on performance, scalability, accuracy, and complexity, the results were grouped, as shown in Figs. 2, 3, 4 and 5 respectively.

Performance—consists of operations; select, insert, update, delete. The average results can be seen in Fig. 2 as below.

Scalability—In Fig. 3 shows the operations such as data storage (write), retrieval (read), data sharding, data chacing, and cluster management, where virtually all types of NoSQL databases have high values.

Accuracy—includes of operations such as import data, export data, and load data. The average results were illustrated in Fig. 4, where the average score for MongoDB and Redis is the highest.

Complexity—consists of tasks such as questions, functions, combinations, and it is possible to see the results in Fig. 5.

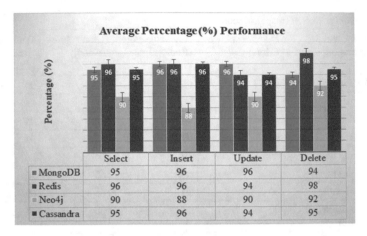

Fig. 2 Comparison of averages based on performance criteria

	Select	Insert	Update	Delete
▪ MongoDB	95	96	96	94
▪ Redis	96	96	94	98
▪ Neo4j	90	88	90	92
▪ Cassandra	95	96	94	95

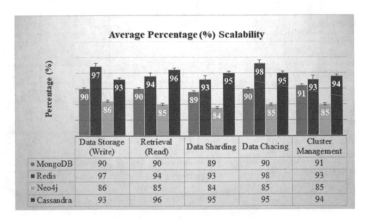

Fig. 3 Comparison of averages based on scalability criteria

	Data Storage (Write)	Retrieval (Read)	Data Sharding	Data Chacing	Cluster Management
▪ MongoDB	90	90	89	90	91
▪ Redis	97	94	93	98	93
▪ Neo4j	86	85	84	85	85
▪ Cassandra	93	96	95	95	94

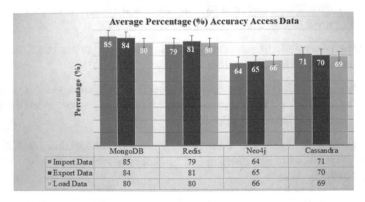

Fig. 4 Comparison of averages based on accuracy criteria

	MongoDB	Redis	Neo4j	Cassandra
▪ Import Data	85	79	64	71
▪ Export Data	84	81	65	70
▪ Load Data	80	80	66	69

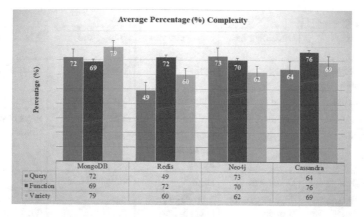

Fig. 5 Comparison of averages based on complexity criteria

The findings showed that MongoDB and Cassandra had the highest results, while Redis and Cassandra owned data scalability, while Neo4j and Cassandra owned middle-class data access accuracy. The highest percentage for complexity issues are MongoDB and Cassandra, while Redis and Neo4j are relatively poor. This demonstrates the difference between NoSQL databases by referring to performance, scability, the accuracy of database access, and complexity as shown in Fig. 6.

The complexity and accuracy given a moderate value is shown by several studies that have been carried out because it is affected by the semi-structured data format of the input. It is also possible to view the results of this study in a Table 2, where all NoSQL databases have high average output criteria. As for the complexity criterion, as shown in Table 2, all NoSQL databases have moderate values.

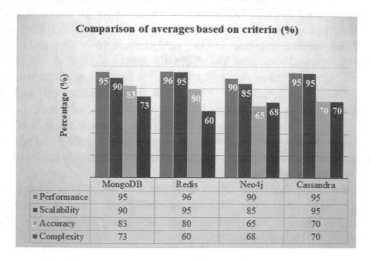

Fig. 6 Comparison of averages based on indicator criteria

Table 2 Implementation of NoSQL comparisons based on items

No.	Items	Document base	Key-value store	Graph store	Wide column store
1	Database Name	MongoDB	Redis	Neo4j	Cassandra
2	License	Open source	Open source	Open source	Open source
3	Database	Database	Database	Database	Keyspace
4	Table	Collection	Hash, List, Set, Sorter Set, and String	Label	Column Family
5	Value	Document	High	Node and edges	Rows
6	Data Source	Web Page data is open	Web Page data is open		Web Page data is open
7	Performance	High	High	High	High
8	Scalability	High	High	High	High
9	Accuracy of Data Access	High	High	Moderate	Moderate
10	Complexity	Moderate	Moderate	Moderate	Moderate

5 Conclusion

This study is to measure and compare the use of the NoSQL databases product for smart city data lake which is includes several data model. The evaluation has been conducted based on quantitative analysis through the experiment. The result of evaluation able to find the appropriate NoSQL databases product based on performance, scalability, accuracy, and complexity. The most important technical characteristics from each NoSQL databases have been studied in selecting the appropriate level of each database within the given features and capabilities. Although the NoSQL database product has the same performance, scalability and complexity scores, the accuracy shows the effect of the difference. MongoDB and Redis have high scores, although there are modest values for Neo4j and Cassandra. NoSQL databases compromise consistency to provide high performance and scalability in order to indicate the requirements that NoSQL is suitable for analyzing and accessing data across agencies based on investigation through experiments. It is in line with the success of the web-scale information system, that availability and speed are of high importance, and accuracy is compromised to some degree by sufficient NoSQL databases to meet these needs.

The future work from this study will involves optimizing algorithms and supporting complex transaction features for NoSQL databases with security and data integrity. In addition, we intend to support more categories of NoSQL databases in future testing and implementation.

Acknowledgements The authors would like to express gratitude to the University Kebangsaan Malaysia (UKM) for providing the opportunity and financial support under the project code ZG-2019-003.

References

1. Lakhe B (2016) Practical Hadoop Migration. Academic Press
2. Memoriam I, Gray J (2018) Database-Centric Scienti fi c Computing
3. Al-mandhari IS, Guan L, Edirisinghe EA (2019) Advances in Information and communication networks, vol 886. Springer
4. Patil MM, Hanni A, Tejeshwar CH, Patil P (2017) A qualitative analysis of the performance of MongoDB vs MySQL database based on insertion and retriewal operations using a web/android application to explore load balancing-Sharding in MongoDB and its advantages. In: Proceedings of international conference IoT in social, mobile, analytics and cloud on I-SMAC, pp 325–330
5. González-Aparicio MT, Ogunyadeka A, Younas M, Tuya J, Casado R (2017) Transaction processing in consistency-aware user's applications deployed on NoSQL databases. Hum Cent Comput Inf Sci 7(1)
6. Patil MM, Hanni A, Tejeshwar CH, Patil P (2017) A qualitative analysis of the performance of MongoDB vs MySQL database based on insertion and retriewal operations using a web/android application to explore load balancing-Sharding in MongoDB and its advantages. In: Proceedings of international conference on I-SMAC (IoT in social, mobile, analytics and cloud) (I-SMAC) 2017, pp 325–330
7. Chen JK, Lee WZ (2019) An introduction of NoSQL databases based on their categories and application industries. Algorithms 12(5)
8. Tool B, Chakraborttii C (2019) Performance evaluation of NoSQL systems using yahoo cloud serving performance evaluation of NoSQL systems using yahoo cloud serving benchmarking tool. February 2015
9. Patil MM, Hanni A, Tejeshwar CH, Patil P (2017) A qualitative analysis of the performance of MongoDB vs MySQL database based on insertion and retriewal operations using a web/android application to explore load balancing-Sharding in MongoDB and its advantages. In: *Proc. Int. Conf. IoT Soc. Mobile, Anal. Cloud, I-SMAC 2017*, pp. 325–330
10. Challenges HD, Gupta S, Giri V (2018) Practical enterprise data lake insights. Academic Press
11. Phyu KP, Shun WZ (2018) Data lake : a new ideology in big data era. In: ITM Web Conference 17, vol 03025, pp 1–11
12. Mathis C (2017) Data lakes. Datenbank-Spektrum
13. I. Nosql, R. Database, B. Data, and CC ((2018)) Transactions "Romulo Alceu Rodrigues, Lineu Alves Lima Filho, Gildarcio Sousa Gonc ¸ alves, Lineu F.S. Mialaret, Adilson Marques da Cunha, and Luiz Alberto Vieira Dias, pp 443–451
14. AE Lotfy, AI Saleh, HA El-Ghareeb, HA Ali A middle layer solution to support ACID properties for NoSQL databases. J King Saud Univ Comput Inf Sci 28(1):133–145
15. Davoudian A, Chen L, Liu M (2018) A Survey on NoSQL Stores. ACM Comput Surv 51(2):1–43
16. Schreiner GA, Duarte D, dos S. Mello R (2019) When relational-based applications go to NoSQL databases: A survey. INF 10(7):1–22
17. Flores A, Ramirez S, Toasa R, Vargas J, Barrionuevo RU, Lavin JM (2018) Performance evaluation of NoSQL and SQL queries in response time for the e-government. In: 2018 5th international conference eDemocracy eGovernment, ICEDEG 2018, pp 257–262
18. Mohan A, Ebrahimi M, Lu S, Kotov A (2016) A NoSQL data model for scalable big data workflow execution. In: Proceedings of 2016 IEEE international congress on big data, big data congress 2016, pp 52–59

19. Li C, Gu J (2019) An integration approach of hybrid databases based on SQL in cloud computing environment. Softw Pract Exp 49(3):401–422
20. Sánchez-de-Madariaga R, Muñoz A, Castro AL, Moreno O, Pascual M (2018) Executing complexity-increasing queries in relational (MySQL) and NoSQL (MongoDB and EXist) size-growing ISO/EN 13606 Standardized EHR databases. J Vis Exp 133:1–11
21. Hajoui O, Dehbi R, Talea M, Batouta ZI (2015) An advanced comparative study of the most promising NoSQL and NewSQL databases with a multi-criteria analysis method. J Theor Appl Inf Technol 81(3):579–588
22. Sánchez-de-Madariaga R, Muñoz A, Castro AL, Moreno O, Pascual M (2018) Executing complexity-increasing queries in relational (MySQL) and NoSQL (MongoDB and EXist) size-growing ISO/EN 13606 standardized EHR databases. J Vis Exp 133
23. Kolonko K (2018) Master of Science in Software Engineering Performance comparison of the most popular relational and non-relational database management systems
24. Kumar MS, Prabhu J (2018) Comparison of NoSQL database and traditional database-an emphatic analysis. JOIV Int J Inf Vis 2(2)51
25. Reniers V, Van Landuyt D, Rafique A, Joosen W (2019) Object to NoSQL database mappers (ONDM): a systematic survey and comparison of frameworks. Inf Syst 85:1–20

Construction of GAN Dataset to Generate Images Similar to Copyrighted Images

Chikazawa Yuta and Miyoshi Tsutomu

Abstract In this paper, we propose a system for improving the convenience of using images on the Internet and SNS, and for preventing copyright infringement that people take without notice. In this system, many similar images are collected from one image on the Internet, and an image is generated from the collected images using deep learning. As a study of image collection means, we will conduct an experiment using Content-Based Image Retrieval (CBIR), which performs image retrieval using image features. In the experiment, images are searched using CBIR and similar images are collected to construct a dataset.

Keywords Generative adversarial network · Content-Based image retrieval

1 Introduction

Today, as the number of users of the Internet and SNS increases and develops, the opportunities to use images are increasing. Because you can easily post images, you can easily post images of others and images that are protected by copyright. Therefore, in this research, we propose a system that provides freely usable images for the purpose of preventing easy copyright infringement on SNS and the Internet. In this system, it is assumed that images that you want to use on SNS or the Internet will be used as a search query, and images with similarities will be mechanically collected on the Internet. It is generated by Generative Adversarial Network (GAN), which is one of the methods. The problem is solved by synthesizing images that are similar and different from one image automatically by artificial learning, that is, a network generated by learning of Deep Learning. In this experiment, when searching for similar images, we verified whether it is possible to construct a dataset for image generation by using a method that uses image features.

C. Yuta (✉) · M. Tsutomu
Department of Media Infomatics, Ryukoku University Graduate School, 1-5 Yokotani, Seta Oe-Cho, Otsu Shiga 520-2194, Japan
e-mail: t19m059@mail.ryukoku.ac.jp

2 Methods

In the proposed system, a target image is generated by collecting and learning images similar to the query image from the Internet. The following experiments were conducted to investigate whether the image retrieval method using feature quantities can be used for image acquisition. We performed a similar image search using Content Based Image Search (CBIR). The proposed system collects images from the internet, but the experiment used the four images shown in Fig. 1 as a query, consisting of a bird image and one from the Open Images Dataset V4 CUB_200_2011 data. We used 10,000 images from the set. A total of 20,000 images combined with 10,000 image data. Images in JPEG format were used as the search database. In image search, the features of the query image and all images are compared and the image with high similarity is acquired. Three functional values are used: ORB, A-KAZE, and Histogram. First, we acquired the top 5000 most similar images and compared the usefulness of the search methods. An experiment is conducted to verify whether the features of the bird's image can be better captured by combining the local features that capture the features of the object in the image with the global features that capture the hue of the entire image. We examine the images that are common to the images acquired with multiple feature quantities, and examine the proportion of the bird images contained in them. Also, the number of acquired images is divided

Fig. 1 Image used for search

into the top 100, 500, 1000, 2000, 3000, 4000, and 5000, and the number of bird images contained in each is checked.

3 Results and Discussion

Of the 20,000 images, we obtained 5000 with high similarity, and Table 1 shows the number of bird images contained in them.

From Table 1, the histogram retrieval method shows the best results. However, in the histogram, the results differ greatly depending on the image. Since this is a comparison based on the hue of the entire image in Histogram, not on the object appearing in the image, it is considered that when the background is one color such as white or light blue, the background has a greater effect than the color of the object. Therefore, it is considered that in the image like Query 4, many light-colored images were mixed in the whole image, which resulted in such a result. On the other hand, the result of query 3 is good because there are many birds with colors close to black, and images with a green background such as forests and trees are considered to be a feature of bird images, so good results are obtained. It seems to indicate. With ORB and A-KAZE, which are local features, the number of bird images is about 50%. Since 20,000 images of birds included in the 20,000 database are 50% of 10,000 images, it can be seen that the features of the birds could not be obtained and the images were acquired almost randomly.

Next, Table 2 shows the ratio of the number of images of common images and the number of birds in them among the acquired images of multiple feature quantities. From the table, the combination of ORB and histogram has the highest result. In addition, since the combination of local features and global features is higher than the combination of the same local features, it is better to combine local features and

Table 1 Percentage of bird images in 5000 images

	Query1	Query2	Query3	Query4
ORB	2497(50%)	2756(55%)	2698(54%)	2657(53%)
A-KAZE	2748(55%)	2097(42%)	2316(46%)	2473(49%)
Histogram	2979(60%)	3194(64%)	3421(68%)	2156(43%)

Table 2 Bird image common to multiple features

	Query1 (%)	Query2 (%)	Query3 (%)	Query4 (%)
ORB and A-KAZE	50	44	49	54
ORB and Histogram	60	72	72	50
A-KAZE and Histogram	59	58	67	48
All features	63	67	67	53

global features and judge by the two features of the bird image and the hue. It was found that more images of birds could be selected than by increasing the features and selecting the features found in birds. However, the total number of images was reduced to less than half by combining multiple features. The ORB and A-KAZE combination was less than 50% of the bird images, but the number of images in common was greater than the combination of global and local features.

Next, scatter plots of the changes in the percentage of bird images are shown in Figs. 2, 3, 4, and 5 divided by total number of images. In the graph, the horizontal axis shows the number of images acquired and the vertical axis shows the ratio. From the graph, it can be seen from Experiment 2 that the percentage of bird images contained up to 2000 is high, and the percentage of bird images is constant thereafter. However, in most of the search results, among the top 1000 acquired images, the number of common images was less than 10%.

Fig. 2 ORB and A-KAZE

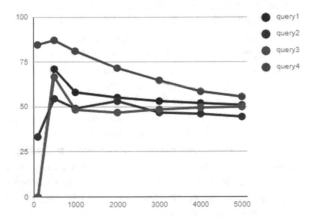

Fig. 3 ORB and Histogram

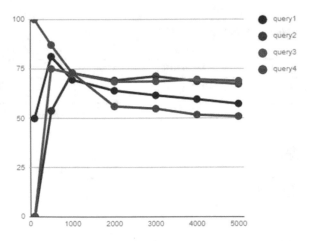

Fig. 4 A-KAZE and
Histogram

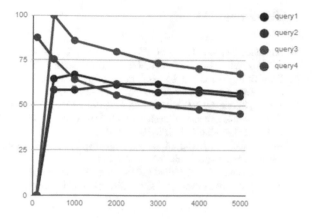

Fig. 5 ORB and A-KAZE
and Histogram

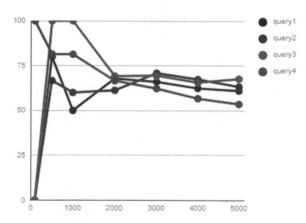

4   Conclusions

We conducted an experiment for the purpose of collecting data for image generation. As a result of acquiring 5000 images using a single method, only about 50% of the images of birds were included, and when using multiple images, it was less than 70%. It was both quality and quantity would be difficult to use as a dataset for image generation.

This time, we evaluated the quality of the dataset for image generation by the number of bird images included in the collected images. However, even if the images do not show the birds, it is judged that they are similar due to the image features. There is a possibility that it will have a good effect on image generation by learning. In the future, you will need to generate an image in GAN based on all the images you collect and use the generated image to evaluate your dataset.

References

1. Alexander M, Abid K, OpenCV3-python tutorial, 2018–1125. http://whitewell.sakura.ne.jp/OpenCV/py_tutorials/py_tutorials.html (accessed 2018-8-20)
2. Edward R, Tom D (2006) Machine learning for highspeed corner detection. In: 9th European conference on computer vision, vol 1, pp 430–443
3. Wah C, Branson S, Welinder P, Perona P, Belongie S, The CaltechUCSD birds-200-2011 dataset. Computation and neural systems technical report, CNS-TR-2011-001. http://www.vision.caltech.edu/visipedia/CUB-200-2011.html (accessed 2018-8-22)
4. Ian JG, Jean P-A, Mehdi M, Bing X, David W-F, Sherjil O, Aaron C, Yoshua B (2014) Generative adversarial networks. arXiv:1406.2661
5. Alec R, Luke M, Soumith C (2015) Unsupervised representation learning with deep convolutional generative adversarial networks. arXiv:1511.06434

Printed in the United States
by Baker & Taylor Publisher Services